Handbook of
Ecological Indicators for Assessment of Ecosystem Health

Second Edition

T0203765

Applied Ecology
and Environmental Management

A SERIES

Series Editor
Sven E. Jørgensen

Copenhagen University, Denmark

Handbook of Ecological Indicators for Assessment of
Ecosystem Health, Second Edition
Sven E. Jørgensen, Fu-Liu Xu, and Robert Costanza

ADDITIONAL VOLUMES IN PREPARATION

Handbook of
Ecological
Indicators for
Assessment of
Ecosystem Health

Second Edition

Edited by
Sven E. Jørgensen
Fu-Liu Xu
Robert Costanza

CRC Press
Taylor & Francis Group
Boca Raton London New York

CRC Press is an imprint of the
Taylor & Francis Group, an **informa** business

CRC Press
Taylor & Francis Group
6000 Broken Sound Parkway NW, Suite 300
Boca Raton, FL 33487-2742

First issued in paperback 2019

© 2010 by Taylor & Francis Group, LLC
CRC Press is an imprint of Taylor & Francis Group, an Informa business

No claim to original U.S. Government works

ISBN-13: 978-0-4398-0936-5 (hbk)
ISBN-13: 978-0-367-86442-2 (pbk)

Library of Congress Cataloging-in-Publication Data

Handbook of ecological indicators for assessment of ecosystem health / edited by Sven E. Jørgensen, Fu-Liu Xu, Robert Costanza. -- 2nd ed.
    p. cm. -- (Applied ecology and environmental management)
  Includes bibliographical references and index.
  ISBN 978-1-4398-0936-5 (hard back : alk. paper)
  1. Ecosystem health. 2. Environmental indicators. I. Jørgensen, Sven Erik, 1934- II. Xu, Fu-Liu. III. Costanza, Robert.

  QH541.15.E265H36 2010
  577.27--dc22   2009051368

Visit the Taylor & Francis Web site at
http://www.taylorandfrancis.com

and the CRC Press Web site at
http://www.crcpress.com

Contents

Section II   Assessment of Ecosystem Health

Preface

Application of ecological indicators has had increasing importance in environmental management since the beginning of the century. The editors of the first edition found it necessary, therefore, to cover the recent developments in the field through publication of a second edition.

The second edition of *Handbook of Ecological Indicators for Assessment of Ecosystem Health* has two sections. The first gives an overview of the applicable indicators. The second covers the application of the indicators in concrete illustrative cases. A wide spectrum of ecosystems is covered by these generally applicable illustrations: wetlands, estuaries, coastal zones, transitional waters, lakes, coastal lagoons, marine ecosystems, landscapes, agricultural systems, and forest.

It is our hope that the second edition will be as well received as the first edition. The organization of the chapters in two sections, one for an overview of the indicators and one for the assessment of ecosystem health, should facilitate the use of the book as a handbook that gives a brief and useful overview of the field.

S. E. Jørgensen
F.-L. Xu
R. Costanza

Editors

Sven Erik Jørgensen is a professor of environmental chemistry at Copenhagen University. He received a doctorate of engineering in environmental technology and a doctorate of science in ecological modeling. He is an honorable doctor of science at Coimbra University, Portugal, and at Dar es Salaam University, Tanzania. He was editor in chief of *Ecological Modelling* from the journal's inception in 1975 to 2009. He has also been the editor in chief of *Encyclopedia of Ecology*.

In 2004 Dr. Jørgensen was awarded the Stockholm Water Prize and the Prigogine Prize. He was awarded the Einstein Professorship by the Chinese Academy of Science in 2005. In 2007 he received the Pascal medal and was elected a member of the European Academy of Science. He has written close to 350 papers, most of which have been published in international peer-reviewed journals. He has edited or written 64 books. Dr. Jørgensen has given lectures and courses in ecological modeling, ecosystem theory, and ecological engineering worldwide.

Fu-Liu Xu is a professor at the College of Urban and Environmental Sciences, Peking University, China. He received his Ph.D. from the Department of Pharmaceutics and Analytical Chemistry, Faculty of Pharmaceutical Sciences, University of Copenhagen, Denmark. His research interests include ecosystem health and ecological indicators, system ecology and ecological modeling, ecotoxicology and risk assessment of POPs.

Dr. Robert Costanza is the Gordon and Lulie Gund Professor of Ecological Economics and founding director of the Gund Institute for Ecological Economics at the University of Vermont. His transdisciplinary research integrates the study of humans and the rest of nature to address research, policy and management issues on multiple scales, from small watersheds to the global system. He is co-founder and past president of the International Society for Ecological Economics, and was founding chief editor of the society's journal, *Ecological Economics*. He has published over 400 papers and 20 books. His awards include a Kellogg National Fellowship, the Society for Conservation Biology Distinguished Achievement Award, and a Pew Scholarship in Conservation and the Environment.

Contributors

Kossi Adjonou
Laboratoire de Botanique et
    Ecologie Végétale
Faculté des Sciences
Université de Lomé
Lomé, Togo

M. Austoni
Department of Environmental
    Sciences
University of Parma
Parma, Italy

S. Bastianoni
Department of Chemistry
University of Siena
Siena, Italy

Mark T. Brown
Department of Environmental
    Engineering Sciences
University of Florida
Gainesville, Florida

Benjamin Burkhard
Ecology Center
University of Kiel
Kiel, Germany

Nuno Caiola
Aquatic Ecosystems Program, IRTA
Catalonia, Spain

Villy Christensen
Fisheries Centre
University of British Columbia
Vancouver, British Columbia, Canada

Robert Costanza
Gund Institute for Ecological
    Economics
Rubenstein School of Environment
    and Natural Resources
University of Vermont
Burlington, Vermont

Philippe Cury
Centre de Recherche Halieutique
    Méditerranéenne et Tropicale
Sète, France

G. A. De Leo
Department of Environmental
    Sciences
University of Parma
Parma, Italy

S. Focardi
Department of Chemistry
University of Siena
Siena, Italy

C. Gaggi
Department of Environmental
    Sciences "G. Sarfatti"
University of Siena
Siena, Italy

G. Giordani
Department of Environmental
    Sciences
University of Parma
Parma, Italy

Carles Ibáñez
Aquatic Ecosystems Program, IRTA
Catalonia, Spain

S. E. Jørgensen
Section of Environmental Chemistry
Copenhagen University
Copenhagen, Denmark

Kouami Kokou
Laboratoire de Botanique et
    Ecologie Végétale
Faculté des Sciences
Université de Lomé
Lomé, Togo

Adzo Dzifa Kokutse
Laboratoire de Botanique et
    Ecologie Végétale
Faculté des Sciences
Université de Lomé
Lomé, Togo

João C. Marques
Department of Zoology
Faculty of Sciences and Technology
Institute of Marine Research (IMAR)
University of Coimbra
Coimbra, Portugal

Felix Müller
Ecology Center
University of Kiel
Kiel, Germany

J. Neto
Department of Zoology
Faculty of Sciences and Technology
Institute of Marine Research (IMAR)
University of Coimbra
Coimbra, Portugal

V. Niccolucci
Department of Chemistry
University of Siena
Siena, Italy

Georg Niedrist
European Academy of Bolzano
Bolzano, Italy

J. Patrício
Department of Zoology
Faculty of Sciences and Technology
Institute of Marine Research (IMAR)
University of Coimbra
Coimbra, Portugal

Irene Petrosillo
Landscape Ecology Laboratory
Department of Biological and
    Environmental Sciences and
    Technologies, Ecotekne
University of Salento
Lecce, Italy

R. Pinto
Department of Zoology
Faculty of Sciences and Technology
Institute of Marine Research (IMAR)
University of Coimbra
Coimbra, Portugal

M. Plus
Département Environnement
    Littoral
Ifremer-Station d'Arcachon
Arcachon, France

F. M. Pulselli
Department of Chemistry
University of Siena
Siena, Italy

R. M. Pulselli
Department of Chemistry
University of Siena
Siena, Italy

Kelly Chinners Reiss
Department of Environmental
    Engineering Sciences
University of Florida
Gainesville, Florida

Fuensanta Salas
Department of Zoology
Faculty of Sciences and Technology
Institute of Marine Research (IMAR)
University of Coimbra
Coimbra, Portugal

Peter Sharpe
Aquatic Ecosystems Program, IRTA
Catalonia, Spain

Ulrike Tappeiner
European Academy of Bolzano
Bolzano, Italy

University of Innsbruck
Innsbruck, Austria

Erich Tasser
European Academy of Bolzano
Bolzano, Italy

H. Teixeira
Department of Zoology
Faculty of Sciences and Technology
Institute of Marine Research (IMAR)
University of Coimbra
Coimbra, Portugal

Rosa Trobajo
Aquatic Ecosystems Program, IRTA
Catalonia, Spain

Sergio Ulgiati
Department of Chemistry
University of Siena
Siena, Italy

P. Viaroli
Department of Environmental
    Sciences
University of Parma
Parma, Italy

Fu-Liu Xu
MOE Laboratory for Earth Surface
    Process
College of Urban and
    Environmental Sciences
Beijing, China

Nicola Zaccarelli
Landscape Ecology Laboratory
Department of Biological and
    Environmental Sciences and
    Technologies, Ecotekne
University of Salento
Lecce, Italy

J. M. Zaldívar
European Commission
Joint Research Centre
Institute for Health and Consumer
    Protection
Ispra, Italy

Patrick Zimmermann
European Academy of Bolzano
Bolzano, Italy

Giovanni Zurlini
Landscape Ecology Laboratory
Department of Biological and
    Environmental Sciences and
    Technologies, Ecotekne
University of Salento
Lecce, Italy

Section I

Ecological Indicators

1

Introduction

S. E. Jørgensen

CONTENTS

1.1 The Role of Ecosystem Health Assessment in Environmental Management

The idea to apply an assessment of ecosystem health in environmental management emerged in the late 1980s. The parallel to the assessment of human health is very obvious. We go to a doctor to get a diagnosis (What is wrong? What causes me to not feel completely healthy?) and hopefully initiate a cure to bring us back to normal (or healthy conditions). Doctors will apply several indicators/examinations (pulse, blood pressure, sugar in the blood, and urine, etc.) before they will come up with a diagnosis and a proper cure. The idea behind the assessment of ecosystem health is similar (Figure 1.1). We observe that an ecosystem is not healthy and want a diagnosis: What is wrong? What caused this unhealthy condition? And what can we do to bring the ecosystem back to normal? To answer these questions, and also to follow the results of the "cure," ecological indicators are applied.

Since ecosystem health assessment emerged in the late 1980s, numerous attempts have been made to use the idea in practice, and again and again environmental managers and ecologists have asked the question: Which ecological indicators should we apply? It is clear today that it is not possible to find one indicator or even a few indicators that can be used generally, as some naïvely thought when ecosystem health assessment (EHA) was introduced. Of course, there are general ecological indicators that are used

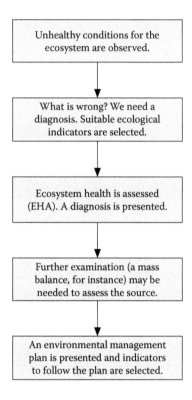

FIGURE 1.1
The figure illustrates how ecological indicators are used for EHA and to follow the effect of the environmental management plan.

almost every time we have to assess ecosystem health, but they are never sufficient to present a complete diagnosis—the general indicators always have to be supplemented by other indicators. Doctors also have general indicators. They will always take your pulse, temperature, and blood pressure—which are very good general indicators—but doctors also always have to supplement these general indicators with other indicators that they select according to the description of the symptoms given by the patient. It's the same with ecological doctors. If they observe dead fish but clear water, they will suspect the presence of a toxic substance in the ecosystem, and they will associate dead fish and very muddy water with oxygen depletion. In these two cases they will use two different sets of indicators, although some general indicators may be used in both cases.

The first international conference on the application of ecological indicators for the assessment of ecosystem health was held in Fort Lauderdale, Florida, in October 1990. Since then, there have been several international and national conferences on ecological indicators and on EHA. In 1992, a book titled *Ecosystem Health*, edited by Costanza, Norton, and Haskell, was published by Island Press. Blackwell published a book with the same title in

1998, edited by Rapport, Costanza, Epstein, Gaudet, and Levins. Blackwell launched a journal with the title *Ecosystem Health* in the mid-1990s with Rapport as the editor in chief. Elsevier launched a journal with the title *Ecological Indicators* in 2000 with Felix Mueller as editor in chief. As can be seen from this short overview of the development of the use of EHA and ecological indicators to perform EHA, there has been significant interest in EHA and ecological indicators.

Some may have expected that EHA, to a certain extent, would replace ecological modeling as it was a new method to quantify the disease of an ecosystem. It is also possible, as discussed in the next chapter, to assess ecosystem health based on observations only. On the other hand, EHA cannot be used to make prognoses and does not provide the overview of ecological components and their interactions that a model does. EHA and ecological modeling are two different and complementary tools that together give a better image of the environmental management possibilities than if EHA or ecological modeling are used independently. Today models are increasingly used, as also demonstrated in this volume, as a tool to perform an EHA. The models are also used to give a prognosis of the development of the EHA's applied ecological indicators when a well-defined environmental management plan is followed.

A number of ecological indicators have been applied during the last 20 years to assess ecosystem health. As already stressed, general ecological indicators do not exist, or at least have not been found yet. A review of the literature published in the last 20 years about EHA and selection of ecological indicators reveals that it is also not possible to indicate a set of indicators to be used for specific problems or specific ecosystems. There are general indicators and there are problem- and ecosystem-specific indicators that will be used again and again for the same problems or the same type of ecosystems. But because all ecosystems are different, even ecosystems of the same type, there are always some very case-specific indicators that are selected on the basis of sound theoretical considerations. We can therefore not just give, say, 300 lists of ecological indicators—each list being valid for a specific problem in a specific ecosystem (we could presume, for instance, 20 different problems and 15 different types of ecosystems, totaling 300 combinations). Our knowledge about human health is much more developed than our knowledge about ecosystem health, and still there is no general procedure on how to assess a diagnosis for each of the several hundred different possible cases doctors will meet in their practices. We will, however, attempt in the next chapter to give an overview of the most applied ecological indicators for different ecosystems and their classification. It is possible to give such an overview, but not to give a general applicable procedure with a general valid list of indicators. This does not mean, of course, that we have nothing to learn from case studies. Because the selection of indicators is difficult and varies from case to case, it is, of course, possible to expand one's experience by learning about as many case studies as possible. This is the general idea behind this volume.

By presenting different types of ecological indicators and a number of different case studies representing different ecosystems and different problems, an overview of the applicable indicators should be achieved.

1.2 The Conceptual Flow in This Volume

The first part of the book, Section I, covering Chapters 1–9, focuses on presenting and reviewing all the applicable indicators. Chapter 2 gives an overview of the applicable ecological indicators, which can be classified in eight levels. Six of the eight levels are straightforward, namely, the application of indicator species; use of ratio of species; use of chemical concentrations, for instance, a limiting nutrient or a toxic substance; use of an entire tropic level; rates for primary, secondary, or tertiary production; and the ecologically well known E. P. Odum's attributes, which are used to distinguish between an early state and a mature state. These six levels are discussed in some detail in Chapter 2.

The last two levels—seven and eight—cover indicators that are able to capture the holistic properties or the system characteristics of a focal ecosystem. These indicators require a more comprehensive explanation and, in particular, a more profound understanding of how they can be used in environmental management and which health aspects they are able to cover. These indicators are presented in Chapters 3–9, with the three thermodynamic indicators—exergy, emergy, and the ratio of exergy to emergy—covered in Chapters 3–5. The holistic indicator buffer capacity—the ability to meet impacts by changes that are as small as possible—is covered by exergy and therefore discussed in Chapter 3, while resilience is covered in Chapter 6. Biodiversity and species richness, genuine holistic indicators, are discussed and reviewed in Chapter 7. Chapter 8 presents a Landscape Development Intensity index (LDI) that integrates human disturbances on the level of landscape. LDI attempts to go holistic on the hierarchical level above ecosystems, namely, on the level of landscapes. The last chapter in this part, Chapter 9, presents the concept of ecosystem services and its linkages to ecological indicators and ecosystem health.

Section II of the book, covering Chapters 10–18, shows how the indicators can be used to assess ecosystem health for different ecosystems: wetlands, estuaries, coastal zones, lakes, forests, marine ecosystems, lagoons, agricultural systems, landscapes, and rivers. These chapters present illustrations on how the indicators are generally applied in practice to support environmental management. Chapter 10 illustrates how the indicators are selected based on use of the ecosystem. Wetlands are used to treat drainage water and wastewater. In this case the selection of indicators is straightforward, as the indicators should answer the question: Are the treatment results satisfactory? If not, what can we do to improve the results? Chapters 11 and 12 illustrate the use of a wide spectrum of indicators and compare the results of the ecosystem health assessment obtained by the different indicators. The conclusion is that different indicators view the ecosystem health from different aspects, which environmental management should have in mind by the selection of indicators in practice. Chapter 13 illustrates the particular difficulties that environmental management has in an African country and touches on the political complications by selection of ecological indicators. Chapter 14 gives an illustration of the particular marine aspects by environmental management, and Chapter 15 illustrates the role of economics in the selection of indicators and the assessment of ecosystem health in environmental management. Chapter 16 looks into the particular considerations that are needed when landscapes consisting of several ecosystems are the focus. Chapter 17 illustrates the ecosystem health assessment when a completely man-controlled ecosystem is the issue. The last chapter, Chapter 18, is similar to Chapter 10 in the sense that the ecosystem problem presented defines the selection of indicators.

2

Application of Indicators for the Assessment of Ecosystem Health

S. E. Jørgensen, Fu-Liu Xu, João C. Marques, and Fuensanta Salas

CONTENTS

2.1 Criteria for the Selection of Ecological Indicators for EHA

Von Bertalanffy (1952) characterized the evolution of complex systems in terms of four major attributes:

1. Progressive integration (entails the development of integrative linkages between different species of biota and between biota, habitat, and climate)

2. Progressive differentiation (progressive specialization as systems evolve biotic diversity to take advantage of abilities to partition resources more finely, and so forth)

3. Progressive mechanization (covers the growing number of feedbacks and regulation mechanisms)

4. Progressive centralization (probably does not refer to a centralization in the political meaning, as ecosystems are characterized by short and fast feedbacks and decentralized control, but to the more developed cooperation among the organisms [the Gaia effect] and the growing adaptation to all other components in the ecosystem)

Costanza et al. (1992) summarizes the concept definition of ecosystem health as follows: (1) homeostasis, (2) absence of disease, (3) diversity or complexity, (4) stability or resilience, (5) vigor or scope for growth, and (6) balance between system components. He emphasizes that it is necessary to consider all or at least most of the definitions simultaneously. Consequently, he proposes an overall system health index, $HI = V*O*R$, where V is system vigor, O is the system organization index, and R is the resilience index. With this proposal Costanza touches on probably the most crucial ecosystem properties to cover ecosystem health.

Kay and Schneider (1992) uses the term "ecosystem integrity" to refer to the ability of an ecosystem to maintain its organization. Measures of integrity should therefore reflect the two aspects of the organizational state of an ecosystem: functional and structural. Function refers to the overall activities of the ecosystem. Structure refers to the interconnection between the components of the system. Measures of function would indicate the amount of energy being captured by the system. It could be covered by measuring the exergy captured by the system. Measures of structure would indicate the way in which energy is moving through the system. The exergy stored in the ecosystem could be a reasonable indicator of the structure.

When using ecological indicators for ecosystem health assessment from a practical environmental management point of view, five criteria could be proposed:

1. Simple to apply and easily understood by laymen
2. Relevant in the context
3. Scientifically justifiable
4. Quantitative
5. Acceptable in terms of costs

On the other hand, from a more scientific point of view, we may say that the characteristics defining a good ecological indicator are

1. Ease in handling
2. Sensitivity to small variations of environmental stress
3. Independence of reference states
4. Applicability in extensive geographical areas and in the greatest possible number of communities or ecological environments
5. A possible quantification

It is not easy to fulfill all these "two times five" requirements. In fact, despite the panoply of bio-indicators and ecological indicators that can be found in the literature, very often the selected indicators are more or less specific for a given stress or they are applicable only to a particular type of community or ecosystem and/or scale of observation. Rarely has their wider validity in fact been utterly proven. As will be seen throughout this volume, the generality of the applied ecological indicators is limited.

2.2 Classification of Ecosystem Health Indicators

The ecological indicators applied today in different contexts, for different ecosystems, and for different problems can be classified on eight levels from the most reductionistic to the most holistic indicators. Ecological indicators for ecosystem health assessment (EHA) do not include indicators of the climatic conditions, which in this context are considered entirely natural conditions.

Level 1 covers the presence or absence of specific species. The best known application of this type of indicator is the saprobien system (Hynes 1971), which classifies streams in four classes according to their pollution by organic matter causing oxygen depletion: oligosaprobic water (unpolluted or almost unpolluted), beta-mesosaprobic (slightly polluted), alpha-mesosaprobic (polluted), and polysaprobic (very polluted). This classification was originally based on observations of species that were either present or absent. The species that were applied to assess the class of pollution were divided into four groups: organisms characteristic of unpolluted water, species dominating in polluted water, pollution indicators, and indifferent species. Records of fish in European rivers have been used to find by Artificial Neural Network (ANN) a relationship between water quality and presence (and absence) of fish species. The result of this examination has shown that the presence or absence of fish species can be used as strong ecological indicators for the quality of river water.

Level 2 uses the ratio between classes of organisms. A characteristic example is the Nyggard Algae Index.

Level 3 is based on concentrations of chemical compounds. Examples are assessment of the level of eutrophication on the basis of the total phosphorus concentration, assuming that phosphorus is the limiting factor for eutrophication. When the ecosystem is unhealthy due to too-high concentrations of specific toxic substances, the concentration of one or more focal toxic compounds is, of course, a very relevant indicator. PCB contamination of the Great North American Lakes has been applied as an ecological indicator by recording the concentrations of PCB in birds and in water. It is often important to find a concentration in a medium or in organisms where the concentration can be easily determined and has a sufficiently high value that is magnitudes higher than the detection limit, which facilitates a clear indication.

Level 4 applies concentration of entire trophic levels as indicators; for instance, the concentration of phytoplankton (as chlorophyll a or as biomass per m³) is used as an indicator for the eutrophication of lakes. A high fish concentration has also been applied as an indicator for good water quality or birds as an indicator for a healthy forest ecosystem.

Level 5 uses process rates as indication; for instance, primary production determinations are used as an indicator for eutrophication either as maximum gC/m^2 per day or gC/m^3 per day, or gC/m^2 per year or gC/m^3 per year. A high annual growth of trees in a forest is used as an indicator for a healthy forest ecosystem, and a high annual growth of a selected population may be used as an indicator for a healthy environment. On the other hand, a high mortality in a population can be used as an indication of an unhealthy environment. High respiration may indicate that an aquatic ecosystem has a tendency for oxygen depletion.

Level 6 covers composite indicators, for instance, as represented by many of E. P. Odum's attributes (see Table 2.1). Examples are biomass, respiration/

TABLE 2.1

Differences between Initial Stage and Mature Stage

Properties		Early Stages	Late or Mature Stage
A	Energetic		
	P/R	>>1 <<1	Close to 1
	P/B	High	Low
	Yield	High	Low
	Specific entropy	High	Low
	Entropy production per unit of time	Low	High
	Exergy	Low	High
	Information	Low	High
B	Structure		
	Total biomass	Small	Large
	Inorganic nutrients	Extrabiotic	Intrabiotic
	Diversity, ecological	Low	High
	Diversity, biological	Low	High
	Patterns	Poorly organized	Well organized
	Niche specialization	Broad	Narrow
	Size of organisms	Small	Large
	Life cycles	Simple	Complex
	Mineral cycles	Open	Closed
	Nutrient exchange rate	Rapid	Slow
	Life span	Short	Long
C	Selection and homeostatis		
	Internal symbiosis	Undeveloped	Developed
	Stability (resistance to external perturbations)	Poor	Good
	Ecological buffer capacity	Low	High
	Feedback control	Poor	Good
	Growth form	Rapid growth	Feedback controlled
	Growth types	r-strategists	k-strategists

Note: A few attributes are added to those published by E. P. Odum (1969, 1971).

biomass, respiration/production, production/biomass, and ratio primary producer/consumers. E. P. Odum uses these composite indicators to assess whether an ecosystem is at an early stage of development or is a mature ecosystem. It is presumed that a mature ecosystem has more resistance toward changes due to impacts.

Level 7 encompasses holistic indicators such as resistance; resilience; buffer capacity; biodiversity; all forms of diversity; size and connectivity of the ecological network; turnover rate of carbon, nitrogen, etc., and of energy. As will be discussed in the next section, high resistance, high resilience, high buffer capacity, high diversity, big ecological network with a medium connectivity, and normal turnover rates are all indications of a healthy ecosystem.

Level 8 indicators are thermodynamic variables, which we may call superholistic indicators as they try to see the forest through the trees and capture the total image of the ecosystem without inclusion of details. Such indicators are exergy, emergy, exergy destruction, entropy production, power, mass, and/or energy system retention time. The economic indicator cost/benefit (which includes all ecological benefits—not only the economic benefits of the society) also belongs to this level.

Section 2.4 gives an overview of the application of the eight levels discussed in Chapters 3–15.

2.3 Indices Based on Indicator Species

When talking about indicator species, we must distinguish two cases, as using either indicator species or bioaccumulative species (the latter being more appropriate in toxicological studies) leads to confusion.

The first case refers to those species whose appearance and dominance are associated with an environmental deterioration, as being favored for such fact, or for its tolerance of that type of pollution in comparison to other less resistant species. In a sense, the possibility of assigning a certain grade of pollution to an area in terms of the present species has been pointed out by a number of researchers, such as Bellan (1967) or Glemarec and Hily (1981), mainly in organic pollution studies.

Following the same policy, some authors have focused on the presence/absence of such species to formulate biological indices, as detailed below.

Indices such as the Bellan (based on polychaetes) or the Bellan-Santini (based on amphipods) attempt to characterize environmental conditions by analyzing the dominance of species indicating some type of pollution in relation to the species considered to indicate an optimal environmental situation (Bellan 1980; Bellan-Santini 1980). Several authors consider the use of these indicators inadvisable because often such indicator species may occur naturally in relatively high densities. The point is that there is no reliable methodology to know at which level one of these indicator species can be well represented in a community that is not really affected by any kind of pollution, which leads to a significant exercise of subjectivity (Warwick 1993). Despite these criticisms, even recently, the AMBI index (Borja et al. 2000), based on the Glemarec and Hily (1981) species classification regarding pollution, as well as the one proposed by Simboura and Zenetos (2002), both of which apply the same principles, have gone back to update such pollution detecting tools. Moreover, Roberts et al. (1998) also proposed an index based on macrofauna species that accounts for the ratio of each species abundance in control samples versus samples procured from stressed areas. It is, however, semi-quantitative as well as site- and pollution-type specific. In the same way, the Benthic Response Index (Smith et al. 2001) is based on the type (pollution tolerance) of species in a sample, but its applicability is complex as it is calculated using a two-step process in which ordination analysis is employed to quantify a pollution gradient within a calibration data set.

The AMBI index, for instance, which accounts for the presence of species indicating a type of pollution and of species indicating a not polluted situation, has been considered useful in terms of the application of the European Water Framework Directive in coastal ecosystems and estuaries. In fact, although this index is very much based on the paradigm of Pearson and Rosenberg (1978), which emphasizes the influence of organic matter enrichment on benthic communities, it was shown to be useful for the assessment of other anthropogenic impacts, such as physical alterations in the habitat, heavy metal inputs, etc. And what's more, it has been successfully applied in Atlantic (North Sea, Bay of Biscay, and south of Spain) and Mediterranean (Spain and Greece) European coasts (Borja et al. 2003).

Regarding submarine vegetation, there is a series of genera that universally appear when pollution situations occur. Among them are the green algae *Ulva, Enteromorpha, Cladophora,* and *Chaetomorpha,* and the red algae *Gracilaria, Porphyra,* and *Corallina.*

High structural complexity species, such as Phaeophyta belonging to *Fucus* and *Laminaria* orders, are seen worldwide as the most sensitive to any kind of pollution, with the exception of the species of the *Fucus* genus that cope with moderate pollution (Niell and Pazo 1978). On the other hand, marine Spermatophytae are considered indicator species of good quality in the water.

In the Mediterranean Sea, for instance, the presence of Phaeophyta *Cystoseira* and *Sargassum* or meadows of *Posidonia oceanica* indicate good quality in the water. Monitoring population density and distribution of such species allows detecting and evaluating the impact of any possible activity (Pérez-Ruzafa and Marcos 2003). *Posidonia oceanica* is possibly the most widely used indicator of water quality in the Mediterranean Sea (Pergent et al. 1995, 1999) and conservation index (Moreno et al. 2001), based on the named marine Spermatophyta, is used in such littoral.

Descriptions of the above-mentioned indices follow.

2.3.1 Pollution Index (Bellan 1980)

$$IP = \sum \frac{Dominance\ of\ pollution\ indicator\ species}{Dominance\ of\ pollution\ clean\ water\ indicators}$$

Species considered as pollution indicators by Bellan are *Platenereis dumerilli, Theosthema oerstedi, Cirratulus cirratus,* and *Dodecaria concharum.*

Species considered as clear water indicators by Bellan are *Syllis gracillis, Typosyllis prolifera, Typosyllis* spp., and *Amphiglena mediterranea.*

Index values over 1 show that the community is pollution disturbed. As organic pollution increases, the value of the index goes higher; that is why, in theory, different pollution grades can be established, although the author does not fix them.

This index, in principle, was designed for application on rocky superficial substrates. Nevertheless, Ros et al. (1990) modified it in terms of the used indicator species in order to be applicable in soft bottoms. In this case, the pollution indicator species are *Capitella capitata, Malococerus fuliginosus,* and *Prionospio malmgremi,* and the clear water indicator species is *Chone duneri.*

2.3.2 Pollution Index (Bellan-Santini 1980)

This index follows the same formulation and interpretation as Bellan's, but is based on the amphipods group.

$$IP = \sum \frac{Dominance\ of\ pollution\ indicator\ species}{Dominance\ of\ pollution\ clean\ water\ indicators}$$

Pollution indicator species: *Caprella acutrifans* and *Podocerus variegatus*

Clear waters indicator species: *Hyale* sp., *Elasmus pocllamunus,* and *Caprella liparotensis*

2.3.3 AMBI (Borja et al. 2000)

For the development of the AMBI the soft bottom macrofauna is divided into five groups according to their sensitivity to an increasing stress:

I. Species very sensitive to organic enrichment and present under unpolluted conditions

II. Species indifferent to enrichment, always in low densities with non-significant variations with time

III. Species tolerant to excess of organic matter enrichment. These species may occur under normal conditions, but their populations are stimulated by organic enrichment.

IV. Second-order opportunist species, mainly small-sized polychaetes

V. First-order opportunist species, essentially deposit-feeders

The formula is as follows:

$$\text{AMBI} = \frac{\{(0 \times \%GI) + (1,5 \times \%GII) + (3 \times \%GIII) + (4,5 \times \%GIV) + (6 \times \%GV)\}}{100}$$

Classification	AMBI
Normal	0–1.2
Slightly polluted	1.2–3.2
Moderately polluted	3.2–5
Highly polluted	5–6
Very highly polluted	6–7

For the application of this index, nearly 2,000 taxa have been classified, which are representative of the most important soft bottom communities present in European estuarine and coastal systems. The Marine Biotic Index can be applied using the AMBI© software (Borja et al. 2003, and www.azti.es, where the software is freely available).

2.3.4 BENTIX (Simboura and Zenetos 2002)

This index is based on the AMBI index but lies in the reduction of the ecological groups involved in the formulae in order to avoid errors in the grouping of the species, and reduce effort in calculating the index.

$$\text{BENTIX} = \frac{\{(6 \times \%GI) + 2 \times (\%GII + \%GIII)\}}{100}$$

Group I: This group includes species sensitive to disturbance in general.

Group II: This group includes species tolerant to disturbance or stress whose populations may respond to enrichment or other sources of pollution.

Group III: This group includes the first order opportunistic species (pronounced unbalanced situation), pioneers, colonizers, or species tolerant to hypoxia.

A complete list of indicator species in the Mediterranean Sea was made and scores were assigned ranging from 1–3 corresponding to each of the three ecological groups.

Classification	BENTIX
Normal	4.5–6.0
Slightly polluted	3.5–4.5
Moderately polluted	2.5–3.5
Highly polluted	2.0–2.5
Very highly polluted	0

2.3.5 Macrofauna Monitoring Index (Roberts et al. 1998)

The authors developed an index for biological monitoring of dredge spoil disposal. Each of 12 indicator species is assigned a score, based primarily on the ratio of its abundance in control versus impacted samples. The index value is the average score of those indicator species present in the sample.

Index values of <2, 2–6, and >6 are indicative of severe, patchy, and no impact, respectively.

The index is site- and impact-specific, but the process of developing efficient monitoring tools from an initial impact study should be widely applicable (Roberts et al. 1998).

2.3.6 Benthic Response Index (Smith et al. 2001)

The Benthic Response Index (BRI) is the abundance weighted average pollution tolerance of species occurring in a sample, and is similar to the weighted average approach used in gradient analysis (Goff and Cottam 1967; Gauch 1982). The index formula is:

$$I_s = \frac{\sum_{i=1}^{n} p_i \sqrt[3]{a_{si}}}{\sum_{i=1}^{n} \sqrt[3]{a_{si}}}$$

where I_s is the index value for sample s, n is the number of species for sample s, p_i is the position for species i on the pollution gradient (pollution tolerance score), and a_{si} is the abundance of species i in sample s.

According to the authors, determining the pollutant score (p_i) for the species involves four steps: (1) assembling a calibration infaunal data set, (2) conducting an ordination analysis to place each sample in the calibration set on a pollution gradient, (3) computing the average position of each species along the gradient, and (4) standardizing and scaling the position to achieve comparability across depth zones.

The average position of species I (p_i) on the pollution gradient defined in the ordination is computed as:

$$P_i = \frac{\sum_{j=1}^{t} g_j}{t}$$

where t is the number of samples to be used in the sum, with only the highest t species abundance values included in the sum. The g_j is the position on the pollution gradient in the ordination space for sample j.

This index has only been applied for assessing benthic infaunal communities on the Maryland Shelf of Southern California using a 717-sample calibration data set.

2.3.7 Conservation Index (Moreno et al. 2001)

$$CI = \frac{L}{L+D}$$

where L is the meadow of living *Posidonia oceanica* and D is the dead meadow coverage.

The authors applied the index near chemical industries. The results led them to establish four grades of *Posidonia* meadow conservation, which allow identification of increasing impact zones, as changes in the industry's activity can be detected by the conservation status in a certain location.

There are also species classified as bioaccumulative, defined as those capable of resisting and accumulating diverse pollutant substances in their tissues, facilitating, then, their detection when they are found in very low levels in the environment, making them difficult to detect through analytical techniques (Philips 1977).

The disadvantage of using accumulator indicator species in the detection of pollutants arises from the fact that a number of biotic and abiotic variables may affect the rate at which the pollutant is accumulated, and therefore both laboratory and field tests need to be undertaken so that the effects of extraneous parameters can be identified.

The mollusks group, precisely the bivalve class, has been one of the most used to determine the existence and quantity of a toxic substance.

Individuals of the genera *Mytilus* (De Wolf 1975; Goldberg et al. 1978; Dabbas et al. 1984; Cossa and Rondeau 1985; Miller 1986; Renberg et al. 1986; Carrell et al. 1987; Lauenstein et al. 1990; Viarengo and Canesi 1991; Regoli and Orlando 1993), *Cerastoderma* (Riisgard et al. 1985; Mohlenberg and Riisgard 1988; Brock 1992), *Ostrea* (Lauenstein et al. 1990; Mo and Neilson 1991), and *Donax* (Marina and Enzo 1983; Romeo and Gnassia-Barelli 1988) have been considered ideal in many works when detecting the concentration of a toxic substance in the environment, due to their sessile nature, wide geographical distribution, and capability to detoxify when pollution ceases. In that sense, Goldberg et al. (1978) introduced the concept of "Mussel Watch" when referring to the use of the mollusks group in the detection of polluting substances, due to their wide geographical distribution and their capability of accumulating those substances in their tissues. In 1980, the National Oceanic and Atmospheric Agency (NOAA) in the United States developed the "Mussel Watch Program," focused on pollution control along the North American coasts. There are also programs, similar to the North American one, in Canada (Cossa et al. 1983; Picard-Berube and Cossa 1983), the Mediterranean Sea (Leonzio et al. 1981), the North Sea (Golovenko et al. 1981), and in the Australian coasts (Cooper et al. 1982; Ritz et al. 1982; Richardson and Waid 1983).

Likewise, certain species of the amphipods group are considered capable of accumulating toxic substances (Albrecht et al. 1981; Reish 1993), as well as species of the polychaetes group, like *Nereis diversicolor* (Langston et al. 1987; McElroy 1988), *Neanthes arenaceodentata* (Reish and Gerlinger 1984), *Glycera alba*, *Tharix marioni* (Gibbs et al. 1983), or *Nephtys hombergi* (Bryan and Gibbs 1987).

Some fish species have also been used in various works focused on the effects of toxic pollution of the marine environment, due to their bioaccumulative capability (Eadie et al. 1982; Gosset et al. 1983; Varanasi et al. 1989) and the existing relationship among pathologies suffered by any benthic fishes and the presence of polluting substances (Malins et al. 1984; Couch and Harshbarger 1985; Myers et al. 1987).

Other authors such as Levine (1984), Maeda and Sakaguchi (1990), Newmann et al. (1991), and Storelli and Marcotrigiano (2001) have looked into algae as optimal detectors of heavy metals, pesticides, and radionuclides, *Fucus*, *Ascophyllum*, and *Enteromorpha* being the most utilized.

For reasons of comparison, the concentrations of substances in organisms must be translated to uniform and comparable units. This is done through the Ecologic Reference Index (ERI), which represents a potential for environmental effects. This index has only been applied using blue mussels.

$$\text{ERI} = \frac{measured\ concentration}{\text{BCR}}$$

where BCR is the value of the background/reference concentration (Table 2.2).

Few indices like the latter, based on the use of bioaccumulative species, have been formulated. More common is the simple measurement of the effects (e.g., % incidence, % mortality) of a certain pollutant on those species, or the use of biomarkers by scientists to evaluate the specificity of the responses to natural or anthropogenic changes, but it is very difficult for the environmental manager to interpret increasing or decreasing changes in biomarker data.

The Working Group on Biological Effects of Contaminants (WGBEC) in 2002 recommended different techniques for biological monitoring programs (Table 2.3).

TABLE 2.2

Upper Limit of BCR for Hazardous Substances in Blue Mussel According to OSPAR/MON (1998)

Substance	Upper Limit of BCR Value (ng/g dry weight)
Cadmium	550
Mercury	50
Lead	959
Zinc	150000

TABLE 2.3

Review of Different Techniques for Biological Monitoring

Method	Organism	Issues Addressed	Biological Significance	Threshold Value
Bulky DNA adduct formation	Fish	PAHs, other synthetic organics	Measures genotoxic effects. Sensitive indicator of past and present exposure	2 × reference site or 20% change
AChE	Fish	Organophosphates and carbonates or similar molecules	Measures exposures	Minus 2.5 × reference site
	Bivalve mollusks			
Metallothionein induction	Fish	Measures induction of metallothionein protein by certain metals	Measures exposure and disturbance of copper and zinc metabolism.	2.0 × reference site
	Mytilus spp.			
EROD or P4501A induction	Fish	Measures induction of enzymes with metabolized planar organic contaminants		2.5 × reference site
ALA-D inhibition	Fish	Lead	Index of exposure	2.0 × reference site
PAH bile metabolites	Fish	PAHs	Measures exposure to and metabolism PAHs	2.0 × reference site

(continued)

TABLE 2.3

Review of Different Techniques for Biological Monitoring (continued)

Method	Organism	Issues Addressed	Biological Significance	Threshold Value
Lysososmal stability	Fish	Not contaminant specific but responds to a wide variety of xenobiotics, contaminants, and metals	Provides a link between exposure and pathological end points	2.5 × reference site
Lysosomal neutral red retention	*Mytilus* spp.	Not contaminant specific but responds to a wide variety of xenobiotics, contaminants, and metals	Provides a link between exposure and pathological end points	2.5 × reference site
Early toxicopathic lesions, preneoplastic and neoplastic liver histopathology	Fish	PAHs	Measures pathological changes associated with exposure to genotoxic and nongenotoxic carcinogens	2.0 × reference site or 20% change
Scope for growth	Bivalve mollusks	Responds to a wide variety of contaminants	Integrative response that is a sensitive and sublethal measure of energy available for growth	
Shell thickening	*Crassostea gigas*	Specific to organotins	Disruption to pattern of shell growth	
Vitellogenin induction	Male and juvenile fish	Estrogenic substances	Measures feminization of male fish and reproductive impairment	
Imposex	Neogastropod mollusks	Specific to organotins	Reproductive interference	2.0 × reference site or 20% change
Intersex	*Littorina littorina*	Specific to reproductive effects of organotins	Reproductive interference in coastal waters	2.0 × reference site or 20% change
Reproductive success in fish	*Zoarces viviparus*	Not contaminant specific; will respond to a wide of environmental contaminants	Measures reproductive output and survival of eggs and fry in relation to contaminants	

2.4 Indices Based on Ecological Strategies

Some indices try to assess environmental stress effects accounting for the ecological strategies followed by different organisms. That is the case of trophic indices such as the infaunal index proposed by Word (1979), which are based on the different feeding strategies of the organisms. Another example is the nematodes/copepods index (Rafaelli and Mason 1981), which accounts for the different behavior of two taxonomic groups under environmental stress situations, but several authors have rejected them due to their dependence on parameters like depth and sediment particle size, as well as because of their unpredictable pattern of variation depending on the type of pollution (Gee et al. 1985; Lambshead and Platt 1985). More recently, other proposals appeared, such as the polychaetes/amphipods ratio index (Gomez-Gesteira and Dauvin 2000), or the index of r/K strategies, which considers all benthic taxa, although it emphasizes the difficulty of exactly scoring each species through the biological trait analysis.

The R/P index of Feldman, based on marine vegetation, is often used in the Mediterranean Sea. It was established as a biogeographical index, and it is based on the fact that *Rodophyceae* spp. number decreases from the tropics to the poles. Its application as an indicator rests on the higher or lower sensitivity of Phaeophyceae and Rhodophyceae to disturbances.

2.4.1 Nematodes/Copepods Index (Rafaelli and Mason 1981)

This index is based on the ratio between nematodes and copepods abundances.

$$I = \frac{nematodes\ abundance}{copepods\ abundance}$$

Values of such ratio can increase or decrease according to high or low organic pollution. This happens by means of a different response of those groups to the input of organic matter to the system. Values over 100 show high organic pollution.

According to the authors, the index application should be limited to certain intertidal zones. In infralittoral areas, at certain depth, despite the absence of pollution, the values obtained were very high. The explanation for this is the absence of copepods in such depths, maybe due to a change in the optimal interstitial habitat for that taxonomic group (see Rafaelli and Mason 1981).

2.4.2 Polychaetes/Amphipods Index

This index is similar to the nematodes/copepods, but now it is applied to the macrofauna level using the polychaetes and amphipods groups. The index was formerly designed to measure the effects of crude pollution.

$$I = Log_{10}\left(\frac{Polychaetes\ abundance}{Amphipods\ abundance} + 1\right)$$

$I \leq 1$: nonpolluted

$I > 1$: polluted

2.4.3 Infaunal Index (Word 1980)

The macrozoobenthos species can be divided into: (1) suspension feeders; (2) interface feeders; (3) surface deposit feeders; and (4) subsurface deposit feeders.

Based on this division, the trophic structure of macrozoobenthos can be determined using the formula:

$$ITI = 100 - 100/3 \times (0n_1 + 1n_2 + 2n_3 + 3n_4)/(n_1 + n_2 + n_3 + n_4)$$

in which n_1, n_2, n_3, and n_4 are the number of individuals sampled in each of the above-mentioned groups.

ITI values near 100 mean that suspension feeders are dominant and that the environment is not disturbed.

Near a value of 0 subsurface, feeders are dominant, meaning that the environment is probably strongly disturbed due to human activities.

One of the disadvantages to attribute to a trophic index is the determination of the diet of the organisms, which can be developed through the study of the stomach contents or in laboratory experiments. Generally, the real diet, that is, the one studied observing the stomach contents is difficult to establish and can vary from one population to another among the same taxonomic entity. *Nereis virens*, for instance, considered as an omnivore species found along the European coast, behaves as an herbivore on the North American coasts (Fauchald and Jumars 1979).

Another aspect to consider at the time of determining the trophic category of many polychaetes species is their alternative feeding behavior that can appear under certain circumstances. Buhr (1976) determined through laboratory experiments that the terebellid *Lanice conchylega*, considered as a detritivore, changes into a filterer when a certain concentration of phytoplankton occurs in the water column. Taghon et al. (1980) observed that some species of the Spionidae family, usually taken for detritivores, could change into filterers, modifying the palps into a characteristic helicoidal shape.

On the other hand, some species of the Sabellidae and Owenidae families can shift from filterers to detritivores. And we can consider some limnivore and detritivore species changed into carnivores when they consume the remains of other animals (Dauer et al. 1981).

Those facts now lead to doubts about the existence of a clear separation among the diverse feeding strategies. That is why other characteristics such as the grade of an individual's mobility and the morphology of the mouth apparatus intervene in the definition of the trophic category of polychaetes (Gambi and Giangrande 1985). The different combinations of that set of characteristics are what Fauchald and Jumars, in 1979, named "feeding guilds."

Some authors such as Maurer et al. (1981), Dauer (1984), and Pires and Múniz (1999) have tried using the classification of the different polychaetes species in feeding guilds when studying the structure of the benthic system and when identifying the different impacts, both with good results.

The problem of using such a classification is, no doubt, the difficulty that carries the determination of each one of those combinations for each species, as according to a study by Dauer (1984), many families hold more than one combination depending on the type of feeding they follow, their grade of mobility, and the morphology of their mouth apparatus, therefore every combination being monospecific. Which leads us to think that such a classification very often is not practical.

2.4.4 Feldman Index

$$I = \frac{N^{\underline{o}} \text{ species of } Rhodophyceae}{N^{\underline{o}} \text{ species of } Phaeophyceae}$$

Cormaci and Furnari (1991) detected values over eight in polluted areas in southern Italy, when normal values in a balanced community oscillate between 2.5 and 4.5. Verlaque (1977) studied the effects of a thermal power station, and also found higher values than the index, but considers those due to the presence of communities of warmth affinity.

However, Belsher and Boudouresque (1976) analyzed vegetation in small harbors and found out that when quality importance of Phaeophyceae increases, the index decreases. Therefore, knowledge of the index behavior does not seem to be enough to consider it, by itself, a pollution symptom.

2.5 Indices Based on the Diversity Value

Diversity is the other most used concept focusing on the fact that the relationship between diversity and disturbances can be seen as a decrease in the first one as stress increases.

Magurran (1989) divides the diversity measurements into three main categories:

- Indices that measure the enrichment of the species, such as Margalef, which are, in essence, a measurement of the number of species in a defined sampling unit.

- Models of the abundance of species, such as the K-dominance curves (Lambshead and Platt 1985) or the log normal model (Gray 1979), which describe the distribution of their abundance, going from those that represent situations in which there is a high uniformity to those that characterize cases in which the abundance of the species is very unequal. However, the log normal model deviation was rejected once by several authors due to the impossibility of finding any benthic marine sample that clearly responded to such a log normal distribution model (Shaw et al. 1983; Hughes 1984; Lambshead and Platt 1985)

- Indices based on the proportional abundance of species that aim at integrating enrichment and uniformity in a simple expression. Such indices can also be divided into those based on statistics, information theory, and dominance indices. Indices derived from the information theory, such as the Shannon–Wiener, are based on something logical: diversity, or information, in a natural system can be measured in a similar way as information contained in a code or message. On the other hand, dominance indices such as the Simpson or Berger–Parker are referred to as measurements that ponder the abundance of the most common species, instead of the enrichment of the species.

Meanwhile, average taxonomic diversity and distinctness measures have been used in some researches (e.g., Warwick and Clarke 1995, 1998; Clarke and Warwick 1999) to evaluate biodiversity in the marine environment, as it takes into account taxonomical, numerical, ecological, genetic, and phylogenetic aspects of diversity. These measures address some of the problems identified with species richness and the other diversity indices (Warwick and Clarke 1995).

2.5.1 Shannon–Wiener Index (Shannon and Wiener 1963)

This index is based on the information theory. It assumes that individuals are sampled at random, out of an "indefinitely large" community, and that all the species are represented in the sample.

The index takes this shape:

$$H' = -\sum p_i \log_2 p_i$$

where p_i is the proportion of individuals found in the species i. In the sample, the real value of p_i is unknown, but it is estimated through the ratio N_i/N, for N_i = number of individuals of the species i and N = total number of individuals.

The units for the index depend on the log used. So, for $\log_2$, the unit is bits/individual; "natural bels" and "nat" for $\log_e$; and "decimal digits" and "decits" for $\log_{10}$.

The index can take values between 0 and 5. Maximal values are rarely over five bits/individual. Diversity is a logarithmic measurement that has, to a certain extent, asyntonic character, which makes the index a little sensitive in the range of values next to the upper limit (Margalef 1978).

As an ordinary basis, in the literature, index low values are considered an indication of pollution (Stirn et al. 1971; Anger 1975; Hong 1983; Zabala et al. 1983; Encalada and Millan 1990; Calderón-Aguilera 1992; Pocklington et al. 1994; Engle et al. 1994; Mendez-Ubach et al. 1997; Yokoyama 1997).

But one of the problems arising with its use is the lack of objectivity when establishing in a precise manner from what value it should start detecting the effects of such pollution.

Molvær et al. (1997) established the following relation between the indices and the different ecological levels according to what is recommended by the Water Framework Directive:

High status: >4 bits/indv

Good status: 4–3 bits/indv

Moderate status: 3–2 bits/indv

Poor status: 2–1 bits/indv

Bad status: 1–0 bits/indv

Detractors of the Shannon index base their criticisms on its lack of sensitivity when it comes to detecting the initial stages of pollution (Leppakoski 1975; Pearson and Rosenberg 1978; Rygg 1985).

Gray (1979), in a study on the effects of a cellulose paste factory's waste, set out the uselessness of this index as it responds to such obvious changes that there is no need of a tool to detect them.

Ros and Cardell (1991), in their study about the effects of great industrial and human domestic pollution, consider the index as a partial approach to the knowledge of pollution effects on marine benthic communities and, without any explanation for that statement, set out a new structural index proposal, for which lack of applicability has already been shown (Salas 1996).

2.5.2 Pielou Evenness Index

$$J' = H'/H'_{max} = H'/log\ S$$

where H'_{max} is the maximum possible value of Shannon diversity.
The index oscillates from 0 to 1.

2.5.3 Margalef Index

The Margalef index quantifies the diversity by relating specific richness to the total number of individuals.

$$D = (S-1)/log_2 N$$

where S is number of species and N is total number of individuals.
The author did not establish reference values.
The main problem that arises when applying this index is the absence of a limit value; therefore, it is difficult to establish reference values. Ros and Cardell (1991) consider values below four as typical of polluted. Bellan-Santini (1980), on the contrary, established that limit when the index takes values below 2.05.

2.5.4 Berger–Parker Index

This index expresses the proportional importance of the most abundant species, and takes this shape:

$$D = n_{max}/N$$

where n_{max} is the number of individuals of the one most abundant species and N is the total number of individuals. The index oscillates from 0 to 1 and, contrary to the other diversity indices, the high values show a low diversity.

2.5.5 Simpson Index (Simpson 1949)

This index was defined on the probability that any two individuals randomly extracted from an infinitely large community could belong to the same species:

$$D = \Sigma\, p_i^2$$

where p_i is the individuals proportion of the species i. To calculate the index for a finite community use:

$$D = \Sigma\, [n_i(n_i-1)/N(N-1)]$$

where n_i is the number of individuals in the species i, and N is the total number of individuals.

Like the Berger–Parker index, this one oscillates from 0 to 1, it has no dimensions and, similarly, the high values imply a low diversity.

2.5.6 Deviation from the Log Normal Distribution (Gray and Mirza 1979)

This method, proposed by Gray and Mirza in 1979, is based on the assumption that when a sample is taken from a community, the distribution of the individuals tends to follow a log-normal model.

The adjustment to a logarithmic normal distribution assumes that the population is ruled by a certain number of factors and it constitutes a community in a steady equilibrium; meanwhile, the deviation from such distribution implies that any perturbation is affecting it.

2.5.7 K-Dominance Curves (Lambshead et al. 1983)

The k-dominance curve is the representation of the accumulated percentage of abundance versus the logarithm of the sequence of species ordered in a decreasing order. The slope of the straight line obtained allows the valuation of the pollution grade. The higher the slope is, the higher the diversity is as well.

2.5.8 Average Taxonomic Diversity (Warwick and Clarke 1995)

This measure, equal to taxonomic distinctness, is based on the species abundances (denoted by x_i, the number of individuals of species i in the sample) and on the taxonomic distance (ω_{ij}) through the classification tree, between every pair of individuals (the first from species i and the second from species j).

It is the average taxonomic distance apart of every pair of individuals in the sample, or the expected path length between any two individuals chosen at random.

$$\Delta = [\Sigma\Sigma_{\,i<j}\,\omega_{ij} x_i x_j]/[N(N-1)/2]$$

where the double summation is over all pairs of species i and j ($i,j = 1.2,\ldots,S; i<j$), and $N = \Sigma_i x_i$, the total number of individuals in the sample.

2.5.9 Average Taxonomic Distinctness (Warwick and Clarke 1995)

To remove the dominating effect of the species abundance distribution, Warwick and Clarke (1995) proposed to divide the average taxonomic diversity index by the Simpson index (Simpson 1949), giving the average taxonomic distinctness index.

$$\Delta^* = [\Sigma\Sigma_{\,i<j}\,\omega_{ij}x_ix_j]/[\Sigma\Sigma_{\,i<j}\,x_ix_j]$$

When quantitative data are not available and the sample consists simply of a species list (presence/absence data) the average taxonomic distinctness takes the following form:

$$\Delta^+ = [\Sigma\Sigma_{\,i<j}\,\omega_{ij}]/[S(S-1)/2]$$

where S, as usual, is the observed number of species in the sample and the double summation ranges over all pairs i and j of the species ($i < j$).

Taxonomic distinctness is reduced in respect to increasing environmental stress and this response of the community lies at the base of this index concept. Nevertheless, it is most often very complicated to meet certain requirements to apply it, like having a complete list of the species present in the study area corresponding to pristine situations. Moreover, some works have shown that in fact taxonomic distinctness is not more sensitive than other diversity indices usually applied when detecting disturbances (Sommerfield and Clarke 1997), and consequently this measure has not been widely used on marine environment quality assessment and management studies.

2.6 Indicators Based on Species Biomass and Abundance

Other approaches account for the variation of organisms' biomass as a measure of environmental disturbances. Along these lines, we have methods such as SAB (Pearson and Rosenberg 1978), consisting of a comparison between the curves resulting from ranking the species as a function of their representativeness in terms of both their abundance and biomass. The use of this method is not advisable because it is purely graphical, which leads to a high degree of subjectivity that impedes relating it quantitatively with the different environmental factors. The ABC method (Warwick 1986) also involves the comparison between the cumulative curves of species biomass and abundance, from which Warwick and Clarke (1998) derived the W-statistic index.

2.6.1 ABC Method (Warwick 1986)

This method is based on the thesis that the distribution of number of individuals for the different species in the macrobenthos communities behaves in a different way than the biomass distribution.

It is adapted from the k-dominance curve already mentioned, showing in one graphic the k-dominance and biomass curves. The graphics are made up comparing the interval of species (in the abscise axis), decreasingly arranged and in logarithmical scale, to the accumulated dominance (in the ordinate axis).

- According to the range of disturbance, three different situations can be given: In a system with no disturbances, a relatively low number of individuals contribute with the major part of the biomass, and at the same time, the distribution of the individuals among the different species is more balanced. The representations would show the biomass curve above the dominance one, indicating higher numeric diversity than biomass.

- Under moderate disturbances, there is a decrease in the dominance with regard to biomass; however, abundances increase. The graphic shows both curves intersected.

- In the case of intense disturbances, the situation is totally the opposite, and only a few species monopolize the greater part of the individuals, which are of a small size; that is why the biomass is low and it is more equivalently shared. It can be seen in the representation how the curve of the number of individuals is placed above the biomass curve, indicating a higher diversity in the biomass distribution.

Some authors like Beukema (1988), Clarke (1990), McManus and Pauly (1990), and Meire and Dereu (1990) have tried to lead this method into a measurable index, the Clarke (1990) approach being the most commonly accepted one.

$$W = \sum_{i=1}^{s} (B_i - A_i) \Big/ 50(S-1)$$

where B_i is the biomass of species i, A_i the abundance of species i, and S is the number of species.

The index can take values from +1, indicating a nondisturbed system (high status) to –1, which defines a polluted situation (bad status). Values close to 0 indicate moderate pollution (moderate status).

The method is specific to organic pollution and it has been applied, with satisfactory results, to soft bottom tropical communities (Anderlini and Wear 1992; Agard et al. 1993), to experiments (Gray et al. 1988), to fish factoring disturbed areas (Ritz et al. 1989), and on coastal lagoons (Reizopoulou et al. 1996; Salas 2002).

However, Ibanez and Dauvin (1988), Beukema (1988), Weston (1990), Craeymeersch (1991), and Salas et al. (2004) obtained confusing results after applying this technique to estuarine zones, induced by the appearance of dominant species in normal conditions, favored by different environmental factors.

Despite being a method designed for application to benthic macrofauna, Abou-Aisha et al. (1995) used it to detect the impact of phosphorus waste in macroalgae, in three areas of the Red Sea. In spite of that, the problem when applying it to marine vegetation lies in the difficulty of counting the number of individuals in the vegetal species.

2.7 Indicators Integrating All Environment Information

From a more holistic point of view, some authors proposed indices that at least tried to integrate all the environmental information. A first approach for application in coastal areas was developed by Satsmadjis (1982), relating sediment particle size to benthic organism diversity. Vollenweider et al. (1998) developed a trophic index (TRIX) integrating chlorophyll a, oxygen saturation, total nitrogen, and phosphorus to characterize the trophic state of coastal waters

In a progressively more complex way, other indices such as the Index of Biotic Integrity (IBI) for coastal systems (Nelson 1990), the Benthic Index of Environmental Condition (Engle et al. 1994), or the Chesapeake Bay B-BI Index (Weisberg et al. 1997) included physicochemical factors, diversity measures, specific richness, taxonomical composition, and the trophic structure of the system.

Similarly, a set of specific indices of fish communities has been developed to measure the ecological status of estuarine areas. The Estuarine Biological Health Index (BHI) combines two separate measures (health and importance) into a single index. The Estuarine Fish Health Index (FHI) (Cooper et al. 1993) is based on both qualitative and quantitative comparisons with a reference fish community. The Estuarine Biotic Integrity Index (EBI) (Deegan et al. 1993) reflects the relationship between anthropogenic alterations in the ecosystem and the status of higher trophic levels, and the Estuarine Fish Importance Rating (FIR) is based on a scoring system of seven criteria that reflect the potential importance of estuaries to the associated fish species. This index is able to provide a ranking based on the importance of each estuary, and helps to identify the systems of major importance for fish conservation.

Nevertheless, these indicators are rarely used in a generalized way because they were usually developed for application in a particular system or area, which makes them dependent on the type of habitat and seasonality. On the other hand, they are difficult to apply as they need a large amount of data of different natures.

2.7.1 Trophic Index (Vollenweider et al. 1998)

$$TRIX = \frac{k}{n} \times \sum (M_i - L_i) / (U_i - L_i)$$

in which K = 10 (scaling the result between 0 and 10), $n = 4$ (number of variables are integrated), M_i = measured value of variable i, U_i = upper limit of variable i, L_i = lower limit of value i.

The resulting TRIX values are dependent on the upper and the lower limit chosen and indicate how close the current state is to the natural state. However, comparing TRIX values of different areas becomes more difficult. When a wide, more general range is used for the limits, TRIX values for different areas are more easily compared with each other.

2.7.2 Coefficient of Pollution (Satsmadjis 1982)

Calculation of the index is based on several integrated equations. These equations are:

$$S' = s + t/(5 + 0.2s)$$
$$i_0 = (-0.0187s'2 + 2.63s' - 4)(2.20 - 0.0166h)$$
$$g' = i/(0.0124i + 1.63)$$
$$P = g'/[g(i/i_0)^{1/2}]$$

where

P = coefficient of pollution

S' = sand equivalent
s = percent sand
t = percent silt
i_0 = theoretical number of individuals
i = number of individuals
h = station depth

g' = theoretical number of species
g = number of species

2.7.3 Benthic Index of Environmental Condition (Engle et al. 1994)

Benthic index = (2.3841 × Proportion of expected diversity) +
(–1.6728 × Proportion of total abundance as tubifids) + (0.6683 ×
Proportion of total abundance as bivalves)

The expected diversity is calculated throughout Shannon–Wiener Index adjusted for salinity.

Expected Diversity = 0.75411 + (0.00078 × salinity) +
(0.00157 × salinity²) + (–0.00030 × salinity³)

This index was developed for estuarine macrobenthos in the Gulf of Mexico in order to discriminate between areas with degraded environmental conditions and areas with nondegraded or reference conditions.

The final development of the index involved calculating discriminating scores for all sample sites and normalizing calculated scores to a scale of 0 to 10, setting the break point between degraded and nondegraded reference sites at 4.1. So the index values lower than 4.1 indicate degraded conditions, higher values than 6.1 indicate nondegraded situations, and values between 6.1 and 4.1 reveal moderate disturbance.

2.7.4 B-IBI (Weisberg et al. 1997)

Eleven metrics are used to calculate the B-IBI (Weisberg et al. 1997):

1. Shannon–Wiener species diversity index
2. Total species abundance
3. Total species biomass
4. Percent abundance of pollution-indicative taxa
5. Percent abundance of pollution-sensitive taxa
6. Percent biomass of pollution-indicative taxa
7. Percent biomass of pollution-sensitive taxa
8. Percent abundance of carnivores and omnivores
9. Percent abundance of deep-deposit feeders
10. Tolerance score
11. Tanypodinae to Chironomidae percent abundance ratio

The scoring of metrics to calculate the B-IBI is done by comparing the value of a metric from the sample of unknown sediment quality to thresholds established from reference data distributions.

This index was developed to establish the ecologic status of Chesapeake Bay, and it is specific to habitat type and seasonality, its use being advisable only in spring.

2.7.5 Index of Biotic Integrity (IBI) for Fishes

A fish Index of Biotic Integrity (IBI) was developed for tidal fish communities of several small tributaries to the Chesapeake Bay (Vaas and Jordan 1990; Carmichael et al. 1992).

Nine metrics are used to calculate the index, taking into account species richness, trophic structure, and abundance:

- Number of species
- Number of species comprising 90% of the catch
- Number of species in the bottom trawl
- Proportion of carnivores
- Proportion of planktivores
- Proportion of benthivores
- Number of estuarine fish
- Number of anadromous fish
- Total fish with Atlantic menhaden removed.

The scoring of the metrics to calculate the index is done by comparing the value of a metric from the sample of unknown water quality to thresholds established from reference data distributions.

2.7.6 Fish Health Index (FHI) (Cooper et al. 1993)

This index is based on the Community Degradation Index (CDI) developed by Ramm (1988, 1990), which measures the degree of dissimilarity (degradation) between a potential fish assemblage and the actual measured fish assemblage.

FHI provides a measure of the similarity (health) between the potential and actual fish assemblages and is calculated using the formula:

$$\text{FHI} = 10\ (J)[\text{Ln}\ (P)/\text{Ln}\ (P_{max})]$$

where J = the number of species in the system divided by the number of species in the reference community; P = the potential species richness (number of species) of each reference community; and P_{max} = the maximum potential species richness from all the reference communities.

The index ranges from 0 (poor) to 10 (good).

The FHI was used to assess the state of South Africa's estuaries (Cooper et al. 1993; Harrison et al. 1994, 1995, 1997, 1999). Although the index has proved to be a useful tool in condensing information of estuarine fish assemblages into a single numerical value, the index is only based on presence/absence data and does not take into account the relative proportions of the various species present.

2.7.7 Estuarine Ecological Index (EBI) (Deegan et al. 1993)

The EBI includes the following eight metrics:

- Total number of species
- Dominance
- Fish abundance
- Number of nurseries
- Number of estuarine spawning species
- Number of resident species
- Proportion of benthic-associated species
- Proportion of abnormal or diseased fishes

The utility of this index requires it to not only reflect the current status of fish communities, but also be applicable over a wide range of estuaries, although this is not entirely achieved (Bettencourt et al. 2004).

2.7.8 Estuarine Fish Importance Rating (FIR) (Maree et al. 2000)

This index is constructed from seven weighted measures of species and estuarine importance and is designed to work on a presence-absence data set where species are only considered to be present if they constitute more than 1% of any catch by number.

Measures of species importance:

- Number of exploitable species
- Number of estuarine-dependent species
- Number of endemic species

Measures of estuarine importance:

- Type
- Size
- Condition
- Isolation

This index can provide a ranking, based on the importance of each estuary, and helps to identify the systems of major importance for fish conservation.

2.8 Presentation and Definition of Level 7 and 8 Indicators—Holistic Indicators

An ecological network is often drawn as a conceptual diagram that is used as the first step in a modeling development procedure. Figure 2.1 shows a nitrogen cycle in a lake and represents a conceptual diagram and the ecological network for a model of the nitrogen cycle. The complexity of the ecological network in Figure 2.1 cannot be used as an ecological indicator because the real network is simplified too much in the figure. But if observations of the real network make it possible to draw close to the real network, we obtain a similar figure. This is much more complicated, of course, and the complexity of the network in this figure could be used as an indicator for the function of the real ecosystem, even if the network was still a simplification of the real ecosystem.

Gardner and Ashby (1970) examined the influence on stability of connectivity (defined as the number of food links in the food web as a fraction of the number of topologically possible links) of large dynamic systems. They suggest that all large, complex, dynamic systems may show the property of being stable up to a critical level of connectivity and then as the connectivity

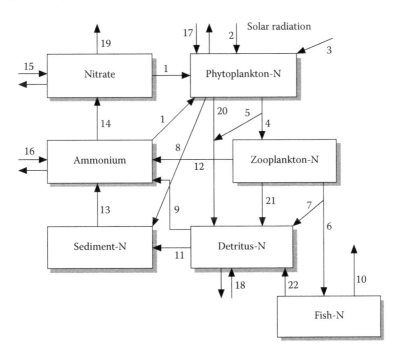

FIGURE 2.1
A conceptual diagram of the nitrogen cycle in a lake. The figure gives an illustration of the ecological network, but the real network is much more complex and the figure can therefore hardly be applied as an ecological indicator.

Adjacency Matrix

From/to	1	2	3	4	5	6	7
1		1	0	0	0	0	0
2	0		1	0	1	1	0
3	0	0		1	1	0	1
4	0	0	0		1	0	0
5	0	0	0	0		0	1
6	0	0	0	0	0		1
7	1	1	0	0	0	0	

FIGURE 2.2
The adjencency matrix for the diagram in Figure 2.1. Twelve connections are realized out of 42, which gives a connectivity index of 28.6.

increases further, the system suddenly gets unstable. A connectivity of about 0.3 to 0.5 seems to provide the highest stability. Although the network in Figure 2.1 is too simple, it can be used to illustrate the calculation of the connectivity. An adjencency matrix is set up as shown in Figure 2.2. The state variables in Figure 2.1 are numbered 1 to 7 in the following sequence: (1) nitrate, (2) phytoplankton-N, (3) zooplankton-N, (4) fish-N, (5) detritus-N, (6) sediment-N, and (7) ammonium-N. The links from – to are indicated by 1, and no links are indicated by 0. The connectivity is the ratio of realized links to the number of possible links. As see in Figure 2.2, by counting the number of 1s relative to all the possible connections, the connectivity is $12/42 = 28.6$. Thus, the connectivity of the diagram in Figure 2.1 is on the low side.

O'Neill (1976) examined the role of heterotrophs on resistance and resilience and found that only small changes in heterotroph biomass could reestablish system equilibrium and counteract perturbations. He suggests that the many regulation mechanisms and spatial heterogeneity should be accounted for when the stability concepts are applied to explain ecosystem responses.

These observations explain why it has been very difficult to find a relationship between ecosystem stability in its broadest sense and species diversity. Compare this with Rosenzweig (1971), where almost the same conclusions are drawn.

It is observed that increased phosphorus loading causes decreased diversity (Ahl and Weiderholm 1977; Weiderholm 1980), but very eutrophic lakes *are* very stable. Figure 2.3 shows the result of a statistical analysis from a number of Swedish lakes. The relationship shows a correlation between number of species and the eutrophication, measured as chlorophyll a in μg/L. A similar relationship is obtained between the diversity of the benthic fauna and the phosphorus concentration relative to the depth of the lakes.

Therefore, it seems appropriate to introduce another but similar concept named buffer capacity, β. It is defined as follows (Jørgensen 1994, 2002):

$$\beta = 1/[\partial \text{ (State variable)}/\partial \text{ (Forcing function)}]$$

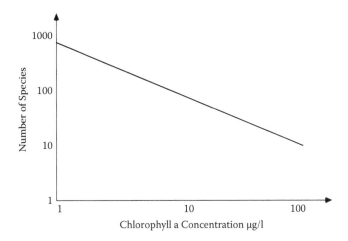

FIGURE 2.3
Weiderholm (1980) obtained the relationship shown for a number of Swedish lakes between the number of species and eutrophication, expressed as chlorophyll a in µg/L.

Forcing functions are the external variables that are driving the system such as discharge of wastewater, precipitation, wind, and so on, while state variables are the internal variables that determine the system, for instance, the concentration of soluble phosphorus, the concentration of zooplankton, and so on.

The concept of buffer capacity has a definition that allows us to quantify, for instance, in modeling, and it is furthermore applicable to real ecosystems, as it acknowledges that *some* changes will always take place in the ecosystem in response to changed forcing functions. The question is how large these changes are relative to changes in the conditions (the external variables or forcing functions).

The concept should be considered multidimensionally, as we may consider all combinations of state variables and forcing functions. It implies that even for one type of change there are many buffer capacities corresponding to each of the state variables. Rutledge (1974) defines ecological stability as the ability of the system to resist changes in the presence of perturbations. It is a definition very close to buffer capacity, but it is lacking the multidimensionality of ecological buffer capacity.

The relation between forcing functions (impacts on the system) and state variables indicating the conditions of the system are rarely linear, and buffer capacities are therefore not constant. In environmental management, it may therefore be important to reveal the relationships between forcing functions and state variables to observe under which conditions buffer capacities are small or large, as compared with Figure 2.4.

Model studies (Jørgensen and Mejer 1979, 1981; Jørgensen 2002) have revealed that in lakes with a high eutrophication level, a high buffer capacity

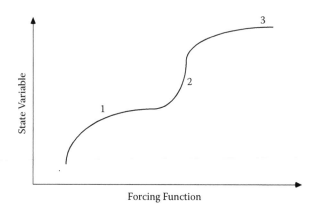

FIGURE 2.4
The relation between state variables and forcing functions is shown. At points 1 and 3 the buffer capacity is high; at point 2 it is low.

to nutrient inputs is obtained by a relatively small diversity. The low diversity in eutrophic lakes is consistent with the above-mentioned results by Ahl and Weiderholm (1977) and Weiderholm (1980). High nutrient concentrations = large phytoplankton species. The specific surface does not need to be large, because there are plenty of nutrients. The selection or competition is not on the uptake of nutrients but rather on escaping the grazing by zooplankton, and here greater size is an advantage. In other words, the spectrum of selection becomes more narrow, which means reduced diversity. It demonstrates that a high buffer capacity may be accompanied by low diversity.

If a toxic substance is discharged to an ecosystem, the diversity will be reduced. The species most susceptible to the toxic substance will be extinguished, while other species, the survivors, will metabolize, transform, isolate, excrete, etc., the toxic substance and thereby decrease its concentration. We observe a reduced diversity, but simultaneously maintain a high buffer capacity to input of toxic compounds, which means that only small changes, caused by the toxic substance, will be observed. Model studies of toxic substance discharge to a lake (Jørgensen and Mejer 1979, 1981) demonstrate the same inverse relationship between the buffer capacity to the considered toxic substance and diversity.

Ecosystem stability is therefore a very complex concept (May 1977) and it seems impossible to find a simple relationship between ecosystem stability and ecosystem properties. Buffer capacity seems to be an applicable stability concept, as it is based

1. On an acceptance of the ecological complexity—it is a multidimensional concept—and

2. On reality, i.e., that an ecosystem will never return to exactly the same situation again.

Another consequence of the complexity of ecosystems mentioned above should be considered here. For mathematical ease, the emphasis—particularly in population dynamics—has been on equilibrium models. The dynamic equilibrium conditions (steady state, not thermodynamic equilibrium) may be used as an attractor (in the mathematical sense, the ecological attractor is the thermodynamic equilibrium) for the system, but the equilibrium will never be attained. Before the equilibrium should have been reached, the conditions, determined by the external factors and all ecosystem components, have changed and a new dynamic equilibrium, and thereby a new attractor, is effective. Before this attractor point has been reached, new conditions will again emerge, and so on. A model based upon the equilibrium state will therefore give a wrong picture of ecosystem reactions. The reactions are determined by the present values of the state variables, and they are different from those in the equilibrium state. We know from many modeling exercises that the model is sensitive to the initial values of the state variables. These initial values are a part of the conditions for further reactions and development. Consequently, the steady-state models may give other results than the dynamic models, and it is therefore recommended to be very careful when drawing conclusions on the basis of equilibrium models. We must accept the complication that ecosystems are dynamic systems and will never attain equilibrium. We therefore need to apply dynamic models as widely as possible, and it can easily be shown that dynamic models give different results from static ones.

The thermodynamic variable exergy has been widely used as a very holistic indicator that is able to capture the system properties of ecosystems. Exergy covers the work capacity of a system. It is strictly defined as the amount of work the system can perform when it is brought into thermodynamic equilibrium with its environment. Exergy is, as can be seen from the definition, dependent on the environment and the system and not entirely on the system (see Figure 2.5). Exergy is therefore not a state variable, such as, for instance, free energy and entropy.

If we choose the same ecosystem as a homogeneous "inorganic soup" that is at same temperature and pressure as the reference state (the environment), exergy will measure the thermodynamic distance from the "inorganic soup" in energy terms. This form for exergy is not strictly in accordance with the exergy introduced to calculate the efficiency of technological processes. With the same system as thermodynamic equilibrium at the same temperature and pressure as the reference state, we can, however, calculate the exergy content (work capacity) of the system as coming entirely from the biochemical energy and from the information embodied in the organisms (see Figure 2.6). The exergy of the system measures the contrast—the difference in work capacity—against the surrounding environment. To distinguish this exergy from the technological exergy, we may call the exergy applied here eco-exergy. Whereever the expression exergy is used in this volume, it is assumed that it is eco-exergy.

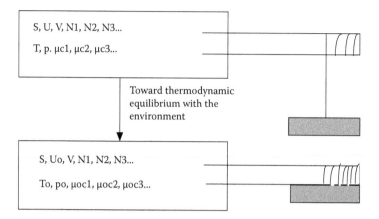

FIGURE 2.5
The definition of "technological" exergy is illustrated.

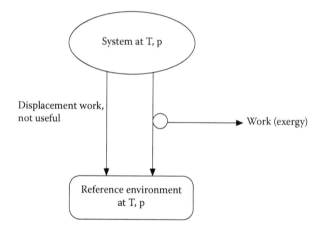

FIGURE 2.6
The definition of eco-exergy is illustrated. Eco-exergy is the amount of work that a system can perform when it is brought into equilibrium with the same system, but with all the chemical compounds in the form of inorganic decomposition products at the highest possible oxidation state. The reference system is an inorganic soup without life and without gradients. The reference state therefore has no eco-exergy.

If the system is in equilibrium with the surrounding environment the exergy is zero. The only way to move systems away from thermodynamic equilibrium is to perform work on them, and the available work in a system is a measure of the ability we have to distinguish between the system and its environment, or thermodynamic equilibrium, also known as the inorganic soup.

Survival implies maintenance of the biomass, and growth means increase of biomass. It costs exergy to construct biomass and obtain/store information. Survival and growth can therefore be measured by use of the thermodynamic concept exergy. Darwin's theory may therefore be reformulated in thermo-dynamic terms and expanded to the system level, as follows: The prevailing conditions of an ecosystem steadily change and the system will continuously select the species that can contribute most to the maintenance or even growth of the exergy of the system.

Notice that the thermodynamic translation of Darwin's theory requires that populations have the properties of reproduction, inheritance, and varia-tion. The selection of the species that contribute most to the exergy of the system under the prevailing conditions requires that there are enough indi-viduals with different properties that a selection can take place. This means that the reproduction and the variation must be high and that once a change has taken place due to a combination of properties giving better fitness, it can be conveyed to the next generation.

If we presume, as proposed above, a reference environment that represents the system (ecosystem) at thermodynamic equilibrium, we can calculate the approximate exergy content of the system as coming entirely from the chem-ical energy:

$$\Sigma(\mu c - \mu co) \, Ni$$

Only what is called ca chemical exergy is therefore included in the com-putation of exergy. The physical exergy is omitted in these calculations as there are no temperature and pressure differences between the system and the reference system. We find by these calculations the exergy of the system compared with the same system at the same temperature and pressure but in the form of an inorganic soup without any life, biological structure, infor-mation, or organic molecules. As $(\mu c - \mu co)$ can be found from the definition of the chemical potential replacing activities by concentrations, we get the following expressions for the eco-exergy:

$$Ex = RT \sum_{i=0}^{i=n} c_i \ln c_i / c_{ieq} \qquad [ML^2T^{-2}] \qquad (2.1)$$

where R is the gas constant, T is the temperature of the environment, and c_i is the concentration of the ith component expressed in a suitable unit (e.g., for phytoplankton in a lake ci could be expressed as mg/L or as mg/L of a focal nutrient. c_{ieq} is the concentration of the ith component at thermodynamic equilibrium and n is the number of components. $c_{i,eq}$ is of course a very small concentration (except for $i = 0$, which is considered to cover the inorganic compounds), but is not zero, corresponding to a very low probability of forming complex organic compounds spontaneously in an inorganic soup at thermodynamic equilibrium.

The problem related to the assessment of c_{ieq} has been discussed and a possible solution proposed in Jørgensen et al. (1995), but the most essential arguments should be repeated here. For dead organic matter, detritus, which is given the index 1, it can be found from classical thermodynamics:

$$\mu 1 = \mu_{1eq} + RT \ln c_1 / c_{1eq} \qquad [ML^2T^{-2}\text{moles}^{-1}] \qquad (2.2)$$

where μ indicates the chemical potential. The difference $\mu_1 - \mu_{1eq}$ is known for organic matter, e.g., detritus, which is a mixture of carbohydrates, fats, and proteins. We find that detritus has approximately 18.7 kJ/g corresponding to the free energy of the mixture of carbohydrates, fats, and proteins.

Generally, c_{ieq} can be calculated from the definition of the probability P_{ieq} to find component i at thermodynamic equilibrium:

$$P_{ieq} \equiv c_{ieq} / \sum_{i=0}^{N} c_{ieq} \qquad [-] \qquad (2.3)$$

If we can find the probability, P_i, to produce the considered component i at thermodynamic equilibrium, we have determined the ratio of c_{ieq} to the total concentration. As the inorganic component, c_0, is very dominant by the thermodynamic equilibrium, equation (2.3) may be rewritten as:

$$P_{ieq} \approx c_{ieq} / c_{0eq} \qquad [-] \qquad (2.4)$$

By combining equations, we get:

$$P_{1eq} = [c_1 / c_{0eq}] \exp[-(\mu_1 - \mu_{1eq}) / RT] \qquad [-] \qquad (2.5)$$

For the biological components, 2, 3, 4, ... N, the probability, P_{ieq}, consists of the probability of producing the organic matter (detritus), i.e., P_{1eq}, and the probability, $P_{i,a}$, to obtain the information embodied in the genes, which determine the amino acid sequence. Living organisms use 20 different amino acids and each gene determines the sequence of about 700 amino acids. $P_{i,a}$ can be found from the number of permutations among which the characteristic amino acid sequence for the considered organism has been selected. It means that we have the following two equations available to calculate P_i:

$$P_{ieq} = P_{1eq} P_{i,a} \qquad (2.5a)$$

($i \geq 2$; 0 covers inorganic compounds and 1 detritus) and

$$P_{i,a} = 20^{-700g} \qquad [-] \qquad (2.6)$$

where g is the number of genes.

Equation (2.4) is reformulated to:

$$c_{ieq} \approx P_{ieq} c_{0eq} \qquad [\text{Moles L}^{-3}] \qquad (2.7)$$

Equations (2.7) and (2.2) are combined:

$$Ex \approx RT \sum_{i=0}^{N} \left[c_i \cdot \ln\left(c_i / (P_{ieq} c_{0eq}) \right) \right] \qquad [ML^2 T^{-2}] \qquad (2.8)$$

This equation may be simplified by using the following approximations (based upon $P_{ieq} \ll c_i$, $P_{ieq} \ll P_0$ and $1/P_{ieq} \gg c_i$, $1/P_{ieq} \gg c_{0eq}/c_i$): $c_i/c_{0eq} \approx 1$, $c_i \approx 0$, $P_i c_{0eq} \approx 0$ and the inorganic component can be omitted. The significant contribution is coming from $1/P_{ieq}$; see Equation (2.8). We obtain:

$$Ex \approx -RT \sum_{i=1}^{N} c_i \cdot \ln(P_{ieq}) \qquad [ML^2 T^{-2}] \qquad (2.9)$$

where the sum starts from 1, because P0,eq $\approx$ 1.

Expressing P_{ieq} as in Equation (2.5a) and P_{1eq} as in Equation (2.5), we obtain the following expression for the calculation of an exergy index:

$$Ex / RT = \sum_{i=1}^{N} \left[c_i \cdot \ln\left(c_i / (c_{0eq}) \right) \right] - (\mu_1 - \mu_{1eq}) \sum_{i=1}^{N} c_i / RT - \sum_{i=2}^{N} c_i \ln P_{i,a} \quad [\text{Moles L}^{-3}]$$

As the first sum is minor compared with the following two sums (use, for instance, $c_i/c_{0eq} \approx 1$), we can write:

$$Ex / RT = (\mu_1 - \mu_{1eq}) \sum_{i=1}^{N} c_i / RT - \sum_{i=2}^{N} c_i \ln P_{i,a} \qquad [\text{Moles L}^{-3}] \qquad (2.10)$$

This equation can now be applied to calculate contributions to the exergy index by important ecosystem components. If we consider only detritus, we know that the free energy released per gram of organic matter is about 18.7 kJ/g. R is 8.4 J/mole and the average molecular weight of detritus is assumed to be 100,000. We get the following contribution of exergy by detritus per liter of water, when we use the unit g detritus exergy equivalent/L:

$$Ex1 = 18.7 \, c_i \, \text{kJ/L} \text{ or } Ex_1/RT = 7.34*105 \, c_i \, [\text{ML-3}] \qquad (2.11)$$

A typical unicell alga has on average 850 genes. Previously, we have purposely used the number of genes and not the amount of DNA per cell, which would include unstructured and nonsense DNA. In addition, a clear correlation between the number of genes and the complexity has been shown (Li and Grauer 1991). However, recently it was discussed that the nonsense genes play an important role; for instance, that they may be considered as spare parts that are able to repair genes when they are damaged or be exposed to mutations. If it is assumed that only the informative genes contribute to the embodied information in organisms, an alga has a total of 850 information genes, i.e., they determine the sequence of $850.700 = 595,000$ amino acids, the contribution of exergy per liter of water, using g detritus equivalent/L as concentration unit would be:

$$Exalgae/RT = 7.34*10^5 \, c_i - c_i \ln 20^{-595000} = 25.2*10^5 \, c_i \, g/L \qquad (2.12)$$

The contribution to exergy from a simple prokaryotic cell can be calculated similarly as:

$$Exprokar/RT = 7.34*10^5 \, c_i + c_i \ln 20329\,000 = 17.2*10^5 \, c_i \, g/L \qquad (2.13)$$

Organisms with more than one cell will have DNA in all cells determined by the first cell. The number of possible microstates therefore becomes proportional to the number of cells. Zooplankton has approximately 100,000 cells and 15,000 genes per cell, each determining the sequence of approximately 700 amino acids. $\ln P_{zoo}$ can therefore be found as:

$$-\ln P_{zoo} = -\ln (20^{-15000*700} * 10^{-5}) \approx 315*105 \qquad (2.14)$$

As seen, the contribution from the numbers of cells is insignificant. Similarly, $P_{fish,a}$ and the P values for other organisms can be found.

The contributions from phytoplankton, zooplankton, and fish to the exergy of the entire ecosystem are significant and far more than corresponding to the biomass. Notice that the unit of Ex/RT is g/L. Exergy can always be expressed in joules per liter, provided that the right units for R and T are used. Equations (2.12)–(2.14) can be rewritten by converting g/L to g detritus/L by dividing by $(7.34*10^5)$.

The exergy index can be found as the concentrations of the various components, c_i, multiplied by weighting factors, β_i, reflecting the exergy that the various components possess due to their chemical energy and to the information embodied in DNA:

$$Ex = \sum_{i=n}^{i=0} \beta_i c_i \qquad (2.15)$$

β_i values based on exergy detritus equivalents have been found for various species. The unit exergy detritus equivalents expressed in g/L can be converted to kJ/L by multiplication by 18.7 corresponding to the approximate average energy content of 1 g detritus. Appendix A, Table A.1, shows the β-values calculated from the number of information genes by the above presented equations.

The index 0 covers the inorganic components, which of course in principle should be included in the calculations of exergy, but in most cases they can be neglected, as the contributions from detritus and even to a higher extent from the biological components are much higher due to an extremely low concentration of these components in the reference system (the ecosystem converted to an inorganic dead system). The calculation of exergy index accounts for use of this equation for the chemical energy in the organic matter as well as for the information embodied in the living organisms. It is measured by the extremely small probability of forming the living components, for instance, algae, zooplankton, fish, mammals, and so on, spontaneously from inorganic matter. The weighting factors may also be considered quality factors reflecting how developed the various groups are and to which extent they contribute to the exergy because of their content of information, which is reflected in the computation. This is completely according to Boltzmann (1905), who gave the following relationship for the work, W, that is embodied in the thermodynamic information:

$$W = RT \ln N \qquad (ML^2T^{-2}) \qquad (2.16)$$

where N is the number of possible states, among which the information has been selected. N is as seen for species the inverse of the probability to obtain the valid amino acid sequence spontaneously.

It is furthermore consistent with the following reformulation of Reeves (1991): "information appears in nature when a source of energy (exergy) becomes available but the corresponding (entire) entropy production is not emitted immediately, but is held back for some time (as exergy)."

The total eco-exergy of an ecosystem *cannot* be calculated exactly, as we cannot measure the concentrations of all the components or determine all possible contributions to exergy in an ecosystem. If we calculate the exergy of a fox, for instance, the above shown calculations will only give the contributions coming from the biomass and the information embodied in the genes, but what is the contribution from the blood pressure, the sexual hormones, and so on? These properties are at least partially covered by the genes, but is that the entire story? We can calculate the contributions from the dominant components, for instance, by using a model or measurements that cover the most essential components for a focal problem.

Exergy calculated using the above shown equations has some shortcomings; it is therefore proposed to consider the exergy found by these calculations as a *relative exergy index*:

1. We account only for the contributions from the organisms' bio-mass and information in the genes. Although these contributions most probably are the most important ones, it cannot be completely excluded that other important contributions are omitted.

2. We don't account for the information embodied in the network—the relations between organisms. The information in the model network that we use to describe ecosystems is negligible compared with the information in the genes, but we cannot exclude that the real, much more complex network may contribute considerably to the total exergy of a natural ecosystem.

3. We have made approximations in our thermodynamic calculations. They are all indicated in the calculations and are in most cases negligible.

4. We can never know all the components in a natural (complex) eco-system. Therefore, we will only be able to use these calculations to determine exergy indices of our simplified images of ecosystems, for instance, of models.

5. The exergy indices are, however, useful as they have been success-fully used as goal function (orientor) to develop structural dynamic models. The *difference* in exergy by *comparison* of two different possible structures (species composition) is here decisive. Moreover, exergy computations always give only relative values, as the exergy is calculated relative to the reference system.

As already stressed, the presented calculations do not include the infor-mation embodied in the structure of the ecosystem, i.e., the relationships between the various components, which is represented by the network. The information of the network encompasses the information of the components and the relationships of the components. The latter contribution has been calculated by Ulanowicz (1986, 1997; see also Ulanowicz and Puccia 1990), as a part of the concept of ascendancy. In principle, the information embodied in the network should be included in the calculation of the exergy index of structural dynamic models, as the network is also dynamically changed. It may, however, often be omitted in most dynamic model calculations because the contributions from the network relationships of models (not from the components of the network, of course) are minor, compared with the con-tributions from the components. This is due to the extreme simplifications made in the models compared with the networks in real ecosystems. It can therefore not be excluded that networks of real ecosystems may contribute considerably to the total exergy of the ecosystems, but for the type of models that we are using at present, we can probably omit the exergy of the informa-tion embodied in the network.

Specific exergy is defined as the exergy, or rather exergy index, divided by the biomass. Specific exergy expresses, in other words, the dominance of the higher organisms as they carry more information per unit of biomass, i.e., have higher β values. A very eutrophic ecosystem has a very high exergy due to the high concentration of biomass, but the specific exergy is low, as the biomass is dominated by algae with low β values.

The combination of the exergy index and the specific exergy index usually gives a more satisfactory description of the health of an ecosystem than the exergy index alone, because it considers the diversity and the life conditions for higher organisms (see Jørgensen 1995). The combination of exergy, specific exergy, and buffer capacities, defined as the change in a forcing function relative to the corresponding change in a state variable, has been used as an ecological indicator for lakes. Exergy or eco-exergy and specific exergy will be presented in more detail in Chapter 3.

H. T. Odum (1983) defines the maximum power principle as a maximization of *useful* power. It implies that the contributions to the *total* power that are useful are summarized. It means that non-useful power is not included in the summation. The difference between useful and non-useful power is perhaps the key to understanding Odum's principle and to using it to interpret ecosystem properties.

According to H. T. Odum, it is the transformation of energy into work (consistent with the term *useful power*) that determines success and fitness. Many ecologists have incorrectly assumed that natural selection tends to increase efficiency. If this were true, endothermy could never have evolved. Endothermic birds and mammals are extremely inefficient compared with reptiles and amphibians. They expend energy at high rates in order to maintain a high, constant body temperature, which, however, gives high levels of activities independent of environmental temperature. Fitness may be defined as reproductive power, dW/dt, the rate at which energy can be transformed into work to produce offspring. This interpretation of the maximum power principle is consistent with the maximum exergy.

In a book titled *Maximum Power—The Ideas and Applications of H. T. Odum*, Hall (1995) has presented a clear interpretation of the maximum power principle, as it has been applied in ecology by H. T. Odum. The principle claims that power or output of useful work is maximized—not the efficiency and not the rate but the trade-off between a high rate and high efficiency yielding most useful energy = useful work. It is illustrated in Figure 2.7. Hall is using an interesting seminatural experiment to illustrate the application of the principle in ecology. Streams were stocked with different levels of predatory cutthroat trout. When predator density was low, there was considerable invertebrate food per predator, and the fish used relatively little maintenance of food searching energy per unit of food obtained. With a higher fish-stocking rate, food became less available per fish, and each fish had to use more energy searching for it. Maximum production occurred at intermediate fish-stocking rates, which means intermediate rates at which the fish utilized their food.

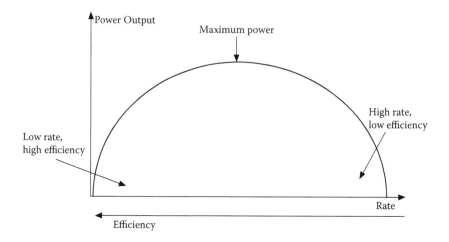

FIGURE 2.7

The maximum power principle claims that the development of an ecosystem is a trade-off (a compromise) between the rate and the efficiency, i.e., the maximum power output per unit of time.

Hall (1995) mentions another example. Deciduous forests in moist and wet climates tend to have a leaf area index of about 6. Such an index is predicted from the maximum power hypothesis applied to the net energy derived from photosynthesis. Higher leaf area index values produce more photosynthate, but do so less efficiently because of the respirational demand of the additional leaf. Lower leaf area indices are more efficient per leaf, but draw less power than the observed intermediate values of roughly 6.

According to Gilliland (1982) and Andreasen (1985) the same concept applies for regular fossil fuel power generation. The upper limit of efficiency for any thermal machine such as a turbine is determined by the Carnot efficiency. A steam turbine could run at 80% efficiency, but it would need to operate at a nearly infinitely slow rate. Obviously, we are not interested in a machine that generates revenues infinitely slowly, no matter how efficiently. Actual operating efficiencies for modern steam-powered generators are therefore closer to 40%, roughly half the Carnot efficiency. The example in Figure 2.7 shows that the maximum power principle is embedded in the irreversibility of the world. The highest process efficiency can be obtained by endoreversible conditions, meaning that all irreversibilities are located in the coupling of the system to its surroundings—there are no internal irreversibilities. Such systems will, however, operate too slowly. Power is zero for any endoreversible system. If we want to increase the process rate, it will imply that we also increase the irreversibility and thereby decrease the efficiency. The maximum power is the compromise between endoreversible processes and very fast completely irreversible processes.

Emergy was introduced by H. T. Odum (1983) and attempts to account for the energy required in the formation of organisms in different trophic levels. The idea is to correct energy flows for their quality. Energies of different types are converted into equivalents of the same type by multiplying by the energy transformation ratio. For example, fish, zooplankton, and phytoplankton can be compared by multiplying their actual energy content with their solar energy transformation ratios. The more transformation steps there are between two kinds of energy, the greater the quality and the greater the solar energy required to produce a unit of energy (J) of that type. When one calculates the energy of one type, that generates a flow of another; this is sometimes referred to as the embodied energy of that type.

Figure 2.8 presents the concept of embodied energy in a hierarchical chain of energy transformation, and Table 2.4 gives embodied energy equivalents for various types of energy.

H. T. Odum (1983) reasons that surviving systems develop designs that receive as much energy amplifier action as possible. The energy amplifier ratio is defined in Figure 2.18 as the ratio of output B to control flow C. H. T. Odum (1983) suggests that in surviving systems the amplifier effects are proportional to embodied energy, but full empirical testing of this theory still needs to be carried out in the future.

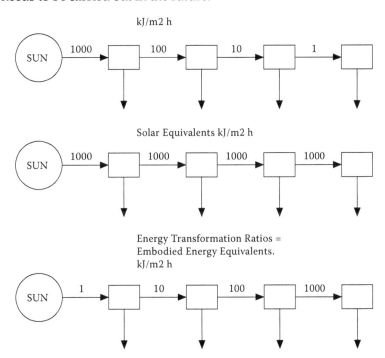

FIGURE 2.8
Energy flow, solar equivalents, and energy transformation ratios equal embodied energy equivalents in a food chain.

TABLE 2.4

Embodied Energy Equivalents for Various Types of Energy

Type of Energy	Embodied Energy Equivalents
Solar energy	1.0
Winds	315
Gross photosynthesis	920
Coal	6800
Tide	11,560
Electricity	27,200

One of the properties of high-quality energies is their flexibility. Whereas low-quality products tend to be special, requiring special uses, the higher-quality part of a web is of a form that can be fed back as an amplifier to many different units throughout the web. For example, the biochemistry at the bottom of the food chain in algae and microbes is diverse and specialized, whereas the biochemistry of top animal consumer units tends to be similar and general, with services, recycles, and chemical compositions usable throughout.

Hannon (1973) and Hannon and Ruth (1997) applied energy intensity coefficients as the ratios of assigned embodied energy to actual energy to compare systems with different efficiencies. The difference between embodied energy flows and power [see Equation (2.17)] simply seems to be a conversion to solar energy equivalents of the free energy ΔF. The increase in biomass, in Equation (2.17), is a conversion to the free energy flow, and the definition of embodied energy is a further conversion to solar energy equivalents.

Embodied energy is, as seen from these definitions, determined by the biogeochemical energy *flow* into an ecosystem component, measured in solar energy equivalents. The stored emergy, Em, per unit of area or volume to be distinguished from the emergy flows can be found from:

$$Em = \sum_{i=1}^{i=n} \Omega_i * c_i \qquad (2.17)$$

where Ω_i is the quality factor, which is the conversion to solar equivalents, as illustrated in Table 2.2 and Figure 2.8, and c_i is the concentration expressed per unit of area or volume. The calculations by Equation (2.17) reduce the difference between stored emergy (= embodied energy) and stored exergy, which also can be found with good approximations as the sum of concentrations times a quality factor [see Equation (2.15)], to a difference between the applied quality factors. Emergy uses as quality factor the cost in the form of solar energy, while the exergy quality accounts for the information embodied in the biomass. Emergy gives the costs while exergy gives the result. For the ratio emergy paid to resulting exergy, see Chapter 5.

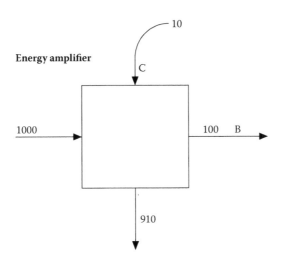

FIGURE 2.9
The Energy Amplifier Ratio, R, is defined as the ratio of output B to control flow C. It means that R = 10 in this case.

Emergy thereby calculates how much solar energy (which is our ultimate energy resource) it has cost to obtain one unit of biomass of various organisms, while exergy accounts for how much "first class" energy (= energy that can do work) the organisms, as a result of the complex interactions in an ecosystem, possess. Both concepts attempt to account for the quality of the energy—emergy by looking into the energy flows in the ecological network to express the energy costs in solar equivalents and exergy by considering the amount of information that the components have embodied.

The differences between the two concepts may be summarized as follows:

1. Emergy has no clear reference state, which is not needed as it is a measure of energy flows, while eco-exergy is defined relative to the same system at thermodynamic equilibrium.

2. The quality factor of exergy is based on the content of information, while the quality factor for emergy is based on the cost in solar equivalents.

3. Exergy is better anchored in thermodynamics and has a wider theoretical basis.

4. The quality factor, Ω, may be different from ecosystem to ecosystem, and in principle it is necessary to assess the quality factor in each case based on an energy flow analysis, which is sometimes cumbersome to make. The quality factors listed in Table 2.2 may be used generally as good approximations. The quality factors used for computation of exergy, β, require a knowledge to the non-nonsense genes of various

organisms, which sometimes is surprisingly difficult to assess (see Chapter 3).

5. In his book *Environmental Accounting—Emergy and Environmental Decision Making*, H. T. Odum (1996) has used calculations of emergy to estimate the sustainability of the economy of various countries. As emergy is based on the cost in solar equivalents, which is the only long-term available energy, it seems to be a sound first estimation of sustainability, although it sometimes is an extremely difficult concept to quantify.

Emergy will be treated in more detail in Chapter 4, and Chapter 5 will use the ratio exergy to emergy as a very powerful and useful indicator.

The diversity index (DI) for an ecosystem is usually represented as:

$$H = -\sum_{i=1}^{s} (P_i * \log_2 P_i) \tag{2.18}$$

which originates from Shannon's theory deriving the average entropy of discrete information. Generally, many kinds of similar indices are proposed and used. P_i in Equation (2.18) originally signified the probability of occurrence of the ith information, which was later replaced by n_i/N in an ecosystem by Margalef (1963), where N is the total number of living elements in the ecosystem and n_i the number of living members of the ith species, i.e., $P_i = n_i/N$. Therefore, the diversity index can be denoted by the relation:

$$DI = -\sum_{i=1}^{s} \left(\frac{n_i}{N}\right) * \log_2 \left(\frac{n_i}{N}\right) \tag{2.19}$$

where N is the total number of living elements; n_i is the living numbers of the ith species.

The use of diversity indices as ecological indicators will be presented in more detail in Chapter 7.

2.9 An Overview of Applicable Ecological Indicators for EHA

Table 2.5 gives an overview of the classes or levels of ecological indicators (see the eight levels in Section 2.2 of this chapter) applied in the Chapters 10–18 in Section II. It is indicated in the table which ecosystems the various chapters consider in the presentation of proposed ecological indicators. It is,

TABLE 2.5

Overview of Applied Ecological Indicators in Chapters 10–18

		Indicator Level							
Chapter	Ecosystem	1	2	3	4	5	6	7	8
10	Wetland			×	×	×	×	×	
11	Coastal, estuary							×	×
12	Lake				×	×	×	×	×
13	Forest	×	×	×	×	×	×		
14	Marine				×	×	×		
15	Coastal			×	×	×	×	×	×
16	Landscape			×	×	×	×	×	×
17	Agroecosystem								×
18	River	×			×	×			

of course, not possible to present all applicable indicators in nine case studies of the use of ecological indicators for EHA. As mentioned in Section 2.2, level 1 indicators have been widely used for EHA of rivers and may also be used in most other ecosystems. Similarly, concentrations of chemical compounds are obvious to use for all unhealthy conditions caused by toxic substances. The experience gained by the use of level 6–8 indicators is usually of more general value in the EHA, because the higher level indicators give an overall (holistic) picture of how far a focal ecosystem on the system level is from healthy conditions. The overview that is a result of this volume is therefore giving information to a higher extent on the applicability of level 6–8 indicators. In practical EHA, these indicators should be supplemented with level 1–5 indicators, which are more specific. The selection of level 1–5 indicators is furthermore obvious in most cases, for instance, the use of PCB and zebra mussels in the EHA of the Great North American Lakes.

2.10 EHA Procedures

2.10.1 Direct Measurement Method

The procedures established for the Direct Measurement Method (DMM) are as follows:

1. Identify the necessary indicators to be applied in the assessment process; use Table 2.5, Section 2.2.
2. Measure directly or calculate indirectly the selected indicators.
3. Assess ecosystem health based on the resulting indicator values.

2.10.2 Ecological Model Method

The procedures established for the Ecological Modeling Method (EMM) for lake ecological health assessment are shown in Figure 2.10.

Five steps are necessary when assessing lake ecosystem health using the EMM procedure:

1. Determine the model's structure and complexity according to the ecosystem structure.
2. Establish an ecological model through designing a conceptual diagram, developing model equations, and estimating model parameters.
3. Calibrate the model as necessary in order to assess its suitability in application to the ecosystem health assessment process.
4. Calculate ecosystem health indicators.
5. Assess ecosystem health based on the values of the indicators.

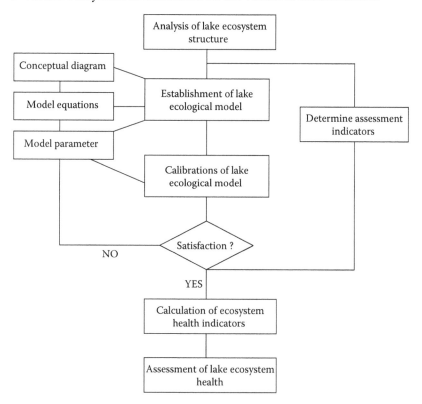

FIGURE 2.10
The procedure of Ecological Model Method (EMM) for ecological health assessment. *Source:* Modified from Xu, F.-L., S. Tao, R. W. Dawson, B. G. Li, and J. Cao. 2001. Lake ecosystem health assessment: Indicators and methods. *Water Research* 35 (13): 3157–167, with permission.

2.10.3 Ecosystem Health Index Method

In order to assess quantitatively the state of ecosystem health, an Ecosystem Health Index (EHI) on a scale of 0 to 100 was developed. It was assumed that when EHI is zero, the healthy state is worst; when EHI is 100, the healthy state is best. In order to facilitate the description of healthy states, EHI is equally divided into five segments or ranges: 0%–20%, 20%–40%, 40%–60%, 60%–80%, 80%–100%, which correspond with the five health states, "worst," "bad," "middle," "good," and "best," respectively.

EHI can be calculated using the following equation:

$$\text{EHI} = \sum_{i=1}^{n} \omega_i \cdot \text{EHI}_i \tag{2.20}$$

where *EHI* is a synthetic ecosystem health index, EHI_i is the *i*th ecosystem health index for the *i*th indicator, and ω_i is the weighting factor for the *i*th indicator.

It can be seen from Equation (2.20) that the synthetic EHI depends on sub-EHIs and weighting factors for each indicator.

The procedure established for lake ecosystem health assessment using the EHI method (EHIM) is shown in Figures 2.11 and 2.12. Five steps are necessary for the EHI method:

1. Select basic and additional indicators.
2. Calculate sub-EHIs for all selected indicators.
3. Determine weighting factors for all selected indicators.
4. Calculate synthetic EHI using sub-EHIs and weighting factors for all selected indicators.
5. Assess ecosystem health based on synthetic EHI values.

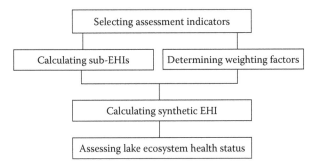

FIGURE 2.11
The procedure of EHI method for lake ecosystem health assessment.

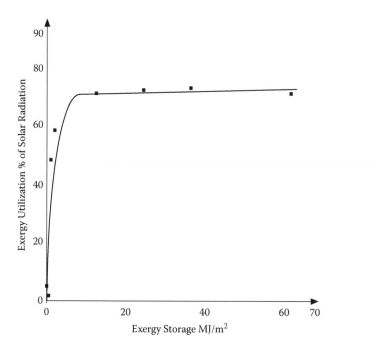

FIGURE 2.12
Exergy storage versus exergy utilization (percentage of solar radiation) for various ecosystems.

2.11 An Integrated, Consistent Ecosystem Theory That Can Be Applied as a Fundament for EHA

Several ecosystem theories have been presented in the scientific literature during the last two to three decades. At first glance they look very different and seem to be inconsistent, but a further examination reveals that they are not very different and that it should be possible to unite them in a consistent pattern. It has been accepted among system ecologists since 1998/1999, but as a result of two important meetings in 2000, one in Porto Venere, Italy, in late May and one in Copenhagen in early June in conjunction with an ASLO meeting, it can be concluded that a consistent pattern of ecosystem theories has been formed. The pattern of ecosystem theories was strongly supported in a brainstorming meeting on the Danish island of Møn in June of 2005. The result of this meeting was published as a book called *A New Ecology—A System Approach* (Jørgensen et al. 2007) in May 2007 at an eco-summit meeting in Beijing. Several system ecologists agreed on the pattern as a working basis for further development in system ecology. This is of utmost importance for the progress in system ecology, because with a theory in hand it will be possible to explain many rules that are published in ecology and applied ecology, which again explain many ecological observations. We should, in other words, be able to attain the same theoretical basis that characterizes physics: a few basic laws, which can be used to deduce rules that explain observations. It has therefore also been agreed that one of the important goals in system ecology would be to demonstrate (prove) the links between ecological rules and ecological laws.

Ten to fifteen years ago the presented theories seemed very inconsistent and chaotic. How could E. P. Odum's attributes (1969), H. T. Odum's maximum power (1983), Ulanowicz's ascendency (1986), Patten's indirect effect (1992), Kay and Schneider's maximum exergy degradation (1992), Jørgensen's maximum exergy principle (Jørgensen and Mejer 1977; Jørgensen 1982, 2002), and Prigogine's (1947) and Mauersberger's minimum entropy dissipation (1983, 1985) be valid at the same time? Everybody insisted that their version of a law for ecosystem development was right, and all the other versions were wrong. New results and an open discussion among the contributing scientists have led to a formation of a pattern, where all the theories contribute to the total picture of ecosystem development.

The first contribution to a clear pattern of the various ecosystem theories came from the network approach often used by Patten. Patten and Fath (see Jørgensen et al. 2007) have shown by a mathematical analysis of networks in steady state (representing, for instance, an average annual situation in an ecosystem with close to balanced inputs and outputs for all components in the network) that the sum of through flows in a network (which is maximum power) is determined by the input and the cycling within the network. The input (the solar radiation) again is determined by the structure of the system (the stored exergy, the biomass). Furthermore, the more structure, the more maintenance is needed and therefore more exergy must be dissipated, the greater the inputs are. Cycling, on the other hand, means that the same energy (exergy) is better utilized in the system, and therefore more biomass (exergy) can be formed without increase of the inputs. It has been shown previously that more cycling means increased ratio of indirect to direct effects, while increased input doesn't change the ratio of indirect to direct effects. Fath and Patten (2001) used these results to determine the development of various variables used as goal functions (exergy, power, entropy, etc.). An ecosystem is, of course, not setting goals, but a goal function is used to describe the direction of development an ecosystem will take in an ecological model. Their results can be summarized as follows:

1. Increased inputs (more solar radiation is captured) mean more bio-mass, more exergy stored, more exergy degraded, and therefore also higher entropy dissipation, more through-flow (power), increased ascendency, but no change in the ratio indirect to direct effect or in the retention time for the energy in the system = total exergy/input exergy per unit of time.

2. Increased cycling implies more biomass, more exergy stored, more through-flow, increased ascendency, increased ratio indirect to direct effect, increased retention, but no change in exergy degradation.

Almost simultaneously Jørgensen et al. (2000) published a paper that claims that ecosystems show three growth forms:

I. Growth of physical structure (biomass), which is able to capture more of the incoming energy in the form of solar radiation, but also requires more energy for maintenance (respiration and evaporation).

II. Growth of network, which means more cycling of energy and matter.

III. Growth of information (more developed plants and animals with more genes), from r-strategists to K-strategists, which waste less energy but also usually carry more information.

These three growth forms may be considered an integration of E. P. Odum's attributes, which describe changes in an ecosystem associated with development from the early stage to the mature stage. Eight of the most applied attributes associated to the three growth forms should be mentioned (for the complete list of attributes, see Table 2.1):

1. Ecosystem biomass (physical structure) increases.

2. More feedback loops (including recycling of energy and matter) are built.

3. Respiration increases.

4. Respiration relative to biomass decreases.

5. Bigger animals and plants (trees) become more dominant.

6. The specific entropy production (relative to biomass) decreases.

7. The total entropy production will first increase and then stabilize on approximately the same level.

8. The amount of information increases (more species, species with more genes; the biochemistry becomes more diverse).

Growth form I covers attributes 1, 3, and 7; growth form II covers 2 and 6; and growth form III covers the attributes 4, 5, 7, and 8.

In the same paper (Jørgensen et al. 2000), Figure 2.13 was presented to illustrate the concomitant development of ecosystems, exergy captured (most of that being degraded), and exergy stored (biomass, structure, information). The points in the figure correspond to different "ecosystems": an asphalt road, bare soil, a desert, grassland, young spruce plantation, older spruce plantation, old temperate forest, and rain forest. Debeljak (2002) has shown that he gets the same shape of the curve when he determines exergy captured and exergy stored in managed forest and virgin forest in different stages of development (see Figure 2.13).

Holling (1986) (see Figure 2.14) has suggested how ecosystems progress through the sequential phases of renewal (mainly growth form I), exploitation (mainly growth form II), conservation (dominant growth form III), and creative destruction. The latter phase also fits into the three growth forms but will require further explanation. The creative destruction phase is a result of either external or internal factors. In the first case (for instance, hurricanes and volcanic activity), further explanation is not needed as an ecosystem has to use the growth forms under the prevailing conditions, which are determined by the external factors.

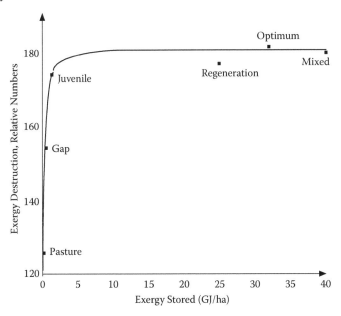

FIGURE 2.13
The plot shows the result by Debeljak (2002). He examined managed and virgin forest in different stages. The gap has no trees, while the virgin forest changes from optimum to mixed to regeneration and back to optimum, although the virgin forest can be destroyed by catastrophic events such as fire or storms. The juvenile stage is a development between the gap and the optimum. Pasture is included for comparison.

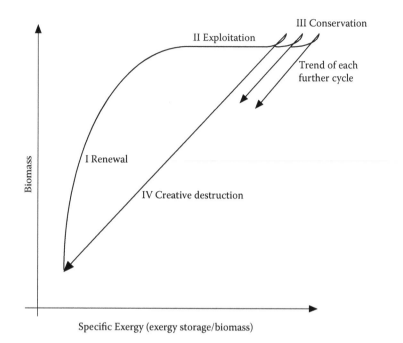

FIGURE 2.14
Holling's four stages are expressed in terms of biomass and specific exergy. Notice that the trend of each further cycle is toward higher exergy storage.

If the destructive phase is a result of internal factors, the question is, "Why would a system be self-destructive?" A possible explanation is that as a result of the conservation phase almost all nutrients will be contained in organisms, which implies that there are no nutrients available to test new and possibly better solutions to move farther away from thermodynamic equilibrium or, expressed in Darwinian terms, to increase the probability of survival. This is also implicitly indicated by Holling (Figure 2.14), as he talks about creative destruction. Therefore, when new solutions are available, in the long run it would be beneficial for the ecosystem to decompose the organic nutrients into inorganic components that can be used to test the new solutions. The creative destruction phase can be considered a method to utilize the three other phases and the three growth forms more effectively in the long run.

Five hypotheses have been proposed to describe ecosystem growth and development, namely:

1. The entropy production tends to be minimum (this is proposed by Prigogine [1947, 1980] for linear systems at steady nonequilibrium state, not for far-from-equilibrium systems). It is applied by

Mauersberger (1983, 1995) to derive expressions for bioprocesses at a stable stationary state.

2. Natural selection tends to make the energy flux through the system at a maximum, as far as is compatible with the constraints to which the system is subject (H. T. Odum 1983). This is also called the maximum power principle (see Section 2.8).

3. Ecosystems will organize themselves to maximize the degradation of exergy (Kay 1984).

4. A system that receives a through-flow of exergy will have a propensity to move away from thermodynamic equilibrium, and if more combinations of components and processes are offered to use the exergy flow, the system has the propensity to select the organization that gives the system as much stored exergy as possible; see Section 2.8 of this chapter, Jørgensen and Mejer (1977, 1979), Jørgensen (1982, 1997), and Mejer and Jørgensen (1979).

5. Ecosystems will have a propensity to develop toward a maximization of the ascendency (Ulanowicz 1986).

The usual description of ecosystem development illustrated, for instance, by the recovery of Yellowstone Park after fire, an island born after a volcanic eruption, reclaimed land, etc., is well covered by E. P. Odum (1969): at first the biomass increases rapidly, which implies that the percentage of captured incoming solar radiation increases but also the energy needed for maintenance. Growth form I is dominant in this first phase, where exergy stored increases (more biomass, more physical structure to capture more solar radiation), but also the through-flow (of useful energy), exergy dissipation, and entropy production increase because of increased need of energy for maintenance. Growth forms II and III become dominant later, although there is an overlap of the three growth forms taking place. When the percentage of solar radiation captured reaches about 80%, it is not possible to increase the amount of captured solar radiation further (in principle because of the second law of thermodynamics). Therefore, further growth of the physical structure (biomass) does not improve the energy balance of the ecosystem. In addition, all or almost all the essential elements are in the form of dead or living organic matter and not inorganic compounds ready to be used for growth. The growth form I will therefore not proceed, but growth forms II and III can still operate. The ecosystem can still improve the ecological network and can still exchange r-strategists with K-strategists, small animals and plants with bigger ones, and less developed with more developed with more information genes. A graphic representation of this description of ecosystem development is already presented in Figures 2.11 and 2.12. The accordance with the five descriptors plus specific entropy production and the three growth forms based on this description of ecosystem development is shown in Table 2.6.

TABLE 2.6

Accordance between Growth Forms and the Proposed Descriptors

	Hypothesis		
	Growth Form I	Growth Form II	Growth Form III
Exergy storage	Up	Up	Up
Power/through-flow	Up	Up	Up
Ascendency	Up	Up	Up
Exergy dissipation	Up	Equal	Equal
Retention time	Equal	Up	Up
Entropy production	Up	Equal	Equal
Exergy/biomass = specific exergy	Equal	Up	Up
Entropy/biomass = specific entropy production	Equal	Down	Down
Ratio indirect/direct effects	Equal	Up	Up

The presented integrated ecosystem theory can be applied in EHA in two ways:

1. As demonstrated, the widely applied E. P. Odum's attributes are covered by the use of several of the presented holistic indicators, for instance, exergy, emergy, ascendency, specific exergy, and entropy production/biomass. The application of the holistic indicators thereby gets wider perspectives.

2. The development of the three growth forms may be used to explain the thermodynamic holistic indicators.

It is mandatory to understand the development of ecosystems and their reactions to stress when the results of an EHA are interpreted in an environmental management context. Therefore, it is important not to consider the indicators just as classification numbers, but to attempt to understand "the story" behind the indicators to be able to answer the questions: Why and where is the ecosystem unhealthy? How did it happen? When will the ecosystem by healthy again? What should we do to recover the ecosystem? This will often require a profound knowledge of ecology and ecosystem theory.

References

Abou-Aisha, K. H., I. A. Kobbia, M. S. Abyad, E. F. Shabana, and F. Schanz. 1995. Impact of phosphorous loadings on macro-algal communities in the Red Sea coast of Egypt. *Water, Air and Soil Pollution* 83:285–97.

Agard, J. B., J. B. R. Gobin, and R. M. Warwick. 1993. Analysis of marine macrobenthic community structure in relation to pollution, natural oil seepage and seasonal disturbance in a tropical environment (Trinidad, West Indies). *Mar Ecol Progr Ser* 92:233–43.

Ahl, T., and T. Weiderholm. 1977. Svenska vattenkvalitetskriterier. Eurofierande ämnen. *SNV PM (Swed)* 918.

Albrecht, W. N., C. A. Woodhouse, and J. N. Miller. 1981. Nearshore dredge-spoil dumping and cadmium, copper and zinc levels in a dermestid shrimp. *Bull Environ Contamin Toxicol* 26:219–23.

Anderlini, V. C., and R. G. Wear. 1992. The effects of sewage and natural seasonal disturbance on benthic macrofaunal communites in Fitzray Bay, Wellington, New Zealand. *Mar Pollut Bull* 24:21–26.

Andreasen, I. 1985. A general ecotoxilogical model for the transport of lead through the system: Air-soil (water)-grass-cow-milk. Thesis, DIA-K, Technical University of Denmark.

Anger, K. 1975. On the influence of sewage pollution on inshore benthic communities in the south of Kiel Bay. 2. Quantitative studies on community structure. *Helgoländer Wiss Meeresunters* 27:408–38.

Bellan, G. 1967. Pollution et peuplements bentiques sur substrat meuble dans la region de Marseille. 1. Le secteur de Cortiou. *Rev Intern Oceanogr Med* 6:53–87.

———. 1980. Annélides polychétes des substrats solids de troits mileux pollués sur les côrtes de Provence (France): Cortiou, Golfe de Fos, Vieux Port de Marseille. *Téthys* 9 (3): 260–78.

Bellan-Santini, D. 1980. Relationship between populations of amphipods and pollution. *Mar Pollut Bull* 11:224–27.

Belsher, T., and C. F. Boudouresque. 1976. L'impact de la pollution sur la fraction algale des peuplements benthiques de Méditerranée. *Atti Tavola Rotonda Internazionale, Livorno* 215–60.

Bettencourt, A., A. Brisker, S. B. Ferreira, J. G. Franco, J. C. Marques, J. J. Melo, J. J. Nobre, A. L. Ramos, C. S. Reis, C. S. Salas, F. Silvaand, and M. C. Simas. 2004. Typology and reference conditions for Portuguese transitional and coastal waters. Development of guidelines for the application of the European Union Water Framework Directive. Ediçào IMAR/INAG.

Beukema, J. J. 1988. An evaluation of the ABC method (abundance/biomass comparison) as applied to macrozoobenthic communities living on tidal flats in the Dutch Wadden Sea. *Marine Biology* 99:425–33.

Boltzmann, L. 1905. The Second Law of Thermodynamics. Reprinted in English in *Theoretical Physics and Philosophical Problems. Selected Writings of L. Boltzmann.* D. Reidel. Drodrecht, the Netherlands.

Borja, A., J. Franco, and V. Perez. 2000. A marine biotic index to establish the ecological quality of soft-bottom benthos within European estuarine and coastal environments. *Mar Pollut Bull* 40 (12): 1100–1114.

Borja, A., I. Muxika and X. Franco, 2003. The application of a Marine Biotic Index to different impact sources affecting soft-bottom communities along European coasts. *Mar Pollut Bull* 46: 835–845.

Brock, V. 1992. Effects of mercury on the biosynthesis of porphyrins in bivalve molluscs (*Cerastoderma edule* and *C. lamarcki*). *J Exp Mar Bio Ecol* 164:17–29.

Bryan, G. W., and P. E. Gibbs. 1987. Polychaetes as indicators of heavy-metal availability in marine deposits. In *Oceanic processes in marine pollution, vol 1: Biological processes and wastes in the ocean*, eds. J. M. Capuzzo and D. R. Kester, 37–49. Melbourne, FL: Krieger Publishing Co., Inc.

Buhr, K. J., 1976. Suspension feeding and assimilation efficiency in *Lanice conchilega* (Polychaeta). *Marine Biology*, 38: 373–383.

Calderón-Aguilera, L. E. 1992. Análisis de la infauna béntica de Bahía de San Quintín, Baja California, con énfasis en su utilidad en la evaluación de impacto ambiental. *Ciencias Marinas* 18 (4): 27–46.

Carmichael, J., B. Richardson, M. Roberts, and S. J. Jordan. 1992. Fish sampling in eight Chesapeake Bay tributaries. Maryland Dept. of Natural Resources, CBRM-HI-92-2. Annapolis.

Carrell, B., S. Forberg, E. Grundelius, L. Henrikson, A. Johnels, U. Lindh, H. Mutvei, M. Olsson, K. Svaerdstroem, and T. Westermark. 1987. Can mussel shells reveal environmental history? *Ambio* 16 (1): 2–10.

Clarke, K. R. 1990. Comparison of dominance curves. *J Exp Mar Biol Ecol* 138 (1–2): 130–43.

Clarke, K. R., and R. M. Warwick. 1999. The taxonomic distinctness measures of biodiversity: Weighting of steps lengths between hierarchical levels. *Mar Ecol Prog Ser* 184:21–29.

Cooper, J. A. G., T. D. Harrison, A. E. L., and R. A. Singh. 1993. Refinement, enhancement and application of the Estuarine Health Index to Natal's estuaries, Tugela—Mtamvuna. Unpublished technical report. CSIR, Durban.

Cooper, R. J., D. Langlois, and J. Olley. 1982. Heavy metals in Tasmania shell fish. I. Monitoring heavy metals contamination in the Dermwert estuary: Use of oysters and mussels. *Jat* 2 (2): 99–109.

Cormaci, M., and G. Furnari. 1991. Phytobenthic communities as monitor of the environmental conditions of the Brindisi coast-line. *Oebalia* XVII-I suppl:177–98.

Cossa, A., and J. G. Rondeau. 1985. Seasonal geographical and size-induced variability in mercury content of *Mytilus edulis* in an estuarine environment: A re-assessment of mercury pollution level in the estuary and gulf of St. Lawrence. *Mar Biol* 88:43–49.

Cossa, F., M. Picard, and J. P. Gouygou. 1983. Polynuclear aromatic hydrocarbons in mussels from the estuary and northwestern gulf of St. Lawrence, Canada. *Bull Environ Contam Toxicol* 31:41–47.

Costanza, R., B. G. Norton, and B. D. Haskell, eds. 1992. *Ecosystem health: New goals for environmental management*. Washington, DC: Island Press.

Couch, J. A., and J. C. Harshbarger. 1985. Effects of carcinogenic agents on aquatic animals: An environmental and experimental overview. *Environ Carcinogenesis Revs* 3C (1): 63–105.

Craeymeersch, J. A. 1991. Applicability of the abundance/biomass comparison method to detect pollution effects on intertidal macrobenthic communities. *Hydrobiol Bull* 24 (2): 133–40.

Dabbas, M., F. Hubbard, and J. McManus. 1984. The shell of Mytilus as an indicator of zonal variatons of water quality within an estuary. *Estuar Coast Shelf Sci* 18 (3): 263–70.

Dauer, D. M., 1984. The use of polychaete feeding guilds as biological variables. *Mar Pollut Bull* 15 (8): 301–304.

Dauer, D. M., M. W. Luckenbach, and A. J. Rodi. 1993. Abundance-biomass comparison ABC method: Effects of an estuarine gradient, anoxic/hypoxic events and contaminated sediments. *Mar Biol* 116:507–18.

Dauer, D. M., C. A. Maybury, and R. M. Ewing. 1981. Feeding behavior and general ecology of several spionid polychaetes from the Chesapeake Bay. *J Exp Mar Biol Ecol* 54 (1): 21–38.

De Wolf, P. 1975. Mercury content of mussel from West European Coasts. *Mar Pollut Bull* 6 (4): 61–63.

Debeljak, M. 2002. Application of exergy degradation and exergy storage as indicators for the development of managed and virgin forest. PhD thesis, University of Ljubljana University.

Deegan, L. A., J. T. Finn, S. G. Ayvazian, and C. Ruyder. 1993. Feasibility and application of the Index of Biotic Integrity to Massachusetts Estuaries (EBI). Final report to Massachusetts Executive Office of Environmental Affairs. North Grafton, MA: Department of Environmental Protection.

Eadie, J. B., W. Faust, W. S. Gardner, and T. Nalepa. 1982. Polycyclic aromatic hydrocarbons in sediments and associated benthos in Lake Erie. *Chemosphere* 11:185–91.

Encalada, R. R., and E. Millan. 1990. Impacto de las aguas residuales industriales y domesticas sobre las comunidades bentónicas de la Bahía de Todos Santos, Baja California, Mexico. *Cinc Mar* 16 (4): 121–39.

Engle, V., J. K. Summers, and G. T. R. Gaston. 1994. A benthic index of environmental condition of Gulf of Mexico. *Estuaries* 17 (2): 372–84.

Fath, B., and B. C. Patten. 2001. A progressive definition of network aggradation. In *Advances in energy studies*, ed. S. Ulgiati, 551–62. Padua: SGE Editoriali.

Fauchald, K. 1977. The polychaete worms. Definitions and keys to the orders, families and genera. *Sci Ser* 28:21–30.

Fauchald, K., and P. Jumars. 1979. The diet of the worms: A study of Polychaete feeding guilds. *Oceanogr Mar Biol Ann Rev* 17:193–284.

Fu-Liu, X. 1997. Exergy and estructural exergy as ecological indicators for the state of the Lake Chalou ecosystem. *Ecological Modelling* 99:41–49.

Gambi, M. C., and A. Giangrande. 1985. Characterization and distribution of polychaete trophic groups in the soft-bottoms of the Gulf of Salerno. *Oebalia* 11 (1): 223–40.

Gardner, M. R., and W. R. Ashby. 1970. Connectance of large synamical (cybernetic) systems: Critical values for stability. *Nature* 288:784.

Gauch, H. G. 1982. *Multivariate analysis in community ecology*. Cambridge Studies in Ecology. New York: Cambridge University Press.

Gee, J. M., R. M. Warwick, M. Schaanning, J. A. Berge, and W. G. Ambrose. 1985. Effects of organic enrichment on meiofaunal abundance and community structure in sublittoral soft sediments. *J Exp Mar Biol Ecol* 91 (3): 247–62.

Gibbs, P. E., W. J. Langston, W. J. Burt, G. R., and P. L. Pascoe. 1983. *Tharyx marioni* (Polychaeta): A remarkable accumulator of arsenic. *J Mar Biol Assoc UK* 63 (2): 313–25.

Gilliland, M.W. 1982. *Embodied energy studies of metal and fuel minerals.* Report to National Science Foundation.

Glemarec, M., and C. Hily. 1981. Perturbations apportées à la macrofaune benthique de la Baie de Concarneau par les effluents urbains et portuaries. *Acta Oecol Appl* 2 (2): 139–50.

Goff, F. G., and G. Cottam. 1967. Gradient analysis: The use of species and synthetic indices. *Ecology* 48:793–806.

Goldberg, E., V. T. Bowen, J. W. Farrington, G. Harvey, P. L. Martin, P. L. Parker, R. W. Risebrough, W. Robertson, E. Schnider, and E. Gamble. 1978. The mussel watch. *Environ Conserv* 5:101–25.

Golovenko, V. K., A. A. Schepinsky, and V. A. Shevchenko. 1981. Accumulation of DDT and its metabolism in Black Sea Mussels. *Izv Akad Nauk SSR Ser Biol* 4:453–550.

Gomez-Gesteira, J.L., J.C. Dauvin. 2000. Amphipods are good bioindicators of the impact of oil spills on soft-bottom macrobenthic communities. *Mar Pollut Bull* 40 (11): 1017–1027.

Gosset, R. W., D. A. Brown, and D. R. Young. 1983. Predicting the bioaccumulation of organic compounds in marine organisms using octanol/water partition coefficients. *Mar Pollut Bull* 14:387–92.

Gray, J. S. 1979. Pollution-induced changes in populations. *Phil Trans R Soc London* 286:545–61.

Gray, J. S., M. Aschan, M. R. Carr, K. R. Clarke, R. H. Green, T. H. Pearson, R. Rosenberg, R. M. Warwick, and B. L. Bayne. 1988. Analysis of community attributes of the benthic macrofauna of Frierfjord/Langesundfjord and in a mesocosm experiment. In *Biological effects of pollutants. Results of a practical workshop*, eds. R. M. Warwick and K. R. Clarke, 151–65. 46 (1–3).

Gray, J. S., and F. B. Mirza. 1979. A possible method for the detection of pollution induced disturbance on marine benthic communities. *Mar Pollut Bull* 10:142–46.

Hall, C. A. S. 1995. *Maximum Power: The Ideas and Applications of H.T. Odum*. University Press of Colorado. Niwot, CO.

Hannon, B. 1973. The structure of ecosystems. *J Theor Biol* 41:535–46.

Hannon, B., and M. Ruth. 1997. *Modeling dynamic biological systems*. Berlin: Springer-Verlag.

Harrison, T. D., J. A. Cooper, A. E. Ramm, and R. A. Singh, 1994. *Application of the Estuarine Health Index to South Africa's estuaries. Orange River – Buffels (Oos)*. Unpublished Technical Report, CSIR, Durban.

Harrison, T. D., J. A. Cooper, A. E. Ramm, and R. A. Singh, 1995. *Application of the Estuarine Health Index to South Africa's estuaries. Palmiet – Sout*. Unpublished Technical Report, CSIR, Durban.

Harrison, T. D., J. A. Cooper, A. E. Ramm, and R. A. Singh, 1997. *Application of the Estuarine Health Index to South Africa's estuaries. Old Woman's – Great Kei*. Unpublished Technical Report, CSIR, Durban.

Harrison, T. D., J. A. Cooper, A. E. Ramm, and R. A. Singh, 1999. *Application of the Estuarine Health Index to South Africa's estuaries. Transkei*. Unpublished Technical Report, CSIR, Durban.

Holling, C. S. 1986. The resilience of terrestrial ecosystems: Local surprise and global change. In *Sustainable development of the biosphere*, eds. W. C. Clark and R. E. Munn, 292–317. Cambridge: Cambridge University Press.

Hong, J. 1983. Impact of the pollution on the benthic community. *Bull Korean Fish Soc* 16:273–90.

Hughes, R. G. 1984. A model of the structure and dynamics of benthic marine invertebrate communities. *Mar Ecol Prog Ser* 15 (1–2): 1–11.

Hynes, H. B. N. 1971. *Ecology of running water*. Liverpool, UK: Liverpool University Press.

Ibanez, F., and J. C. Dauvin. 1988. Long-term changes (1977–1987) in a muddy fine sand Abra alba-Melina palmata community from the Western Channel: Multivariate time series analysis. *Mar Ecol Progr Ser* 19:65–81.

Jørgensen, S. E. 1982. Exergy and buffering capacity in ecological system. In *Energetics and systems*, eds. W. Mitsch et al., 61. Ann Arbor, MI: Ann Arbor Science Publishers.

———. 1994. Review and comparison of goal functions in system ecology. *Vie Mileu* 44 (1): 11–20.

———. 1995. Exergy and ecological buffer capacities as measures of ecosystem health. *Ecosystem Health* 1: 150–160

———. 2002. *Integration of ecosystem theories: A pattern*. Dordrecht, The Netherlands: Kluwer Scientific Publishing Co.

Jørgensen, S. E., B. D. Fath, and S. Bastianoni. 2007. *A New Ecology. A System Approach*. 278 pp. Elsevier. Amsterdam and Oxford.

Jørgensen, S. E. and H. F. Mejer, 1977. Ecological buffer capacity. *Ecological Modelling*: 39–61.

Jørgensen, S. E. and H. F. Mejer. 1979. A holistic approach to ecological modelling. *Ecol Model* 7:169–89.

———. 1981. Exergy as a key function in ecological models. In *Energy and ecological modelling. Developments in environmental modelling, vol. 1*, eds. W. Mitsch, R. W. Bosserman, J. A. Klopatek, 587–590. Amsterdam: Elsevier.

Jørgensen, S. E., S. N. Nielsen, and H. F. Mejer. 1995. Energy, environ, exergy and ecological modelling. *Ecol Model* 77:99–109.

Jørgensen, S. E., B. C. Patten, and M. Straskraba. 2000. Ecosystem emerging IV: Growth. *Ecol Model* (in press).

Kay, J. J. 1984. Self organization in living systems. Thesis, Systems Design Engineering, University of Waterloo, Ontario, Canada.

Kay, J., and E. D. Schneider. 1992. Thermodynamics and measures of ecological integrity. In *Proc. "Ecological Indicators,"* 159–82 Amsterdam: Elsevier.

Lambshead, P. J. D., and H. M. Platt. 1985. Structural patterns of marine benthic assemblages and their relationship with empirical statistical models. In *Proceedings of the Nineteenth European Marine Biology Symposium*, ed. P. E. Gibbs, 16–21. Plymouth.

Lambshead, P. J. D., H. M. Platt, and K. M. Shaw. 1983. The detection of differences among assemblages of marine benthic species based on an assessment of dominance and diversity. *Journal of Natural History* 17:859–74.

Langston, W. J., G. R. Burt, and M. J. A. F. Zhou. 1987. Tin and organotin in water, sediments and benthic organisms of Poole Harbour. *Mar Pollut Bull* 18 (12): 634–39.

Lauenstein, G., A. Robertson, and T. O'Connor. 1990. Comparison of trace metal data in mussels and oysters from a Mussel Watch programme of the 1970s with those from a 1980s programme. *Mar Pollut Bull* 21 (9): 440–47.

Leonzio, C., E. Bacci, S. Focardi, and A. Renzoni. 1981. Heavy metals in organisms from the northern Tyrrhenian sea. *Science of the Total Environment* 20 (2): 131–46.

Leppakoski, E. 1975. Assessment of degree of pollution on the basis of macrozoobenthos in marine and brackish water environments. *Acta Acad Aba* 35:1–98.

Levin, S. A. 1994. Patchiness in marine and terrestrial systems: From individuals to populations. *Philosophical Transactions of the Royal Society of London, Series B* 343:99–103.

Levine, H. G. 1984. The use of seaweeds for monitoring coastal waters. In *Algae as ecological indicators*, ed. L. E. Shubert. London: Academic Press.

Li, W.-H. and Grauer, D. 1991. *Fundamentals of Molecular Evolution*. Sunderland, MA: Sinauer.

Maeda, S., and T. Sakaguchi. 1990. Accumulation and detoxification of toxic metal elements by algae. In *Introduction to applied phycology*, ed. I. Akatsuka. SPB Academic Publishing bv.

Magurran, A. E. 1989. *Diversidad ecológica y su medición*. Barcelona: Vedrá.

Malins, D. C., B. B. McCain, M. S. Myers, D. W. Brown, A. K. Sparks, J. F. Morado, and H. O Hodgins. 1984. Toxic chemicals and abnormalities in fish and shellfish from urban bays of Puget Sound. *Responses of Marine Organisms to Pollutants* 14 (1–4): 527–28.

Maree, R. C., A. K. Whitfield, and N. W. Quinn. 2000. Prioritisation of South African estuaries based on their potential importance to estuarine-associated fish species. Unpublished report for Water Research Commission.

Margalef, R. 1978. *Ecología*. Barcelona: Omega.

Marina, M., and O. Enzo. 1983. Variability of zinc and manganese concentrations in relation to sex and season in the bivalve *Donax trunculus*. *Mar Pollut Bull* 14 (9): 342–46.

Maurer, D., and W. Leathem. 1981. Polychaete feeding guilds from Georges Bank, USA. *Mar Biol* 62 (2–3): 161–71.

Maurer, D., W. Leathem, and C. Menzie. 1981. The impact of drilling fluid and well cuttings on Polychaete feeding guilds from the US northeastern continental shelf. *Mar Pollut Bull* 12 (10): 342–47.

Mauersberger, P. 1983. General principles in deterministic water quality modeling. In *Mathematical modeling of water quality: Streams, lakes and reservoirs*. International Series on Applied Systems Analysis 12, ed. G. T. Orlob, 42–115. New York: Wiley.

———. 1985. Optimal control of biological processes in aquatic ecosystem. *Gerlands Beitr Geiophys* 94:141–47.

Mauersberger, P. 1995. Entropy control of complex ecological processes. In *Complex Ecology. The Part Whole Relation in Ecosystems*. B. C. Patten and S. E. Jørgensen (eds.). Englewood Cliffs, NJ: Prentice-Hall. pp. 130–165.

May, R. M. 1977. *Stability and complexity in model ecosystems*. 3rd ed. Princeton, NJ: Princeton University Press.

McElroy, A. E. 1988. Trophic transfer of PAH and metabolites (fish, worm). *Responses of Marine Organisms to Pollutunats* 8 (1–4): 265–69.

McManus, J. W., and D. Pauly. 1990. Measuring ecological stress: Variations on a theme by R. M. Warwick. *Mar Biol* 106 (2): 305–308.

Meire, P. M., and J. Dereu. 1990. Use of the abundance/biomass comparison method for detecting environmental stress: Some considerations based on intertidal macrozoobenthos and bird communities. *J Appl Ecol* 27 (1): 210–23.

Mejer, H. F. and S. E. Jørgensen, 1979. Exergy and ecological buffer capacity. p. 827–846 in *State-of-the-Art of Ecological Modelling*. Proceedings of ISEM's First International Conference, Copenhagen (1978).

Mendez-Ubach, N. 1997. Polychaetes inhabiting soft bottoms subjected to organic enrichment in the Topolobampo lagoon complex, Sinaloa, México. *Océanides* 12 (2): 79–88.

Miller, B. S. 1986. Trace metals in the common mussel *Mytilus edulis* (L.) in the Clyde Estuary. *Proc R Soc Edinb* 90:379–91.

Mo, C., and B. Neilson. 1991. Variability in measurements of zinc in oysters, *C. virginica*. *Mar Pollut Bull* 22 (10): 522–25.

Mohlenberg, F., and H. U. Riisgard. 1988. Partioning of inorganic and organic mercury in cockles *Cardium edule* (L.) and *C. glaucum* (Bruguiere) from a chronically polluted area: Influence of size and age. *Environmental Pollution* 55:137–48.

Molvær, J., J. Knutzen, J. Magnusson, B. Rygg, J. Skei, and J. Sørensen. 1997. Classification of environmental quality in fjords and coastal waters. *SFT Guidelines* 97:3.

Moreno, D., P. A. Anguilera, and H. Castro. 2001. Assessment of the conservation status of seagrass (*Posidonia oceanica*) meadows: Implications for monitoring strategy and the decision-making process. *Biological Conservation* 102:325–32.

Myers, M. S., L. D. Rhodes, and B. B. McCain. 1987. Pathologic anatomy and patterns of occurrence of hepatic neoplasm, putative preneoplastic lesions and other idiopathic hepatic conditions in English sole (*Parophrys vetulus*) from Puget Sound, Washington, U.S.A. *J Natl Cancer Inst* 78 (2): 333–63.

Nelson, W. G. 1990. Prospects for development of an index of biotic integrity for evaluating habitat degradation in coastal systems. *Chemistry and Ecology* 4:197–210.

Newmann, G., M. Notter, and H. Dahlgaard. 1991. Bladder-wrack (*Fucus vesiculosus* L.) as an indicator for radionclides in the environment of Swedish nuclear power plants. *Swedish Environmental Protection Agency* 3931:1–35.

Niell, F. X., and J. P. Pazo. 1978. Incidencia de vertidos industriales en la estructura de poblaciones intermareales. II. Distribución de la biomasa y de la diversidad específica de comunidades de macrófitos de facies rocosa. *Inv Pesq* 42 (2): 231–39.

O'Connor, J. S., and R. T. Dewling. 1986. Indices of marine degradation: Their utility. *Environmental Management* 10:335–43.

Odum, E. P. 1969. The strategy of ecosystem development. *Science* 164:262–70.

———. 1971. *Fundamentals of ecology*. Philadelphia: W. B. Saunders Co.

Odum, H. T. 1983. *System ecology*. New York: Wiley Interscience.

———. 1996. *Environmental Accounting. Emergy and Environmental Decision Making*. New York: Wiley.

O'Neill, R. V. 1976. Ecosystem persistence and heterotrophic regulation. *Ecology* 57:1244–53.

OSPAR/MON, 1998. Report of ad hoc working group on Monitoring (MON), Copenhagen, 23–37 February 1998.

Patten, B. C. 1992. Energy, emergy and environs. *Ecol Model* 62:29–70.

Pearson, T. H., and R. Rosenberg. 1978. Macrobenthic succession in relation to organic enrichment and pollution of the marine environment. *Oceanogr Mar Biol Ann Rev* 16:229–331.

Pérez-Ruzafa, A., and C. Marcos. 2003. Contaminación marina: Enfoques y herramientas para abordar los problemas ambientales del medio marino. In *Perspectivas y herramientas en el estudio de la contaminación marina*, eds. A. Pérez-Ruzafa, C. Marcos, F. Salas, and S. Zamora, Chapter 1, 11–33. Murcia: Servicio de Publicaciones, Universidad de Murcia.

Pergent, G., C. Pergent-Martini, and C. F. Boudoureque. 1995. Utilisation de l'herbier à Posidonie oceanica comme indicateur biologique de la qualité du milieu littoral en Méditerranée: État des connaissances. *Mésogée* 54:3–27.

Pergent, G., C. Pergent-Martini, and V. Pasqualini. 1999. Preliminary data on impact of fish farming facilities on Posidonia oceanica meadows in the Mediterranean. *Oceanologica Acta* 22 (1): 95–107.

Philips, D. H. J. 1977. The use of biological indicator organisms to monitor trace metal pollution in marine and estuarine environments—A review. *Environ Pollut* 13:281–317.

Picard-Berube, M., and D. Cossa. 1983. Teneurs en benzo 3,4 pyréne chz Mytilus edulis. L. de léstuarie et du Golfe du Saint-Laurent. *Marine Environmental Research* 10:63–71.

Pires, S., and P. Múniz. 1999. Trophic structure of polychaetes in the São Sebastião Channel (southeastern Brazil). *Mar Biol* 5:517–28.

Pocklington, P., D. B. Scott, and C. T. Schafer. 1994. Polychaete response to different aquaculture activities. *Proceedings of the 4th International Polychaete Conference* 162:511–20.

Prigogine, I. 1947. *Etude thermodynamique des phénomènes irreversibles*. Liège: Desoer.

———. 1980. *From Being to Becoming. Time and Complexity in the Physical Sciences*. San Francisco, CA: Freeman.

Rafaelli, D. G., and C. F. Mason. 1981. Pollution monitoring with meiofauna using the ratio of nematodes to copepods. *Mar Poll Bull* 12:158–63.

Ramm, A. E. 1988. The community degradation index: A new method for assessing the deterioration of aquatic habitats. *Water Resources*, 22: 293–301.

Ramm, A. E. 1990. Application of the community degradation index to South African estuaries. *Water Resources*, 24: 383–389.

Reeves, H. 1991. *The Hour of our Delight: Cosmic Evolution, Order and Complexity*. New York: Freeman. 246 pp.

Regoli, F., and E. Orlando. 1993. *Mytilus galloprovincialis* as a bioindicator of lead pollution: Biological variables and cellular responses. In *Proceedings of the Second European Conference on Ecotoxicology*, eds. W. Sloof and H. De Kruijf, 1–2 .

Reish, D. J. 1993. Effects of metals and organic compounds on survival and bioaccumulation in two species of marine gammaridean amphipod, together with a summary of toxicological research on this group. *Journal of Natural History* 27:781–94.

Reish, D. J., and T. V. Gerlinger. 1984. The effects of cadmium, lead, and zinc on survival and reproduction in the polychaetous annelid *Neanthes arenaceodentata* (F. Nereididae). *Proceedings of the First International Polychaete Conference*, 383–89.

Reizpoulou, S., M. Thessalou-Legaki, and A. Nicoloaidou. 1996. Assessment of disturbance in Mediterranean lagoons: An evaluation of methods. *Mar Biol* 125:189–97.

Renberg, L., M. Tarpea, and G. Sundstroem, 1986. The use of the bivalve *Mytilus edulis* as a test organism for bioconcentration studies: II. The bioconcentration of two super(14)C-labeled chlorinated paraffins. *Ecotoxicol Environ Saf* 11 (3): 361–72.

Richardson, B. J., and J. S. Waid. 1983. Polychlorinated biphenyls (PCBs) in shellfish from Australian coastal waters. *Ecol Bull (Stockholm)* 35:511–17.

Riisgard, H. U., T. Kierboe, F. Moenberg, I. Drabaeki, and P. Madsen. 1985. Accumulation, elimination and chemical speciation of mercury in the bivalves *Mytilus edulis* and *Macoma balthica*. *Mar Biol* 86:55–62.

Ritz, D. A., M. E. Lewis, and M. Shen. 1989. Response to organic enrichment of infaunal macrobenthic communities under salmonid seacages. *Mar Biol* 103:211–14.

Ritz, D. A., R. Swain, and N. G. Elliot. 1982. Use of the mussel *Mytilus edulis planulatus* (Lamarck) in monitoring heavy metal levels in seawater. *Aust J Mar Freshwater Res* 33 (3): 491–506.

Roberts, R. D., M. G. Gregory, and B. A. Foster. 1998. Developing an efficient macro-fauna monitoring index from an impact study—a dredge spoil example. *Mar Pollut Bull* 36 (3): 231–35.

Roesijadi, G. 1994. Metallothionein induction as a measure of response to metal exposure in aquatic animals. *Genetic and Molecular Ecotoxicology* 102 (12): 91–96.

Romeo, M., and M. Gnassia-Barelli. 1988. *Donax trunculus* and *Venus verrucosa* as bioindicators of trace metal concentrations in Mauritanian coastal waters. *Mar Biol* 99 (2): 223–27.

Ros, J. D., and M. J. Cardell. 1991. La diversidad específica y otros descriptores de contaminación orgánica en comunidades bentónicas marinas. *Actas del Symposium sobre Diversidad Biológica*, 219–23. Madrid: Centro de Estudios Ramón Areces.

Ros, J. D., M. J. Cardell, V. Alva, C. Palacin, and I. Llobet. 1990. Comunidades sobre fondos blandos afectados por un aporte masivo de lodos y aguas residuales (litoral frente a Barcelona, Mediterráneo occidental): Resultados preliminares. *Bentos* 6:407–23.

Rosenzweig, M. L. 1971. Paradox of enrichment: Destabilization of exploitation ecosystems in ecological time. *Science* 171:385–87.

Rutledge R. W. 1974. Ecological Stability: A Systems Theory Viewpoint (thesis). Oklahoma State University, Oklahoma.

Rygg, B. 1985. Distribution of species along a pollution gradient induced diversity gradients in benthic communities in Norwegian Fjords. *Mar Pollut Bull* 16 (12): 469–73.

Salas, F. 1996. Valoración de los indicadores biológicos de contaminación orgánica en la gestión del medio marino. Tesis de Licenciatura, Universidad de Murcia.

———. 2002. Valoración y aplicabilidad de los índices y bioindicadores de contaminación orgánica en la gestión del medio marino. Tesis Doctoral, Universidad de Murcia.

Salas, F., J. M. Neto, A. Borja, and J. C. Marques. 2004. Evaluation of the applicability of a marine biotic index to characterise the status of estuarine ecosystems: The case of Mondego estuary (Portugal). *Ecological Indicators* 4:215–25.

Satsmadjis, J. 1982. Analysis of benthic fauna and measurement of pollution. *Revue Internationale Oceanographie Medicale* 66–67:103–107.

Shannon, C. E., and W. Wiener. 1963. *The mathematical theory of communication*. Urbana: University of Illinois Press.

Shaw, K. M., P. J. D. Lambshead, and H. M. Platt. 1983. Detection of pollution-induced disturbance in marine benthic assemblages with special reference to nematodes. *Mar Ecol Progr Ser* 11:195–202.

Simboura, N., and A. Zenetos. 2002. Benthic indicators to use ecological quality classification of Mediterranean soft bottom marine ecosystems, including a new Biotic Index. *Mediterranean Marine Science* 3 (2).

Simpson, E. H. 1949. Measurement of diversity. *Nature* 163:688.

Smith, R. W., M. Bergen, S. B. Weisberg, D. Cadien, A. Dalkey, D. Montagne, J. K. Stull, and R. G. Velarde. 2001. Benthic response index for assessing infaunal communities on the mainland shelf of southern California. *Ecological Applications* 11:1073–87.

Somerfield, P. J., and K. R. Clarke. 1997. A comparison of some methods commonly used for the collection of sublittoral sediments and their associated fauna. *Mar Environ Res* 43:145–56.

Stirn, J., A. J. Avcin, I. Kerzan, B. M. Marcotte, N. Meith-Avcin, B. Vriser, and S. Vukovic. 1971. Selected biological methods for assessment of pollution. In *Marine pollution and waste disposal*, eds. E. A. Pearson and E. Defraja, 307–28. Oxford, UK: Pergamon Press.

Storelli, M. M., and G. O. Marcotrigiano. 2001. Persistent organochlorine residues and toxic evaluation of polychlorinated biphenyls in sharks from the Mediterranean Sea (Italy). *Mar Pollut Bull* 42 (12): 1323–329.

Taghon, C. L., A. R. M. Nowell, and P. Jumars. 1980. Induction of suspension feeding in Spionid Polychaetes by high particulate fluxes. *Science* 210:562–64.

Ulanowicz, R. E. 1986. *Growth and development ecosystems phenomenology.* New York: Springer-Verlag.

———. 1997. *Ecology, the ascendent perspective.* New York: Columbia University Press.

Ulanowicz, R. E., and C. J. Puccia. 1990. Mixed trophic impacts in ecosystems. *Coenoses* 5 (1): 7–16.

Vaas, P. A., and S. J. Jordan. 1990. Long term trends in abundance indices for 19 species of Chesapeake Bay fishes: Reflection of trends in the Bay ecosystem. In *New perspectives in the Chesapeake system: A research and management partnership. Proceedings of a Conference.* Chesapeake Research Consortium Publ., 137.

Varanasi, U., S. Chan, and R. Clark. 1989. *National Benthic Surveillance Project: Pacific coast.* Washington, DC: National Ocean Service.

Verlaque, M. 1977. Impact du rejet thermique de Martigues-Ponteau sur le macrophytobenthos. *Tethys* 8 (1): 19–46.

Viarengo, A., and L. Canesi. 1991. Mussels as biological indicators of pollution. *Aquaculture* 94:225–43.

Vollenweider, R. A., F. Giovanardi, G. Montanari, and A. Rinaldi. 1998. Characterisation of the trophic conditions of marine coastal waters with special reference to the NW Adriatic Sea: Proposal for a trophic scale, turbidity and generalised water quality index. *Environmetrics* 9:329–57.

Von Bertalanffy, L. 1952. *Problems of life.* New York: Wiley.

Warwick, R. M. 1986. A new method for detecting pollution effects on marine macrobenthic communities. *Mar Biol* 92:557–62.

———. 1993. Environmental impact studies on marine communities: Pragmatical considerations. *Aust J Ecol* 18:63–80.

Warwick, R. M., and K. R. Clarke. 1994. Relearning the ABC: Taxonomic changes and abundance/biomass relationships in disturbed benthic communities. *Mar Biol* 118:739–44.

———. 1995. New biodiversity measures reveal a decrease in taxonomic distinctness with increasing stress. *Mar Ecol Prog Ser* 129:301–305.

———. 1998. Taxonomic distinctness and environmental assessment. *J Appl Ecol* 35:532–43.

Weiderholm, T. 1980. Use of benthos in lake monitoring. *J. Wat. Pollution Control Fed.* 52: 537–557.

Weisberg, S. B., J. A. Ranashinghe, D. M. Dauer, L. C. Schaffner, R. J. Diaz, and J. B. Frithsen. 1997. An estuarine benthic index of biotic integrity (B-IBI) for the Chesapeake Bay. *Estuaries* 20:149–58.

Weston, D. P. 1990. Quantitative examination of macrobenthic community changes along an organic enrichment gradient. *Mar Ecol Progr Ser* 61 (3): 233–44.

Word, J. Q. 1979. The infaunal trophic index. *Calif Coast Wat Res Proj Annu Rep* 19–39.

———. 1980. Classification of benthic invertebrates into infaunal trophic index feeding groups. *Biennial Report*, 1979–80, Coastal Water Research Project, Los Angeles, pp. 103–121.

Xu, F.-L., S. Tao, R. W. Dawson, B. G. Li, and J. Cao. 2001. Lake ecosystem health assessment: Indicators and methods. *Water Research* 35 (13): 3157–167.

Yokoyama, H. 1997. Effects of fish farming on macroinvertebrates. Comparison of three localities suffering from hypoxia. *UJNR Technical Report* 24:17–24.

Zabala, K., A. Romero, and M. Ibanez. 1983. La contaminación marina en Guipuzcoa: I. Estudio de los indicadores biológicos de contaminación en los sedimentos de la Ría de Pasajes. *Lurralde* 3:177–89.

3

Eco-Exergy as Ecological Indicator

S. E. Jørgensen

CONTENTS

3.1 Thermodynamic Concepts as Super-Holistic Indicators

Thermodynamic indicators may be denoted as super-holistic indicators because they attempt to quantify by use of one indicator the holistic properties of an ecosystem. In Section 2.8, we presented the thermodynamic indicators eco-exergy, specific exergy = eco-exergy/biomass, power, and emergy. It was discussed how these thermodynamic indicators holistically express the ecosystem properties. The two latter indicators will be presented in detail in Chapter 4, and Chapter 5 will present the ratio of eco-exergy and emergy as an additional super-holistic thermodynamic indicator.

Eco-exergy measures the work capacity of the system, i.e., the amount of first-class energy = energy that can do work, which the system contains. As all activities require work, the eco-exergy content may also be considered the expression of the sustainability of the system. In this context it has been shown (Svirezhev 1998) that eco-exergy also expresses the amount of energy needed to break down the system to thermodynamic equilibrium. In other words, eco-exergy expresses what the system has been able to gain in sustainability or work capacity. Emergy, on the other hand, expresses how much it has cost to reach the obtained sustainability or work capacity in the ultimate energy source on earth, sunlight. Therefore, the ratio between eco-exergy and emergy expresses how effective the system has been to move away from thermodynamic equilibrium on the basis of the energy input—the captured sunlight. It is not surprising, but will be shown in detail, that natural systems are much better at gaining eco-exergy from a given amount of emergy than man-controlled systems.

This chapter provides further details about the translation of eco-exergy to the assessment of ecosystem health. Furthermore, some examples will be applied to illustrate the use of eco-exergy and specific eco-exergy as holistic indicators.

The theoretical background for the application of eco-exergy as an indicator has already been presented in Section 2.8. In this chapter we will use the equation presented in Section 2.8 to calculate the exergy or eco-exergy index. According to Section 2.8, the exergy index can be found as the concentrations of the various components, c_i, multiplied by weighting factors, β_i, reflecting the exergy that the various components possess due to their chemical energy and to the information embodied in genome:

$$Ex = \sum_{i=n}^{i=0} \beta_i c_i \qquad (3.1)$$

β_i values based on exergy detritus equivalents have been found for various species. The unit exergy detritus equivalents expressed in g/L can be converted to kJ/L by multiplying by 18.7, corresponding to the approximate average energy content of 1 g detritus. Table A.1 in the Appendix shows the β-values calculated from the number of information genes by the above presented equations.

3.2 Eco-Exergy as Ecosystem Health Indicator

The combination of the eco-exergy index and the specific eco-exergy index (= eco-exergy/biomass) usually gives a more satisfactory description of the health of an ecosystem than the exergy index alone, because it considers the diversity and the life conditions for higher organisms (see also Jørgensen 2002). The combination of exergy, specific exergy, and buffer capacities, defined as the change in a forcing function relative to the corresponding change in a state variable, has been used as an ecological indicator for lakes and coastal ecosystems.

The eco-exergy index and the specific eco-exergy index together give an indication of ecosystem health corresponding to the six attributes presented by Costanza (1992). Below are the six attributes presented and explanations for why they are covered by eco-exergy and specific eco-exergy:

1. Homeostasis is required to ensure survival and eco-exergy directly measures survival.
2. Absence of disease is needed to ensure growth. All three growth forms (see Section 2.11) imply increase of eco-exergy.
3. Diversity or complexity is covered, because high exergy must imply that all or almost all ecological niches are occupied and that obviously requires high diversity and complexity. That all ecological niches are occupied implies furthermore that a number of higher organisms with a high β-value are present, i.e., the specific eco-exergy is high.
4. Stability and resilience are partially covered by buffer capacity, which, as discussed in Section 2.8, is proportional to exergy by a statistical analysis of model results.
5. Vigor or scope for growth: as discussed in Section 2.8, the three growth forms all involve increased exergy.
6. Balance between system components means that there are both higher and lower organisms and many different organisms present, which implies that the β-value is relatively high, i.e., the specific eco-exergy is high.

It is possible to conclude that eco-exergy and specific eco-exergy are very informative and can be considered as indicators of health and sustainability for ecosystems. If an ecosystem maintains its eco-exergy and its specific eco-exergy for a longer period of time, we can conclude that the ecosystem is healthy and sustainable. Environmentally sound management of ecosystems should therefore imply that the ecosystems maintain—of course, with seasonal fluctuations—on a long-term basis their levels of eco-exergy and specific eco-exergy.

The relationship between eco-exergy and specific eco-exergy and other relevant indicators has been found through the examination of 12 marine ecosystem models presented by Christensen and Pauly (1993). The results of the examination are presented in Jørgensen (2006) and further support the coverage of Costanza's six attributes by eco-exergy and specific eco-exergy. One of the models used in this examination is shown in Figure 3.1 as an illustration of the steady-state models that can be developed by the downloadable software Ecopath. The following ecological indicators were determined for 12 ecosystems:

A. Biomass, B (g dry weight/m²)
B. Respiration, R [g dry weight/(m² y)]

C. Eco-exergy, Ex (kJ/m^2)

D. Exergy destruction – ΔEx (kJ/m^2 y)

E. Diversity as number of species included in the model (–)

F. Connectivity as number of connections relative to the total number of possible connections (–)

G. Complexity expressed as "diversity" times "connectivity" (–)

H. Respiration/biomass = RE/A (year^{-1}) (RE = respiration, A = biomass)

I. Exergy destruction/exergy = $-\Delta$Ex/Ex (year^{-1})

J. Exergy production (kJ/m^2 year)

K. Specific exergy = Ex/biomass (kJ/g)

It was found by a correlation matrix that only the following 11 indicators were correlated with a correlation coefficient ≥ 0.65:

I. Eco-exergy production to eco-exergy, $r^2 = 0.93$

II. Respiration to exergy, $r^2 = 0.98$

III. Eco-exergy destruction to respiration, $r^2 = 0.87$

IV. Respiration to exergy production, $r^2 = 0.855$

V. Respiration/biomass to specific exergy, $r^2 = 0.86$

VI. Respiration to biomass, $r^2 = 0.68$; notice in this context that respiration is considerably better correlated to exergy than to biomass.

All the correlations are based on a linear relationship, which seems reasonable at least as a first approximation. The six relationships can easily be explained on basis of the presentation of eco-exergy and specific eco-exergy in Section 2.8 and in this chapter.

Higher eco-exergy levels, at least for the examined marine ecosystems, are associated with higher rates of eco-exergy production, which is consistent with the translation of Darwin's theory to thermodynamics by use of eco-exergy (see Section 2.8). The development of an ecosystem is toward increasing biomass, and when almost all the inorganic matter is used to build biomass, a reallocation of the matter in the form of species with more information may take place. Increased information increases the possibility of building even more eco-exergy (information).

Biomass includes plants (algae), which have relatively low eco-exergy and also lower respiration. It explains why eco-exergy with high weighting factors for fish and other higher organisms is better correlated with respiration than biomass. The general relationship between respiration and eco-exergy is not surprising, as more stored eco-exergy means that the ecosystem has more biomass, and becomes more complex and more developed, which implies that it also requires more energy (eco-exergy) for maintenance.

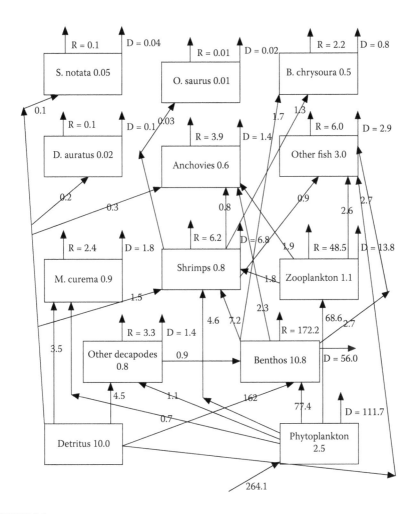

FIGURE 3.1
Ecopath model of Tamahua, a coastal lagoon in Mexico. All biomasses are in g m^{-2} and rates are in g m^{-2} y^{-1}. R indicates the respiration rates and D the rates of detritus formation.

A linear correlation between the respiration level and the rate of eco-exergy destruction is obvious, as the eco-exergy destruction is caused mainly by respiration. It is just two sides of the same coin. Ecosystems with more eco-exergy can also produce more eco-exergy (consider, for instance, that the production is a first-order reaction) and furthermore require more energy for maintenance, which is respiration.

More specific eco-exergy for ecosystems—higher specific eco-exergy means more dominance of higher organisms—is well correlated to the ratio of respiration to biomass. Usually, higher organisms are characteristic for more developed ecosystems, where both the eco-exergy and the specific eco-exergy (the level of information) are high. However, in aquatic ecosystems a high biomass and a low information level (the specific eco-exergy is low) may be characteristic for eutrofied ecosystems, where plants with a relatively low respiration relative to the very high biomass are dominant. Therefore high specific exergy → less biomass and more relative respiration → respiration/biomass increases.

Based upon the overall results in Jørgensen (2006), we can conclude that

A. Exergy measures the distance from thermodynamic equilibrium. Svirezhev (1998) has shown that exergy measures the amount of energy needed to break down the ecosystem. Exergy is therefore a reasonable good measure of (compare with Costanza 1992):
   1. Absence of disease (may be measured by the growth potential)
   2. Stability or resilience (destruction of the ecosystem is more difficult the more eco-exergy the ecosystem has)
   3. Vigor or scope for growth

Eco-exergy is, in other words, a good expression for sustainability.

B. Specific eco-exergy measures organization in the sense that more developed organisms correspond to higher specific eco-exergy. More developed organisms usually represent higher trophic levels. It implies a more complicated food web. Specific eco-exergy is therefore a reasonable good measure of:
   1. Homeostasis (more feedback is present in a more complicated food web)
   2. Diversity or complexity
   3. Balance between system components—the ecosystem is not dominated by the first trophic levels, as is usual for ecosystems at an early stage.

In other words, specific eco-exergy gives information about whether the sustainability is based on biomass or on information. If both eco-exergy and specific eco-exergy are high, the sustainability is based upon both a high biomass and a high information, which, of course, is preferable.

Notice that eco-exergy or specific eco-exergy is not correlated to diversity or complexity determined by the connectivity. Complexity of ecosystems has several dimensions: complexity due to the presence of more complex organisms, complexity due to diversity, and complexity due to a more complex network. These three different complexities may increase independently. As stability and buffer capacities are not related to complexity, biodiversity can hardly be applied as a measure for sustainability directly, but biodiversity can maybe be used to express the probability of maintaining a high sustainability. It should, in this context, not be forgotten that a high biodiversity gives a wider spectrum of buffer capacities (Jørgensen 2002).

3.3 Illustrative Examples of Eco-Exergy as Ecosystem Health Indicator

Zaldívar et al. (2005) have used eco-exergy and specific eco-exergy to assess the ecosystem health of coastal lagoons. They examined three situations for a lagoon with clam production and eutrophication by *Ulva*. They used a lagoon model to examine:

1. The present situation
2. The optimal strategy based on cost-benefit for removal of *Ulva*
3. A significant nutrient loading reduction from watershed by the use of wastewater treatment, constructed wetland, and restrictions for the agricultural use of fertilizers

They calculated for all three scenarios the eco-exergy and the specific eco-exergy. The results are shown in Figures 3.2 and 3.3.

The results show that the cost-benefit optimal solution for removal of *Ulva* has the highest eco-exergy and specific eco-exergy, followed by a significant

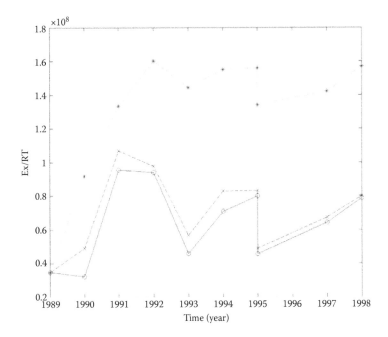

FIGURE 3.2

Exergy mean annual values: Present scenario (*continuous line*), removal of *Ulva*, optimal strategy from cost-benefit point of view (*dotted line*), and nutrient load reduction from watershed (*dashed line*).

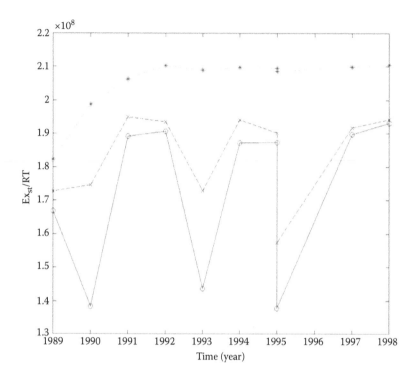

FIGURE 3.3
Specific exergy mean annual values: Present scenario (*continuous line*), removal of *Ulva*, optimal strategy from cost-benefit point of view (*dotted line*), and nutrient load reduction from watershed (*dashed line*).

removal of nutrients from the watershed. The present situation had the lowest eco-exergy and specific eco-exergy. The result shows, in other words, that it is a good sustainability policy to take good care of natural resources (in this case the clams) by optimal ecological management and/or sufficient pollution control. The natural resources always contribute very considerably to the eco-exergy.

In 1963, a new island emerged 33 km off Iceland by volcanic activity. The area of the island, called Surtsey, is 2.8 km^2 and, since the mid-1960s, the development of the flora and fauna has been followed. Figure 3.4 shows the number of plant species found on the island from 1965 to 2000.

The development of the number of plant species is roughly following an exponential increase. In accordance to Moore's Law, complexity of a system should grow in accordance with a first-order reaction as the number of plant species approximately did in the examined period. The eco-exergy has been found for plants and nesting birds, and the result is depicted in Figure 3.5. The development of plants and birds reinforce each other as more plants attract more birds either directly or indirectly due to the presence of more insects, and more birds mean more droppings that will fertilize the soil and

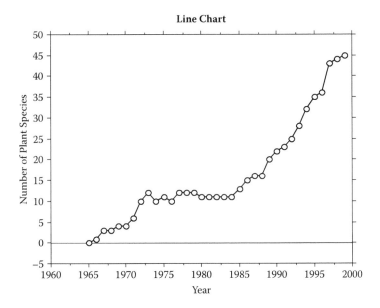

FIGURE 3.4
The number of plant species found on the newly created island of Surtsey from 1965 to 2000.

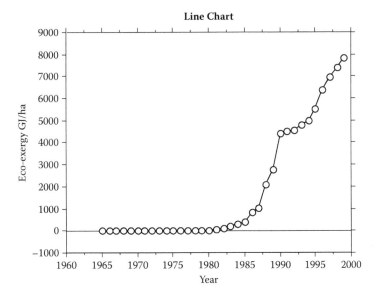

FIGURE 3.5
The development of eco-exergy of plants plus nesting birds over time (1965–2000) is shown. The semi-logarithmic plot in Figure 3.6 indicates that the development is exponential, which is completely in accordance with Moore's Law.

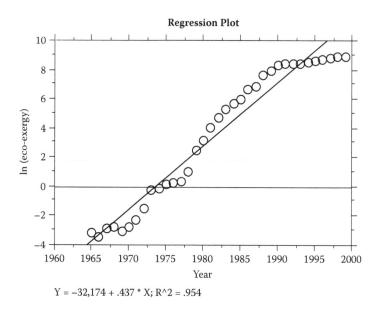

$$Y = -32{,}174 + .437 * X;\ R^2 = .954$$

FIGURE 3.6
A semi-logarithmic plot of eco-exergy versus time (1965–2000).

cause faster plant growth. The eco-exergy development is also following a first-order reaction, exponential growth or Moore's Law. It is seen clearly in Figure 3.6, where a semi-logarithmic plot is applied.

Other similar events show the same pattern. Nature will, after major destruction, have a horizontal (diversity) and vertical (the product of plant biomass and information, i.e., eco-exergy) development, following a first-order reaction, i.e., an exponential increase. The vertical development is beneficially followed by eco-exergy as ecological indicator.

Libralato et al. (2006) have used eco-exergy as an indicator for various marine benthic communities. They found that the eco-exergy declined as a result of disturbances, while the recovery of the ecosystems implied an increase of the eco-exergy. Full recovery of the ecosystems after sufficient time gave approximately the same eco-exergy as after the disturbance. They concluded that eco-exergy is a useful (holistic) indicator for ecosystems, and they recommended a broader application of this indicator.

Jørgensen (2006) has compared different agricultural systems with different complexity, and he finds that eco-exergy is bigger for integrated agriculture or organic agriculture than for industrial agriculture with, for instance, production of pigs and barley only.

3.4 Recommendations and Conclusions

Eco-exergy has been applied in a number of other illustrative examples throughout the book. The conclusions from all the applications are that eco-exergy is a useful holistic indicator, but it must inevitably be supplemented by other indicators to be able to give a more specific diagnosis that can be used to cure the ecosystem. A low eco-exergy can tell us that something is wrong, and that the ecosystem is vulnerable and has little resistance or buffer capacity, but supplementary indicators are needed to tell what is wrong and which medicine we need to use to cure the ecosystem.

References

Christensen, V., and D. Pauly. 1993. *Trophic models of aquatic ecosystems*. ICLARM. Copenhagen: DANIDA.

Costanza, R. 1992. Toward an operational definition of ecosystem health. In *Ecosystem health, new goals for environmental management*, eds. R. Costanza, B. G. Norton, and B. D. Haskell, 239–56. Washington, DC: Island Press.

Jørgensen, S. E. 2002. *Integration of ecosystem theories: A pattern*. 3rd ed. Dordrecht, The Netherlands: Kluwer Academic Publishers.

———. 2006. *Eco-exergy as sustainability*. Southampton, UK: WIT.

Libralato, S., P. Torricelli, and F. Pranovi. 2006. Exergy as ecosystem indicator: An application to the recovery process of marine benthic communities. *Ecological Modelling* 192:571–85.

Svirezhev, Y. M. 1998. Thermodynamic orientors: How to use thermodynamic concepts in ecology. In *Eco targets, goal functions and orientors*, eds. F. Müller and M. Leupelt, 102–22. Heidelberg and Berlin: Springer.

Zaldívar, J. M. 2005. Application of exergy as an ecological indicator in Mediterranean coastal lagoons. In *Handbook of ecological indicators to assess ecosystem health*, eds. S. E. Jørgensen, R. Costanza, and F.-L. Xu. Boca Raton, FL: CRC Press.

4

Emergy Indices of Biodiversity and Ecosystem Dynamics

Mark T. Brown and Sergio Ulgiati

CONTENTS

4.1 Introduction

In this chapter, ecosystems are summarized as energetic systems and ecosystem health is discussed in relation to changes in structure, organization, and functional capacity as explained by changes in emergy, empower, and transformity. The living and nonliving parts and processes of the environment as they operate together are commonly called ecosystems. Examples are forests, wetlands, lakes, prairies, and coral reefs. Ecosystems circulate materials, transform energy, support populations, join components

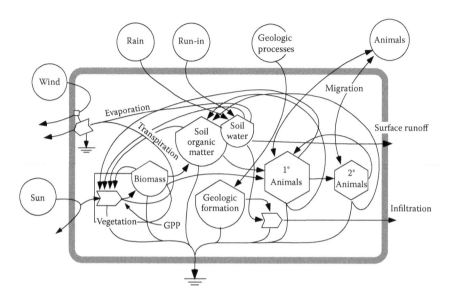

FIGURE 4.1
Generic ecosystem diagram showing driving energies, production, cycling, and the hierarchy of ecological components.

in network interactions, organize hierarchies and spatial centers, evolve and replicate information, and maintain structure in pulsing oscillations. Energy drives all these processes and energetic principles explain much of what is observed.

The living parts of ecosystems are interconnected, each receiving energy and materials from the other, and interacting through feedback mechanisms to self-organize in space, time, and connectivity. Processes of energy transformation throughout the ecosystem build order, cycle materials, and sustain information, degrading energy in the process. The parts are organized in an energy hierarchy as shown in aggregated form in Figure 4.1. As energy flows from driving energy sources on the right to higher and higher order ecosystem components, it is transformed from sunlight to plant biomass, to first-level consumers, to second level, and so forth. At each transformation, second law losses decrease the available energy but the "quality" of energy remaining is increased.

4.2 A Systems View of Ecosystem Health

Conceptually, ecosystem health is related to integrity and sustainability. A healthy ecosystem is one that maintains both system structure and function in the presence of stress. Vigor, resilience, and organization have been suggested as appropriate criteria for judging ecosystem health. Leopold (1949) referred to health of the "land organism" as "the capacity for internal self-renewal." Others have suggested that "[h]ealth is an idea that transcends scientific definition. It contains values, which are not amenable to scientific methods of exploration but are no less important or necessary because of that" (Ehrenfeld 1993). Ecosystem health may be related to the totality of ecosystem structure and function and may only be understood within that framework. Emergy and transformity (Odum 1988; Brown and Ulgiati 2004b) lend some insight into understanding, measuring, and quantifying ecological health (Ulgiati and Brown 2009).

The condition of landscapes and the ecosystems within them is strongly related to levels of human activity. Human-dominated activities and especially the intensity of land uses can affect ecosystems through direct, secondary, and cumulative impacts (Brown and Vivas 2005). Most landscapes are composed of patches of developed land and patches of wild ecosystems. While not directly converted, wild ecosystems often experience cumulative secondary impacts that originate in developed areas and that spread outward into surrounding and adjacent undeveloped lands. The more developed a landscape, the greater the intensity of impacts. The systems diagram in Figure 4.2 illustrates some of the impacts originating in developed lands

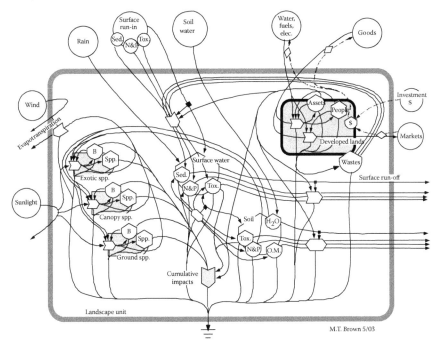

FIGURE 4.2

Landscape unit showing the effects of human activities on ecosystem structure and functions. The more intense the development, the larger the effects. B., biomass; N & P, nitrogen and phosphorus; O.M., organic matter; Sed., sediments; Spp., species; Tox., toxins.

that are experienced by surrounding and adjacent wild ecosystems. These impacts are delivered in the form of air- and water-borne pollutants, physical damage, changes in the suite of environmental conditions (like changes in groundwater levels or increased flooding), or combinations of all of them. Pathways from the developed lands module on the right carry nutrients and toxins that affect surface water and groundwater, which in turn negatively affect terrestrial and marine and aquatic systems. Other pathways interact directly with the biomass and species of wild ecosystems, decreasing the viability and quantity of each. Pathways that affect the inflow and outflow of surface and groundwater may alter hydrologic conditions, which in turn may negatively affect ecological systems. All these pathways of interaction affect ecosystem health.

4.3 Emergy, Transformity, and Hierarchy

Given next are definitions and a brief conceptual framework of emergy theory (Odum 1996; Brown and Ulgiati 2004a) and systems ecology (Odum 1983) that form the basis for understanding ecological systems within the context of ecosystem health.

4.3.1 Emergy and Transformity: Concepts and Definitions

That different forms of energy have different "qualities" is evident from their abilities to do work. While it is true that all energy can be converted to heat, it is not true that one form of energy is substitutable for another in all situations. For instance, plants cannot substitute fossil fuel for sunlight in photosynthetic production, nor can humans substitute sunlight energy for food or water. It should be obvious that the quality that makes an energy flow usable by one set of transformation processes makes it unusable for another set. Thus quality is related to form of energy and to its concentration, where higher quality is somewhat synonymous with higher concentration of energy and results in greater flexibility. So wood is more concentrated than detritus, coal more concentrated than wood, and electricity more concentrated than coal.

The concept of emergy accounts for the environmental-services-supporting process as well as for their convergence through a chain of energy and matter transformations in both space and time. By definition, *emergy* is the amount of energy of one type (usually solar) that is directly or indirectly required to provide a given flow or storage of energy or matter. The units of emergy are emjoules (abbreviated eJ) to distinguish them from energy joules (abbreviated J). Solar emergy is expressed in solar emergy joules (seJ, or solar emjoules). The flow of emergy is *empower*, in units of emjoules per time. Solar empower is solar emjoules per time (e.g., seJ/sec).

When the emergy required to make something is expressed as a ratio to the available energy of the product, the resulting ratio is called a *transformity*.* The solar emergy required to produce a unit flow or storage of available energy is called *solar transformity* and is expressed in solar emergy joules per joule of output flow (seJ/J). The transformity of solar radiation is assumed equal to one (1.0 seJ/J). Transformities of the main natural flows in the biosphere (wind, rain, ocean currents, geological cycles, etc.) are calculated as the ratio of total emergy driving the biosphere, as a whole, to the actual energy of the flow under consideration (Odum 1996). The total emergy driving the biosphere is the sum of solar radiation, deep heat, and tidal momentum and is equivalent to 15.83 E24 seJ/yr, based on a re-evaluation and subsequent recalculation of energy contributions done in the year 2000 (Odum et al. 2000).† This total emergy is used as a driving force for all main biosphere scale processes (winds, rains, ocean currents, and geologic cycles), because these processes and the products they produce are coupled and cannot be generated one without the other (Figure 4.3).

Table 4.1 lists transformities (seJ/J) and specific emergy (seJ/g) of some of the main flows of emergy driving ecological processes. Transformities and specific emergy given in the last column are ratios of the biosphere driving emergy in the second column to the annual production in the third column. Figure 4.3 shows in an aggregated way the emergy of the main biosphere flows that are, in turn, used to account for input flows to processes on smaller space-time scales, like processes in ecosystems as well as in human-dominated systems (Ulgiati and Brown 1999; Brown and Bardi 2001; Brandt-Williams 2002; Kangas 2002). The total emergy driving a process becomes a measure of the self-organization activity of the surrounding environment, converging to make that process possible. It is a measure of the environmental work necessary to provide a given resource. For example, the organic matter in forest soil represents the convergence of solar energy, rain, and winds driving the work processes of the forest over many years that has resulted in layer upon layer of detritus that ever so slowly decomposes into a storage of soil organic matter. It represents part of the past and present ecosystem's work that was necessary to make it available.

---

* The transformity was originally proposed as a measure of energy quality (Odum 1976) and referred to as the energy quality ratio and the energy transformation ratio, but it was renamed transformity in 1983 (Odum et al. 1988). The ratio of emergy to matter produced by a process (i.e., seJ/g) is termed specific emergy. The general term for transformities and specific emergy is Unit Emergy Value (UEV).

† Prior to 2000, the total emergy contribution to the geobiosphere that was used in calculating emergy intensities was 9.44 E24 seJ/yr. The increase in global emergy reference base to 15.83 E24 seJ/yr changes all the emergy intensities, which directly and indirectly were derived from the value of global annual empower. Thus, to be consistent and to allow comparison with older values, emergy intensities calculated prior to the year 2000 are multiplied by 1.68 (the ratio of 15.83/9.44).

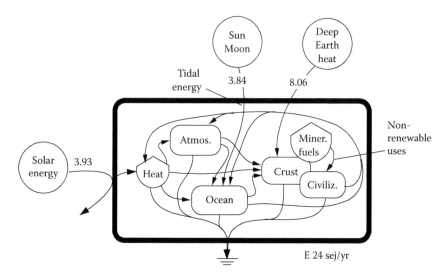

FIGURE 4.3

The main components of the biogeosphere showing the driving energies and the interconnected cycling of energy and matter. The total emergy driving the biogeosphere is the sum of solar, tidal, and deep heat sources totaling 15.83 E24 seJ/yr.

TABLE 4.1

Emergy of Products of the Global Energy System

Product and Units	Emergy[a] E24 seJ/yr	Production Units/yr	Emergy/Unit
Global latent heat, J	15.83	1.26 E24	1.3 E1 seJ/J
Global wind circulation, J	15.83	6.45 E21	2.5 E3 seJ/J
Hurricane, J	15.83	6.10 E20	2.6 E4 seJ/J
Global rain on land, g	15.83	1.09 E20	1.5 E5 seJ/g
Global rain on land (chem. pot.), J	15.83	5.19 E20	3.1 E4 seJ/J
Average river flow, g	15.83	3.96 E19	4.0 E5 seJ/g
Average river geopotential, J	15.83	3.40 E20	4.7 E4 seJ/J
Average river chem. potential, J	15.83	1.96 E20	8.1 E4 seJ/J
Average waves at the shore, J	15.83	3.10 E20	5.1 E4 seJ/J
Average ocean current, J	15.83	8.60 E17	1.8 E7 seJ/J

[a] Main empower of inputs to the geobiospheric system from Figure 4.1 not including nonrenewable consumption (fossil fuel and mineral use).

Source: After Odum, H. T., M. T. Brown, and S. B. Williams. 2000. *Handbook of emergy evaluation: A compendium of data for emergy computation issued in a series of folios. Folio no. 1—Introduction and global budget.* Gainesville, FL: Center for Environmental Policy, Environmental Engineering Sciences, University of Florida. http://www.ees.ufl.edu/cep/

Example transformities of main ecosystem components are given in Tables 4.2 and 4.3. Table 4.2 lists components and processes of terrestrial ecosystems giving several transformities for each. Within each category

TABLE 4.2

Summary of Transformities in Terrestrial Ecosystems

Ecosystem	Transformity (seJ/J)	Reference
Gross primary production		
Subtropical mixed hardwood forest, Florida	1.03E+03	Orrell 1998
Subtropical forest, Florida	1.13E+03	Orrell 1998
Tropical dry savannah, Venezuela	3.15E+03	Prado-Jartar and Brown 1996
Salt marsh, Florida	3.56E+03	Odum 1996
Subtropical depressional forested wetland, Florida	7.04E+03	Bardi and Brown 2001
Subtropical shrub-scrub wetland, Florida	7.14E+03	Bardi and Brown 2001
Subtropical herbaceous wetland, Florida	7.24E+03	Bardi and Brown 2001
Floodplain forest, Florida	9.16E+03	Weber 1994
Net primary production		
Subtropical mixed hardwood forest, Florida	2.59E+03	Orrell 1998
Subtropical forest, Florida	2.84E+03	Orrell 1998
Temperate forest, North Carolina (*Quercus* spp.)	7.88E+03	Tilley 1999
Tropical dry savannah, Venezuela	1.67E+04	Prado-Jartar and Brown 1997
Subtropical shrub-scrub wetland, Florida	4.05E+04	Bardi and Brown 2001
Subtropical depressional forested wetland, Florida	5.29E+04	Bardi and Brown 2001
Subtropical herbaceous wetland, Florida	6.19E+04	Bardi and Brown 2001
Biomass		
Subtropical mixed hardwood forest, Florida	9.23E+03	Orrell 1998
Salt marsh, Florida	1.17E+04	Odum 1996
Tropical dry savannah, Venezuela	1.77E+04	Prado-Jartar and Brown 1997
Subtropical forest, Florida	1.79E+04	Orrell 1998
Tropical mangrove, Ecuador	2.47E+04	Odum and Arding 1991
Subtropical shrub-scrub wetland, Florida	6.91E+04	Bardi and Brown 2001
Subtropical depressional forested wetland, Florida	7.32E+04	Bardi and Brown 2001
Subtropical herbaceous wetland, Florida	7.34E+04	Bardi and Brown 2001
Wood		
Boreal silviculture, Sweden (*Picea aibes*, *Pinus silvestris*)	8.27E+03	Doherty 1995
Subtropical silviculture, Florida (*Pinus elliotti*)	9.78E+03	Doherty 1995
Subtropical plantation, Florida (*Eucalyptus* and *Malaleuca* spp.)	1.89E+04	Doherty 1995
Temperate forest, North Carolina (*Quercus* spp.)	2.68E+04	Tilley 1999
Peat		
Salt marsh, Florida	5.89E+03	Odum 1996
Subtropical depressional forested wetland	2.52E+05	Bardi and Brown 2001
Subtropical shrub-scrub wetland	2.87E+05	Bardi and Brown 2001
Subtropical wetland	3.09E+05	Bardi and Brown 2001

TABLE 4.3

Summary of Transformities in a Marine Ecosystem, Prince William Sound, Alaska

Item	Transformity (seJ/J)
Phytoplankton	1.84E+04
Zooplankton	1.68E+05
Small nekton (molluskans, arthropods, small fishes)	1.84E+06
Small nekton predators (fish)	1.63E+07
Mammals (seal, porpoise, belukha whale, etc.)	6.42E+07
Apex predators (killer whale)	2.85E+08

Source: After Brown, M. T., R. D. Woithe, C. L. Montague, H. T. Odum, and E. C. Odum. 1993. *Emergy analysis perspectives of the* Exxon Valdez *oil spill in Prince William Sound, Alaska.* Final Report to the Cousteau Society. Center for Wetlands, University of Florida, Gainesville, FL.

transformities vary almost one order of magnitude, reflecting the differences in total driving energy of each ecosystem type. The table is arranged in increasing quality of products from gross production to peat. Transformities increase in like fashion. An energy transformation is a conversion of one kind of energy to another kind. As required by the second law, the input energies (sun, wind, rain, etc.) with available potential to do work are partly degraded in the process of generating a lesser quantity of each output energy. With each successive transformation step, a lesser amount of higher quality resources is developed.

When the output energy of a process is expressed as a percent of the input energy, an efficiency results. Lindeman (1942) efficiencies, in ecological systems, are an expression of the efficiency of transfer of energy between trophic levels. Table 4.3 lists transformities of trophic levels in the Prince William Sound of Alaska calculated from a food web and using Lindeman efficiencies of about 10% (Brown et al. 1993). The transformity, which is a ratio of the emergy input to the available energy output, is an expression of quality of the output energy; the higher the transformity, the more emergy is used to make it.

4.3.2 Hierarchy

A hierarchy is a form of organization resembling a pyramid where each level is subordinate to the one above it. Depending on how one views a hierarchy, it can be an organization whose components are arranged in levels from a top level (small in number, but large in influence) down to a bottom level (many in number, but small in influence). Or one can view a hierarchy from the bottom where one observes a partially ordered structure of entities in which every entity but one is successor to at least one other entity, and every entity except the highest entity is a predecessor to at least one other. In general, in ecology we consider hierarchical organization to be a group of processes arranged in order of rank or class in which the nature of function at each higher level becomes more broadly embracing than at the lower level. Thus we often speak of food chains as hierarchical in organization.

Most if not all systems form hierarchical energy transformation series, where the scale of space and time increases along the series of energy transformations. Many small-scale processes contribute to fewer and fewer larger-scale processes (Figure 4.4). Energy is converged from lower- to higher-order processes, and with each transformation step, much energy loses its availability (a consequence of the Second Law of Thermodynamics), while only a small amount is passed along to the next step. In addition, some energy is fed back reinforcing power flows up the hierarchy. Note in Figure 4.4 the reinforcing feedbacks by which each transformed power flow feeds backward so that its special properties can have amplifier actions.

4.3.3 Transformities and Hierarchy

Transformities, by virtue of the fact that they quantify the convergence of energy into products and account for the total amount of energy required to make something, are quality indicators (Ulgiati and Brown 2006). Quality is a system property, which means that an "absolute" scale of quality cannot be made, nor can the usefulness of a measure of quality be assessed without first defining the structure and boundaries of the system. For instance, quality as synonymous with usefulness to the human economy is only one possible definition of quality, a "user-based quality." A second possibile definition

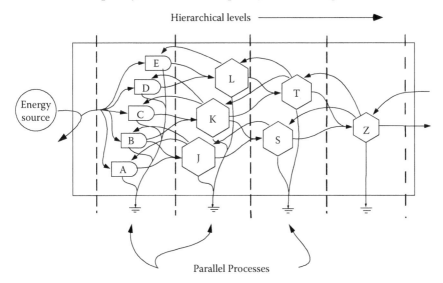

FIGURE 4.4
Diagram of the organization of systems showing the convergence of energy and matter into higher and higher levels via parallel and hierarchical processes.

of quality is one where quality increases with increases of input. That is, the more energy invested in something, the higher its quality. We might describe this type of quality as "donor-based quality."

Self-organizing systems (be they the biosphere or an ecosystem) are organized with hierarchical levels (Figure 4.4) and each level is composed of many parallel processes. This leads to two other properties of quality: (a) *parallel quality,* and (b) *cross quality.* In the first kind, "parallel quality," quality is related to the efficiency of a process that produces a given flow of energy or matter within the same hierarchical level (comparison among units in the same hierarchical level in Figure 4.4). For any given ecological product (organic matter, wood, herbivore, carnivore, etc.) there are an almost infinite number of ways of producing it, depending on surrounding conditions. For example, the same tree species may have different gross production and yield different number and quality of fruit depending on climate, soil quality, rain, etc. Individual processes have their own efficiency, and as a result the output has a distinct transformity. Quality as measured by transformity in this case relates to the emergy required to make like products under differing conditions and processes. Note Table 4.2, where several transformities are given for each of the ecosystem products listed.

The second definition of quality, "cross quality," is related to the hierarchical organization of the system. In this case, transformity is used to compare components or outputs from the different levels of the hierarchy, accounting for the convergence of emergy at higher and higher levels (comparison of transformity between different hierarchical levels, in Figure 4.4). At higher levels, a larger convergence of inputs is required to support the component (a huge amount of grass is needed to support an herbivore, many kilograms of herbivore are required to support a predator, many villages to support a city, etc.). Also, higher feedback and control ability characterize components at higher hierarchical levels, so that higher transformity is linked to higher control ability on lower levels. Therefore, higher transformity, as equated with higher level in the hierarchy, often means greater flexibility and is accompanied by greater spatial and temporal effect.

Figure 4.5 and Table 4.4 give energy and transformity values for an aggregated system diagram of Silver Springs, Florida. The data were taken from H. T. Odum's earlier studies on this ecosystem (Odum 1957). Solar energy drives the system directly (i.e., through photosynthesis) and indirectly through landscape processes that develop aquifer storages, which provide the kinetic energy of the spring. Vegetation in the spring uses solar energy and capitalizes on the kinetic energy of the spring, which brings a constant supply of nutrients. Products of photosynthesis are consumed directly by herbivores and also deposited in detritus. Herbivores are consumed by carnivores who are, in turn, consumed by top carnivores. With each step in the food chain energy is degraded, and with each step some energy is upgraded to a higher quality.

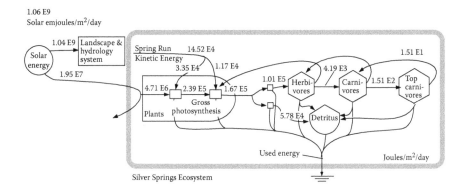

FIGURE 4.5
Aggregated systems diagram of the ecosystem at Silver Springs, Florida, showing decreasing energy with each level in the metabolic chain (after Odum 2004). Table 4.5 gives the transformities that result from the transformations at each level.

TABLE 4.4

Solar Transformities of Ecosystem Components of the Silver Springs

Item	Transformity (seJ/J)
Solar energy	1
Kinetic energy of spring flow	7170
Gross plant production	1620
Net plant production	4660
Detritus	6600
Herbivores	127,000
Carnivores	4,090,000
Top carnivores	40,600,000

4.3.4 Transformity and Efficiency

Transformities can sometimes play the role of efficiency indicators and sometimes the role of hierarchical position indicator (Ulgiati and Brown 2006). This is completely true in systems selected under maximum power principle constraints (Lotka 1922a, 1922b; Odum 1983) and is therefore true in untouched and healthy ecosystems. Things are different in an ecosystem stressed by an excess of outside pressure. Relations among components are likely to change, some components may also disappear, and the whole hierarchy may be altered. The efficiency of a given process may change (no matter if it decreases or increases) and some patterns of hierarchical control of higher to lower levels may diminish or disappear because of a simplified structure of the system. These performance changes translate into different values of the transformities, the variations of which become clear measures of lost or decreased system integrity.

When an ecological network is expressed as a series of energy flows and transformation steps where the transformation steps are represented as Lindeman efficiencies, the resulting transformities represent trophic convergence and a measure of the amount of solar energy required to produce each level in the hierarchy

4.4 Emergy, Transformity, and Biodiversity

In practice, the conservation of biodiversity suggests sustaining the diversity of species in ecosystems as we plan human activities that affect ecosystem health. Biodiversity has no single standard definition. Generally speaking, biodiversity is a *measure of the relative diversity among organisms present in different ecosystems.* "Diversity" in this case includes diversity within species (i.e., genetic diversity), among species, and among ecosystems. Another definition is simply the *totality of genes, species, and ecosystems of a region.* Three levels of biodiversity have been recognized:

- Genetic diversity—diversity of genes within a species
- Species diversity—diversity among species
- Ecosystem diversity—diversity among ecosystems

A fourth level of biodiversity, cultural diversity, has also been recognized.

A main problem with quantifying biodiversity, especially in light of the definition above, is that there is no overall measure of biodiversity since diversity at various levels of an ecological hierarchy cannot be summed. If they were summed, bacteria and other small animals and plants would dominate the resulting diversity to the total neglect of the larger species. It therefore may be possible to develop a quantitative evaluation of total biodiversity within regions or ecosystems by weighting biodiversity at each hierarchical level by typical trophic level transformities. In this way quantitative measures of biodiversity can be compared and changes resulting from species loss can be scaled based on transformities. A more realistic picture of total biodiversity may emerge and allow quantitative comparison of losses and gains that result from changes in ecological health.

As an example of calculating biodiversity at the ecosystem level, we provide data for components of the Everglades ecosystem in south Florida, USA (Table 4.5). Transformities were computed from network flow data (Ulanowicz et al. 2001), using a linear optimization technique that manipulates a set of unknowns (transformity values) to meet a set of constraints (emergy inflow = emergy outflow) (see Brown et al. [2006] for a more detailed discussion of the methods). These values represent an average of all the species of

TABLE 4.5

Number of Species and Average Transformities of Generalized
Compartments in the Everglades Graminoid Marsh Ecosystem

Compartment	Number Species[a]	Avg. Transformity (seJ/J)	Cumulative Emergy (E+06)
Bacteria[b]	?	1–10	?
Invertebrates	48	2.0E+04	1.0
Primary producers[c]	250 (est.)	2.1E+04	5.0
Fishes	24	4.8E+04	1.2
Amphibians	14	1.2E+05	1.7
Reptiles	19	1.5E+05	2.8
Birds	59	2.3E+05	13.6
Mammals	20	2.3E+06	46.0

[a] After Ulanowicz et al. (2001).
[b] Jørgensen, Odum, and Brown 2004.
[c] Lane et al. 2003.

each category and between wet and dry seasons since the driving energy inputs are greater in wet season than in dry season. Transformities increase with increasing complexity of organisms. The final column represents the cumulative emergy associated with each group of organisms obtained by multiplying average transformity by the number of species.

Indices of diversity, well known in the ecological literature, are adaptations of information theory used to describe the organization (and condition) of ecosystems, which began with MacArthur (1955), who suggested the Shannon diversity formulation [Equation (4.1)] to compare flows within ecosystems. Beginning with Margalef (1961), however, the use of physical stocks (biomass, abundance, cover, frequency) of system components instead of flows became common (Peet 1974; Krebs 2000). The typical formulation in ecology is:

$$H = -\sum_{i=1}^{j} p_i * \log[p_i] \qquad (4.1)$$

where H is the diversity, and p_i is the probability of observing component i in a system of j components. Observation probabilities (p_i) are typically measures of relative physical stocks for each ecosystem component. There are two conceptual problems with the standard use of the Shannon equation in ecology: First, it uses stocks instead of flows; and second, one cannot compute whole system diversity since it ignores the hierarchical organization of ecosystems (the equation is maximized when the probability of observing each component is equal). To overcome these limitations, we have proposed (Brown et al. 2006; Brown and Cohen 2007) a quality-adjusted Shannon diversity.

4.5 System-Level Diversity Index

Quality-adjusted Shannon diversity can be computed using Equation (4.1) in the typical manner in ecology, except relative importance value [p_i in Equation (4.1)] is defined as the proportion of total system emergy flow (seJ/yr) allocated to each component. We refer to the relative value of each component calculated in this manner as the Emergy Importance Value (EIV), which is computed as follows:

$$EIV_i = \frac{NP_i * \tau_i}{\sum_i NP_i * \tau_i} \qquad (4.2)$$

where NP_i is the net production (J/yr) and τ_i (seJ/J) is the computed transformity of component i. In this formulation, importance value is the relative contribution of each component to the total emergy flow through all biotic components [i.e., denominator of Equation (4. 2)], computed by summing net production multiplied by transformity over all components.

We then calculate an ecosystem-scale Shannon diversity index [following Equation (4.1)] as follows:

$$Biodiversity = -\sum_{i=1}^{j} EIV_i * \log[EIV_i] \qquad (4.3)$$

The biodiversity is maximized for this index when the emergy on each pathway, and therefore each component's EIV, is equal. When physical flows are adjusted for quality, flow evenness across all ecosystem components is expected; that is, the emergy on all pathways is equal. Odum (1994) postulated flow evenness across all components as the goal condition for network systems that are maximizing power. Actual deviations from this condition could be a useful indicator of system condition. This may be true at the ecosystem scale and at the scale of individual ecosystem components. Our evaluation of the Florida Everglades using this method (Brown et al. 2006) suggested that the system was operating at 42% of its maximum potential diversity (Table 4.6).

TABLE 4.6

System Scale Indices of Biodiversity for the Everglades Graminoid Marsh

Quality-adjusted diversity (bits)	1.73
Theoretical maximum diversity (bits)	4.14
Relative diversity (%)	42

Source: After Brown, M. T., M. J. Cohen, E. Bardi, and W. W. Ingwersen. 2006. Species diversity in the Florida Everglades, USA: A systems approach to calculating biodiversity. *Aquatic Sciences* 68 (3): 254–77.

4.6 Emergy and Information

Ecosystems create, store, and cycle information. The cycles of material, driven by energy, are also cycles of information. Ecosystems, driven by a spectrum of input resources, generate information accordingly and store it in different ways (seeds, structure, biodiversity). The emergy cost of the generated information can be measured by a transformity value and may be a measure of healthy ecosystem dynamics. Odum (1996) suggested transformities for various categories of information within ecosystems given in Table 4.7. In healthy ecosystems (as well as in healthy human-dominated systems, such as a university) suitable emergy input flows contribute to generating, copying, storing, and disseminating information. In stressed ecosystems, such as those where some simplification occurs due to improper loading from outside, the cycle of information is broken or impaired. In this case, the ecosystem exhibits a loss of information, which may manifest itself in simplification of structural complexity, losses of diversity, or decreases in genetic diversity (reduced reproduction).

There are two different concepts of information shown in Table 4.7. The first is the emergy required to maintain information as in the maintenance of DNA in leaves (i.e., copying), or the maintenance of information of the population of trees (emergy in seed DNA, which is the storing and disseminating of information). The second concept is related to generating new information. When a species must be generated anew, the costs are associated with developing one from existing information sources such as trees within the same forest. However, the emergy required to generate biodiversity at the global scale, that is, to generate anew all species, required billions of years and a huge amount of total emergy. Table 4.7 provides very average data

TABLE 4.7

Transformities of Information in Forest Components and the Emergy to Generate Global Biodiversity

Item	Solar Transformity	Units
Forest scale		
DNA in leaves	1.2E+07	seJ/J
DNA in seeds	1.9E+09	seJ/J
DNA in species	1.2E+12	seJ/J
Generate a new species	8.0E+15	seJ/J
Global scale		
Generate global biodiversity	2.1E+25	seJ/species

Sources: After Odum, H. T. 1996. *Environmental accounting. Emergy and environmental decision making.* New York: John Wiley and Sons, Inc.; and Ager, D. U. 1965. *Principles of paleontology.* New York: McGraw-Hill.

for tropical forest ecosystems and, of course, represents only "order of magnitude" estimates of the costs of information generation, copying, storing, testing, and disseminating.

4.7 Measuring Changes in Ecosystem Health

Changes in ecosystem health can result from alterations in driving energy signature, or inflows of a high-quality stressor such as pollutants, or unsustainable activity like overharvesting. In each case there is a consequent change in the pattern of energy flows supporting organization. An energy signature (see Figure 4.6) could change, resulting in ripples that could propagate through the ecosystem. If the change in signature is outside the normal range of fluctuations in the driving energy pattern, the effect is a change in the flows of energy and material throughout the ecosystem. Significant change in system organization might be interpreted as changes in ecosystem health. In general, chemicals, including metals, pollutants, and toxins have high transformity (see Table 4.8) and, as a result of an excess concentration,

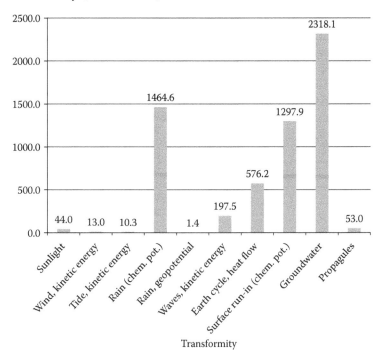

FIGURE 4.6
Emergy signature of driving energies for 1 hectare of typical mangrove ecosystem in Florida.

TABLE 4.8

Transformities of Selected Metals as Global Flows to Atmosphere
and Storages within a River Ecosystem

	Annual Releases to Atmosphere[a] (seJ/J)	River Ecosystem[b] (seJ/J)
Aluminium	9.65E+06	3.30E+07
Iron	8.46E+07	6.19E+07
Chromium	2.59E+10	1.99E+10
Arsenic	8.56E+11	—
Lead	2.39E+12	3.59E+10
Cadmium	1.52E+13	8.78E+10
Mercury	6.85E+14	—

[a] Not including human release.
[b] Genoni et al. 2003

they are capable of initiating significant changes in ecosystem processes, which often result in declines in ecosystem health. As transformities (emergy intensities) increase, their potential effect within the ecosystem increases. Effects can be both positive and negative. Transformity does not suggest the outcome that might result from the interaction of a stressor within an ecosystem, only that with high transformity, the effect is greater.

The ultimate effect of a pollutant or toxin is not only related to its transformity but, more important, to its concentration or empower density (emergy per unit area per unit time [i.e., seJ/m^2*day]) in the ecosystem. Where empower density of a stressor is significantly higher than the average empower density of the ecosystem it is released into, one can expect significant changes in ecosystem function. For instance, because of the very high transformities of most metals like those at the bottom of Table 4.8, their concentrations need only be in the parts per billion range to still have empower densities greater than most natural ecosystems. For instance, using the transformity of mercury in Table 4.8 and the exergy of mercury (Szargut et al. 1988), one can convert the transformity to a specific emergy of 3.7 E17 seJ/g. Using this specific emergy, and a mercury concentration of 0.001 ppb (the level the EPA considers to have chronic effects on aquatic life) the emergy density of the mercury in a lake would be 3.7 E12 seJ/m^2. This emergy density is about two orders of magnitude greater than the empower of renewable sources driving the lake ecosystem. Genoni et al. (2003) measured concentrations of 25 different elements in trophic compartments and in the physical environment of the Steina River in Germany (Table 4.8). They calculated transformities of each element based on global emergy supporting the river ecosystem, which cycles the elements and their Gibbs energy. They suggested that the tendency to bioaccumulate was related to transformity of the elements and the transformity of accumulating compartments (i.e., metals and heavy elements accumulated in high transformity compartments).

Empower density has been used as a predictor of impact of human-dominated activities on ecosystems. In recent studies of the Florida, USA, landscape, Brown and Vivas (2004) showed strong correlations between empower density of urban and agricultural land uses with declines in wetland ecosystem health and pollutant loads in streams. Table 4.9 shows general empower densities of urban and agricultural land uses with natural wildlands for comparison. The empower densities of urban and agricultural land uses are from two to four orders of magnitude greater than the empower density of the natural environment.

A change in ecosystem health is manifested in changes in structural and functional relationships within the system of interest (region, landscape, ecosystem). Often the signs are subtle enough that change is difficult to detect. In other circumstances, indicators are not sensitive enough to detect change or to discern changes in health from "normal variability." Network analysis of the flows of emergy on pathways of ecological systems may add insight into changes in ecosystem health. Using the data from Silver Springs in Figure 4.5, a network analysis of changes in emergy flows and cycling that results from removing the top carnivores (Table 4.10) shows changes in overall cycling emergy of about 15% at the top end of the food chain and diminishing effect cascading back downward toward the bottom. The analysis uses a matrix technique to assign emergy to pathways and includes cycling so that feedbacks within the system are accounted for. Evaluation of the changes in pathway emergy may provide a tool that can help in measuring changes in overall ecosystem health with alterations of components or elimination of trophic levels within the system.

TABLE 4.9

Empower Density of Selected Land Use Categories

Land Use	Empower Density (E14 seJ/ha/yr)
Natural land/open water	7.0
Silviculture and pasture	10–25
High-intensity pasture and agriculture	26–100
Residential and recreational uses	1000–3500
Commercial, transport, and light industrial	3700–5200
High-intensity residential, commercial, and business	8000–30,000

Source: Brown, M. T., and M. B. Vivas. 2004. A landscape development intensity index. *Environmental Monitoring and Assessment* 101:289–309.

TABLE 4.10

The Effect of Changes in System Organization Resulting from Loss of Top Carnivore (Silver Springs, Florida, data)

Item	Transformity (seJ/J)	Pathway Emergy with Top Carnivore[a] (seJ/m²/day)	Pathway Emergy without Top Carnivore[b] (seJ/m²/day)	Percent Change
Solar energy	1	NC	NC	NC
Kinetic energy of spring flow	7170	NC	NC	NC
Gross plant production	1620	3.87E+08	3.84E+08	0.8%
Net plant production	4660	4.71E+08	4.68E+08	0.6%
Detritus	6600	6.67E+08	6.58E+08	1.4%
Herbivores	127000	5.32E+08	5.20E+08	2.3%
Carnivores	4090000	6.13E+08	5.20E+08	15.2%
Top carnivores	40600000	6.13E+08	0	100.0%

NC, no change.

[a] Emergy on pathways of the system depicted in Figure 4.5. Emergy is calculated using a network analysis method (Odum 2001).

[b] Emergy on pathways of the system depicted in Figure 4.5 when the top carnivore is excluded. Emergy is calculated using a network analysis method (Odum 2002).

4.8 Restoring Ecosystem Health

Restoration of ecosystems falls within the sphere of ecological engineering. Ecological engineering is the design and management of self-organizing ecosystems that integrate human society with its natural environment for the benefit of both. The restoration of damaged ecosystems, while resulting in benefits for humanity (increased ecological services) is also necessary to maintain landscape scale information cycles and ultimately biodiversity. The value of active restoration can be measured as the decrease in the time required to restore ecosystem functions to levels characteristic of levels prior to disturbance. The graph in Figure 4.7 illustrates the concept of a net benefit from ecological restoration. The difference between the upper and lower lines in the graph is the benefit of restoration. If the benefit is divided by the costs of restoration a benefit/cost ratio results.

Stressed or damaged ecosystems may be rejuvenated or restored by removal of stresses, or in the case of significant losses, by reconstruction. Table 4.11 gives data for the construction of a forested wetland system in Florida. The data are given for a 50-year time period assuming that 50 years are required to develop a relatively mature forested wetland. While the

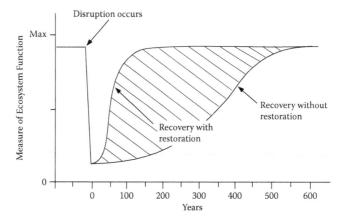

FIGURE 4.7
Graph illustrating the net benefit from ecological restoration. The net benefits can be calculated as the difference between recovery of ecosystem function with and without restoration efforts.

TABLE 4.11

Emergy Costs for Restoration of Forested Wetland in Florida

Item	Data[a]	Units	Unit Emergy Values (seJ/unit)	Emergy (E15 seJ)
Environmental flows				
Sunlight	4.2E+13	J/yr	1	2.10
Wind	3.0E+09	J/yr	2.5E+03	0.38
Rain, chemical potential	6.4E+10	J/yr	3.1E+04	97.60
				97.70
Construction flows				
Planting material	8.4E+07	J	6.7E+04	0.01
Services	8.7E+02	$	1.7E+12	1.48
Fertilizer	6.7E+03	g	4.7E+09	0.03
Services	1.0E+02	$	1.7E+12	0.17
Labor (unskilled)	3.1E+07	J	4.2E+07	1.30
Labor (skilled)	5.4E+07	J	1.2E+08	6.61
Services	4.1E+03	$	1.7E+12	7.01
				16.61
Management				
Chemicals (herbicides)	1.9E+04	g	2.5E+10	0.47
Labor (unskilled)	2.3E+07	J	4.2E+07	0.96
Labor (skilled)	4.6E+07	J	1.2E+08	5.63
				7.06

[a] Based on assumption of 50-year recovery time.

Source: After Bardi, E., and M.T. Brown. 2001. Emergy evaluation of ecosystems: a basis for environmental decision making. In: M.T. Brown (ed) *Emergy Synthesis: Proceedings to the First Biennial Emergy Analysis Research Conference*, Gainesville, FL, Center for Environmental Policy, University of Florida, Gainesville.

inputs of nonrenewable and human-dominated resources are significant, over the 50-year time frame of the restoration effort the renewable emergy dominates.

4.9 Summary and Conclusions

Emergy and transformity are useful measures that may be applied to concepts of ecosystem health. Transformity measures the convergence of biosphere work into processes and products of ecosystems, and as such offers the opportunity to scale ecosystems and their parts based on the energy required to develop and maintain them. Ecosystems are composed of physical structures (i.e., wood, biomass, detritus, animal tissue, etc.) and information found in both its genetic makeup as well as relationships and connections between individuals and groups of individuals. Declines in ecosystem health are manifested in changes in the quality and quantity of relationships and connections between individuals. Stressors may change driving energies pathways, and connections.

When one component in a system is affected, the energy and matter flows in the whole system change, which may translate into declines in ecosystem health. We suggest in this chapter that changes in ecosystem structure and functions are reflected in changes of emergy flows and the corresponding transformities of system components. We suggest that there may be a relationship between the empower density of urban and agricultural lands and their effects on ecosystem health. The effect of a stressor may be predicted by its empower density. Changes in ecosystem structure translate into changes in pathway empower, and thus quantifying changes on networks may provide quantitative evaluation of changes in ecosystem health.

References

Ager, D. U. 1965. *Principles of paleontology*. New York: McGraw-Hill.

Bardi, E., and M.T. Brown. 2001. Emergy evaluation of ecosystems: A basis for environmental decision making. In: M.T. Brown (ed) *Emergy Synthesis: Proceedings to the First Biennial Emergy Analysis Research Conference*, Gainesville, FL, Center for Environmental Policy, University of Florida, Gainesville.

Brandt-Williams, S. 2002. *Handbook of emergy evaluation. Folio no. 4—Emergy of Florida agriculture*. Gainesville FL: The Center for Environmental Policy, University of Florida. (http://www.ees.ufl.edu/cep/)

Brown, M. T., and E. Bardi. 2001. *Handbook of emergy evaluation. Folio no. 3—Emergy of ecosystems.* Gainesville, FL: The Center for Environmental Policy, University of Florida. (http://www.ees.ufl.edu/cep/)

Brown, M. T., and M. J. Cohen. 2007. Emergy and network analysis. In *Encyclopedia of Ecology*, eds. B. D. Fath and S. E. Jørgensen. New York: Elsevier.

Brown, M. T., M. J. Cohen, E. Bardi, and W. W. Ingwersen. 2006. Species diversity in the Florida Everglades, USA: A systems approach to calculating biodiversity. *Aquatic Sciences* 68 (3): 254–77.

Brown, M. T., and S. Ulgiati. 2004a. Emergy and environmental accounting. In *Encyclopedia of energy*, ed. C. Cleveland. New York: Elsevier.

———. 2004b. Energy quality, emergy, and transformity: H. T. Odum's contributions to quantifying and understanding systems. *Ecological Modeling* 78 (1–2):201–13.

Brown, M. T., and M. B. Vivas. 2004. A landscape development intensity index. *Environmental Monitoring and Assessment* 101:289–309.

———. 2005. A landscape development intensity index. *Environmental Monitoring and Assessment* 101:289–309.

Brown. M. T., R. D. Woithe, C. L. Montague, H. T. Odum, and E. C. Odum. 1993. Emergy analysis perspectives of the Exxon Valdez oil spill in Prince William Sound, Alaska. Final Report to the Cousteau Society. Center for Wetlands, University of Florida, Gainesville, FL.

Doherty, S. J. 1995. Emergy evaluations of and limits to forest production. PhD dissertation, Department of Environmental Engineering Sciences, University of Florida. Gainesville, FL.

Ehrenfeld, D. 1993. *From beginning again: People and nature in the new millennium.* New York: Oxford University Press, Inc.

Genoni, G. P., E. I. Mejer, and A. Ulrich. 2003. Energy flow and elemental concentrations in the Steina River ecosystem (Black Forest, Germany). *Aquatic Sciences* 6:143–57.

Jørgensen, S. E., H. T. Odum, and M. T. Brown. 2004. Emergy and exergy stored in genetic information. *Ecological Modeling* 78 (1–2):11–16.

Kangas, P. C. 2002. *Handbook of emergy evaluation. Folio no. 5—Emergy of landforms.* Gainesville, FL: The Center for Environmental Policy, University of Florida. (http://www.ees.ufl.edu/cep/)

Krebs, C. J. 2000. *Ecological methodology.* Menlo Park, CA: Addison Wesley Longman, Inc.

Lane, C., M. T. Brown, M. Murray-Hudson, and B. Vivas. 2003. Florida Wetland Condition Index (FWCI): Biological indicators for wetland condition of herbaceous wetlands in Florida. Final report to Florida Department of Environmental Protection. Center for Wetlands, University of Florida, Gainesville.

Leopold, A. 1949. *A Sand County almanac, and sketches here and there.* New York: Oxford University Press.

Lindeman, R. L. 1942. The trophic—Dynamic aspects of ecology. *Ecology* 23 (4): 399–418.

Lotka, A. J. 1922a. Contribution to the energetics of evolution. *Proceedings of the National Academy of Sciences, U.S.* 8:147–51.

———. 1922b. Natural selection as a physical principle. *Proceedings of the National Academy of Sciences, U.S.* 8:147–51.

MacArthur, R. 1955. Fluctuations of animal populations and a measure of community stability. *Ecology* 36:533–36.

Margalef, R. 1961. Communication of the structure of planktonic populations. *Limnology and Oceanography* 6:124–28.

Odum, H. T. 1957. Trophic structure and productivity of Silver Springs, Florida. *Ecol Monogr* 27:55–112.

———. 1976. Energy quality and carrying capacity of the earth. Response at Prize Ceremony, Institute de la Vie, Paris. *Tropical Ecology* 16(l):1–8.

———. 1983. *Systems ecology: An introduction*. New York: John Wiley and Sons, Inc.

———. 1988. Self organization, transformity, and information. *Science* 242:1132–39.

———. 1994. *Ecological and general systems*, Niwot, CO: University of Colorado Press, 644 pp.

———. 1996. *Environmental accounting. Emergy and environmental decision making*. New York: John Wiley and Sons, Inc.

———. 2001. Simulating emergy and materials in hierarchical steps. pp. 119–127 in *Emergy Synthesis: Theory and Applications of the Emergy Methodology*, Proceedings of the International Workshop on Emergy and Energy Quality, Gainesville, FL, Sept. 1999, ed. by M.T. Brown. Center for Environmental Policy, Univ. of Florida, Gainesville, 328 pp.

———. 2004. *Environment, power and society*, 2nd ed. New York: Columbia University Press.

Odum, H. T., and J. Arding. 1991. *Emergy analysis of shrimp mariculture in Ecuador*. Narragansett, RI: Coastal Resources Center, University of Rhode Island.

Odum, H. T., M. T. Brown, and S. B. Williams. 2000. *Handbook of emergy evaluation: A compendium of data for emergy computation issued in a series of folios. Folio no. 1— Introduction and global budget*. Gainesville, FL: Center for Environmental Policy, Environmental Engineering Sciences, University of Florida. (http://www.ees.ufl.edu/cep/)

Orrell, J. J. 1998. Cross scale comparison of plant production and diversity. M.S. thesis, Department of Environmental Engineering Sciences. University of Florida, Gainesville.

Peet, R. K. 1974. The measurement of biodiversity. *Annual Review of Ecology and Systematics* 5:285–307.

Szargut, J., D. R. Morris, and F. R. Steward. 1988. *Exergy analysis of thermal, chemical and metallurgical processes*. London: Hemisphere Publishing Corporation.

Tilley, D. R. 1999. Emergy basis of forest systems. Ph.D. dissertation, Department of Environmental Engineering Sciences. University of Florida, Gainesville.

Ulanowicz, R., S. Heymans, C. Bondavalli, and M. S. Egnotovich. 2001. *Network analysis of trophic dynamics of south Florida ecosystems*. University of Maryland Center for Environmental Science, Chesapeake Biological Laboratory. (http://www.cbl.umces.edu/~atlss/ATLSS.html)

Ulgiati, S., and M. T. Brown. 1999. Emergy accounting of human-dominated, large-scale ecosystems. In *Thermodynamics and ecology*, eds. S. E. Jørgensen and Kay. New York: Elsevier.

———. 2006. Exploring complexity of landscape and ecosystems. *Ecological Questions* Toruń, Poland: Nicolaus Copernicus University Press. 17:85–102.

———. 2009. Emergy and ecosystem complexity. *Communications in Nonlinear Science and Numerical Simulation* 14 (1): 310–21.

Weber, T. 1994. Spatial and temporal simulation of forest succession with implications for management of bioreserves. M.S. thesis, University of Florida, Gainesville.

5

Eco-Exergy to Emergy Flow Ratio for the Assessment of Ecosystem Health

F. M. Pulselli, C. Gaggi, and S. Bastianoni

CONTENTS

5.1 Introduction

The study of ecosystems from a holistic point of view implies the analysis of the relations between the elements of the entire whole. As pointed out by E. P. Odum, "[T]he old folk wisdom about the forest being more than just a collection of trees is indeed the first working principle for ecology" (Odum 1977). Ecosystems are generally organized hierarchically, and an important consequence of this type of organization is that new properties emerge whenever parts are combined to form a larger entity. Systems characterized by self-organizing behaviors that build gradients and order from thermodynamic equilibrium (disorder) show common patterns: certain collective features emerge and similar attributes can be observed, even between very different environments (Tiezzi 2006). Such behavior is frequent in living systems and ecosystems, which also show enormous creativity in their evolutionary

113

paths. In spite of the wide variability of choices typical of natural systems, such oriented trends are strongly present.

Orientors or goal functions describe systems from a holistic point of view and define their structural and functional features. They are based on certain general principles, such as thermodynamic laws, and they have to reflect the general properties of living dissipative self-organizing systems. These characteristics are difficult to measure, and therefore ecological orientors can indicate some aspects of the degree of naturalness of ecosystems. Goal functions can therefore provide a good basis for finding usable indicators for ecosystem health, ecological integrity, and sustainability (Müller and Leupelt 1998). They can also be used to evaluate the strength of human impact and an ecosystem's structural carrying capacity.

Here a holistic approach has been used in order to explore potentiality and limits of two thermodynamics-based goal functions (emergy and eco-exergy) and their ratio. Emergy can account for the amount of basic energy (solar) required to sustain a process or an ecosystem, and eco-exergy, the level of organization reached by a system. The ratio of eco-exergy to emergy flow is therefore an indicator of efficiency in transforming the basic solar energy available into the structure of an ecosystem. The ratio of the variation (in time or in space) of eco-exergy and emergy flow allows us to study and measure the reaction of the ecosystem structure to the variation of the input flow (Bastianoni 1998). This hypothesis is applied to a case study on a forest ecosystem in Tuscany.

5.2 Eco-Exergy and Emergy

The thermodynamic approach to ecosystems has produced many different functions that try to measure the distance from thermodynamic equilibrium of systems under study with respect to the outside environment, as well as the effort needed to lead the ecosystem to a certain level of organization. Among orientors based on thermodynamics, two are used here to investigate ecosystems and the relations in which they are involved: eco-exergy and emergy.

Under a holistic rather than molecular viewpoint, shared by both ecology and thermodynamics, eco-exergy was first introduced by Jørgensen and Mejer (1977); it is derived from exergy, a thermodynamic potential that measures the distance of an open system from thermodynamic equilibrium, as a function of the gradients of the intensive physical and chemical variables. The further effort due to Jørgensen and coworkers to extend the application of exergy to biological and ecological systems tends to establish a strong relationship between exergy and information content (Jørgensen et al. 1995).

With Jørgensen and his coworkers, exergy is used in ecology to measure complexity. Complexity in ecosystems is expected to be associated with the presence of more complex organisms, which, in principle, correspond to higher information content (in the form of DNA, RNA, and protein sequences) and greater distance from thermodynamic equilibrium (Marques and Nielsen 1998). On the basis of this general framework and of calculations and approximations (see, for example, Bendoricchio and Jørgensen 1997), the mathematical definition of ecological exergy becomes:

$$Ex = \sum \beta_i * c_i,$$

where c_i is the concentration of the ith component of the ecosystem and β_i's are weighting factors related to the probability of forming the organism at the thermodynamic equilibrium. In other words, it represents how much information an organism contains, starting from the genome size. Calculation of beta values (β_i) is based on the number of million bases (Mb) of nucleotides in the genome of the organism, and the percentage of repeating sequences. Calculations of the β values are reported in Jørgensen et al. (2005) and Jørgensen (2008). Eco-exergy, in this way, measures the distance from thermodynamic equilibrium of a *living* organism, while in exergy this biological aspect is absent. In this way, like in a sort of snapshot, any ecosystem can be analyzed and a level of organization can be calculated, as far as a set of affordable data can be collected. As systems develop, they follow the maximum exergy principle, i.e., a system tends to reach the highest possible level of organization that is compatible with the available inputs (see, for example, Jørgensen and Svirezhev 2004 and Jørgensen et al. 2007).

"Emergy is the available energy of one kind previously used up directly and indirectly to make a service or product" (Odum 1996). In particular, solar emergy (measured in solar emergy joules, i.e., sej) is the solar energy required (directly or indirectly) to make a service or product. (Solar) emergy is a measure of convergence of energies, space, and time, both from global environmental work and human services into a product. It is sometimes referred to as "energy memory" and its logic (of "memorization" rather than "conservation") is different from other energy-based analyses, as shown by the emergy "algebra" (see Brown and Herendeen 1996).

Emergy evaluation can be used to assess the sustainability of systems: it accounts for the energy that is needed to sustain anthropic or natural systems. We can view emergy as the work that the biosphere has to do in order to maintain a system far from equilibrium or in order to reproduce an item once it has been used. In fact, the emergy function can be used to express any flow of matter or energy on a common basis, the joule of solar (equivalent) energy. Solar energy is the flow that created, helped develop, and maintains life in the biosphere; all biophysical processes on earth are considered as driven by this energy flow. Formally:

$$Em = \sum_i E_i * Tr_i$$

where E_i is the energy content of the ith independent input flow to the process and it is expressed in joules. Tr_i is the *solar transformity* of the ith input flow, and it is the conversion factor used to transform the energy of a certain flow into emergy.

The total emergy flowing through a system over some unit of time, referenced to its boundary source, is its *empower*, with units sej/s or sej/yr (Odum 1988). If a system, and in particular an ecosystem, can be considered to be in a relatively steady state, the empower (or emergy flow) can be seen as nature's "labor" required for maintaining that state.

Following Lotka (1922), Odum stated a maximum empower principle, saying that "if natural selection has been given time to operate, the higher the emergy flux necessary to sustain a system or a process, the higher is their hierarchical level and the usefulness that can be expected from them" (Odum 1988) or "prevailing systems are those whose designs maximize empower by reinforcing resource intake at the optimum efficiency" (Odum 1996).

Bastianoni et al. (2006) showed how H. T. Odum's maximum empower and Jørgensen's maximum exergy principles can both be valid from a practical viewpoint, given a time order: first the maximization of empower and then the maximization of exergy.

5.3 The Ratio of Eco-Exergy to Emergy Flow

The need to compare the emergy flow that sustains an ecosystem to the consequent ecosystem reaction was already clear to H. T. Odum, who tried to assess the ecosystem response using the emergy/information ratio (Odum 1988) as a measure of information hierarchy.

The relation between emergy and information, used by Keitt (1991) and H. T. Odum (1988), gives a good indication of general character, but information theory has an important problem in the fact that the choice of the basic element of the system under study is arbitrary: one can select an atom or an individual of a species, a letter of an alphabet, or a gene as the basic "symbol." The possibility of joining emergy with another function able to measure the ecosystem structure was proposed by Bastianoni and Marchettini, who introduced a relation between emergy flow and eco-exergy to indicate the solar emergy flow required by the ecosystem to produce or maintain a unit of organization or structure of a complex system (Bastianoni and Marchettini 1997). The role of information and structure is fundamental when we approach the study of complex systems, such as an ecosystem. The use of eco-exergy adds something to the classical exergy approach, which does not take into account information content, in spite of demonstrated connections between it and thermodynamics. In fact, the mixing of two gases previously separated, or a variation in the position of a single component in a PC, are two examples of changes that do not bring a loss of classical exergy, but that give back a loss of usefulness of the product (Susani et al. 2006). The same thing happens in living structures: the difference between, for instance, a living organism and a dead one is not related to this classical exergetic content that is, in fact, the same, but is related to the capability of the living system to use the information content in its DNA (Tiezzi 2006).

Eco-exergy can be considered as a measure of information and a relation can be introduced between emergy flow and eco-exergy. The ratio of eco-exergy to emergy flow represents the state of the system (as eco-exergy) per unit input (as empower).* The eco-exergy/empower ratio can be regarded as a measure of the efficiency of an ecosystem, even though it is not dimensionless, as efficiency usually is, since it has the dimension of time. The higher its value, the higher the efficiency of the system; if the eco-exergy/empower ratio tends to increase (apart from oscillations due to normal biological cycles), it means that natural selection is making the system follow a thermodynamic path that will bring the system to a higher organizational level. As an efficiency indicator the eco-exergy-to-empower ratio enlarges the viewpoint of a pure exergetic approach, where the exergy degraded and the eco-exergy stored for various ecosystems are compared: using emergy there is a recognition of the fact that solar radiation is the driving force of all the energy (and exergy) flows on the biosphere, important when also important "indirect" inputs (of solar energy) are present in a process.

Eco-exergy-to-empower ratio has often been applied in order to assess ecosystem health: in fact, ecosystems different in size, empower, and eco-exergy can be compared with each other as well as their behavior and performances. In general, we can say that in natural systems, where selection has acted undisturbed for a long time, the ratio of eco-exergy to empower is higher and decreases with the progressive introduction of artificial inputs and stress factors that make the emergy flow higher and lower the eco-exergy content of the ecosystem. In the evolutionary process, close to the steady state (climax), the ratio of eco-exergy to empower tends to increase, which means

---

* Bastianoni and Marchettini (1997) first introduced this relation as the ratio of emergy (flow) to eco-exergy. This choice was made in order to maintain coherence with the definition of transformity and point out the differences: transformity is the emergy that contributes to a production system divided by the energy content of a product (or empower divided by power). The emergy flow to eco-exergy ratio instead represents an empower converging to a certain system divided by the eco-exergy of the whole system. Afterward, it seemed more comprehensible to put the effect (eco-exergy) at the numerator and the requirement at the denominator, as in any efficiency indicator.

that the system uses all the materials and energy available to reach a higher eco-exergy content. The same systems, once having reached the climax, will remain in such a state for some time and can grow/develop again only if further energy and/or materials are available. In the latter case, a new source of energy (or better emergy) can be used to build up new biomass and/or complexity of the ecosystem (stored eco-exergy). In terms of eco-exergy-to-empower ratio, when a system is relatively young and acquires new inputs, the ratio tends to be lower; when the system is developing toward the climax stage, the ratio tends to rise (Bastianoni 2008). Fath et al. (2001) identify the ratio of eco-exergy to emergy flow* as one of the possible orientors of an eco-system: they link the emergy flow to the total system throughput (TST) and eco-exergy to the total system storage (TSS), and therefore they connect the maximization of the eco-exergy-to-emergy flow ratio with the maximization of residence time (Fath et al. 2001).

5.4 ΔEx versus ΔEm

A further investigation in this field consists of the analysis of the change in inputs (and emergy flow) to a system and the consequent change in structure of the same system (and in its eco-exergy). We can consider the variation of the emergy flow to the system between two equal and contiguous intervals (these intervals must be significant for the system under study in order to annul the effect of periodical variations like daily and seasonal cycles). The variation of emergy flow is indicated with ΔEm. The change in organization due to the change in emergy input is represented by the variation of the eco-exergy content of the system, ΔEx. The following equation:

$$\sigma = \frac{\Delta Ex}{\Delta Em}$$

as proposed by Bastianoni (1998), has the dimensions of $J \times s \times sej^{-1}$, and represents the change of level of organization (eco-exergy) of the system under study, when it is involved in a change of the emergy flow. It is a quantity that is specific to the inputs that are subtracted or added (Bastianoni 1998).

This kind of analysis confirms the usefulness of the ratio of eco-exergy (or variation in eco-exergy) to emergy flow (or variation in emergy flow) for assessing ecosystem health. Some qualitative implications are added in the reasoning because the inputs to the system (expressed in emergy units) are also evaluated on the basis of virtuous or detrimental consequences of their contribution to the development of the system (represented by eco-exergy or

---

* They actually use the inverse but the rationale is still valid.

eco-exergy variation). "σ" put directly into relation to the change in resource use and the consequence of this change within a system, and the analysis of its value, enables an overall evaluation of both inputs (we can distinguish, for example, nutrients from pollutants) and ecosystem behavior.

In particular, we can identify different possible scenarios depicted in Figure 5.1: σ is positive if the addition of an emergy input gives rise to further organization (case a), or if a lowering of emergy has a negative effect on the system (case c); on the other hand, σ is negative if a higher emergy flow causes a decrease in organization (case b) or if a lower quantity of one or more inputs causes increasing organization (case d). We can say that in cases (b) and (d) the inputs (added or removed) can generally be regarded as pollutants: if we remove them, the system self-organizes; if we add them, the system is damaged. In this way, Bastianoni (1998) provided a definition of pollution based on two orientors, emergy flow and eco-exergy, focusing not on particular aspects of a system, but on the system as a whole. The intensity of the "pollution" is proportional to the absolute value of the slope of the segment connecting the origin to the point that describes the system, since a small increase (decrease) in emergy flow produces a large loss (gain) of organization. The same reasoning can be applied to the cases where σ is positive, namely (a) and (c). The slope of the line connecting the point with the origin represents the benefit that a set of inputs—when added—are able to produce on a system (Figure 5.1): cases (a) and (c) can be seen as an addition or a subtraction of nutrients, respectively.

The points in Figure 5.1 correspond to singular situations that can evolve over time. As an example of the application of this concept, let us consider the change in the composition of rain that falls upon a forest. If the rain becomes more acidic, its emergy content rises, as does the emergy flow through the forest. On the other hand, the eco-exergy of the forest is likely to decrease because of the loss of biomass density and of the consequent loss of biodiversity. In this case, σ would be negative at least until the acidity of the rain

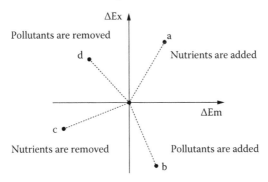

FIGURE 5.1
Diagram of the relationship between the change of emergy (ΔEm) and the change of exergy (ΔEx).

decreases again or the species in the forest learn how to survive in the modified environment or how to use a different input.

Few studies have been conducted on the ΔEx/ΔEm behavior in time. The
step forward in developing this index has been done by moving the variation
of emergy and eco-exergy from time to space. In the case study presented
below, this ratio has been used to analyze a forest area exposed (in the same
time) to a common stress that presents different intensity in space.

5.5 Integration of Analytical and Systems Approaches for the Description of the Influence of Mercury Emissions on Mount Amiata Ecosystem

Here we present the approach of the ratio of ΔEx to ΔEm applied to the ecosystems of Mt. Amiata, located in southern Tuscany, Italy. This mountain is
part of the geologic anomaly of the Mediterranean basin, which contains
about 65% of the world's cinnabar HgS deposits (Figure 5.2). Atmospheric
mercury emissions are from the main sources of geothermal power plants,
abandoned mine structures, and spoil banks of roasted cinnabar ore.

Several studies have been conducted in this area in order to establish
sources of mercury, chemical speciation, concentrations in the environmental matrices, and the effects of mercury, as a pollutant, on the living organisms present on this area of study (Bacci et al. 2000).

In this case study lichens are used as bio-accumulators and bio-indicators
as they are exposed for several months in these areas in order to establish the

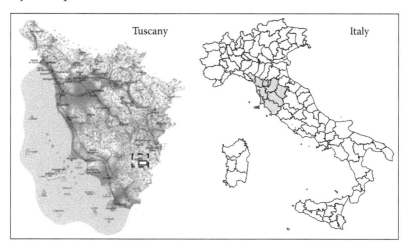

FIGURE 5.2
Mount Amiata (southern Tuscany, *dashed box*) and the area under study (*white box*).

concentration and the effects of mercury. This procedure is used as a basis for calculating the differences in emergy flow to different areas in order to assess the impact of Hg^0 on local vegetable ecosystems (Bosco et al. 2008).

The bioindicators methodology used lichens transplants, *Evernia prunastri*, as bioaccumulators, to measure the spatial distribution of Hg^0. Mercury levels in the aerial parts of vegetation are several orders of magnitude higher than the corresponding concentration in the air and lichen thalli become excellent bioconcentrators. Samples will be taken in at least 20 sample points, representative of squares of 30 × 30 m^2, the center of which is located where the lichen bag is inserted. The emergy flow to the different sampling plots of the Mt. Amiata area will be evaluated including the flow of mercury (as Hg^0) as well as the flows of solar energy, rain, wind, and geothermal energy. In order to estimate the contribution of mercury to the total emergy flow, the concentration of atmospheric mercury revealed by its concentration in lichens (as in Figure 5.3) will be converted in an emergy flow by multiplying the Hg^0 flow by the specific emergy of mercury (Bosco et al. 2007). The biomass of the different species will be multiplied by their β values (see Jørgensen 2008) in order to obtain the value of the eco-exergy in the different areas. Up to now we can only use the β values of angiosperms (β = 393) and gymnosperms (β = 314).

The ΔEx and ΔEm values will be calculated for comparison with a holm oak forest with atmospheric mercury concentration near to the mean Tuscany value, as a reference value. The ratio between the variation of the emergy flow (ΔEm), carried with the geothermal fluid, and the variation of eco-exergy (ΔEx) will enable us to evaluate if there is a correlation between the flows of

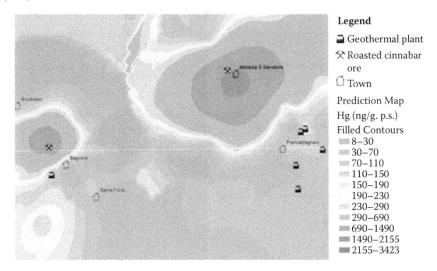

Legend

🔧 Geothermal plant

⚒ Roasted cinnabar ore

⌂ Town

Prediction Map
Hg (ng/g. p.s.)
Filled Contours
▨ 8–30
▨ 30–70
▨ 70–110
▨ 110–150
150–190
190–230
230–290
▨ 290–690
▨ 690–1490
▨ 1490–2155
▨ 2155–3423

FIGURE 5.3
Environmental distribution of atmospheric mercury revealed by means of lichens in summer 2004.

geothermal elements and biomass. In particular, how the vegetable ecosystem organization responds or changes in relation to changes of emergy input related to Hg^0 fallout will be estimated (Bastianoni 1998); it is expected that, at least at high concentrations, a positive variation of ΔEm will cause negative effects on biomass-organization systems, revealed by the value of ΔEx. In this case the variation is calculated not over time but over space, comparing the exergy and emergy values with a blank point, not affected by mercury.

5.6 Preliminary Results

The values of ΔEx and ΔEm obtained from such an analysis will be plotted in a diagram like that in Figure 5.1. From the position of different points in the diagram, a qualitative analysis of the role of resources as "nutrients" or "pollutants" can be performed. In the case presented here, a potential pollutant like mercury is able to affect the environmental performance of the surrounding ecosystem. A negative change in exergy due to an increase in emergy flow (namely, mercury) can be measured and represented on a scale. Preliminary analysis of the area is shown in Figure 5.4.

The points are positioned on the right side of the graph, since the mercury concentration in the sample always exceeds the reference value (in this study the reference value is the mean value of Tuscany). The position of the points corresponding to negative values of the ΔEx axis indicates atmospheric mercury as a pollutant. The graph also shows a low correlation between the

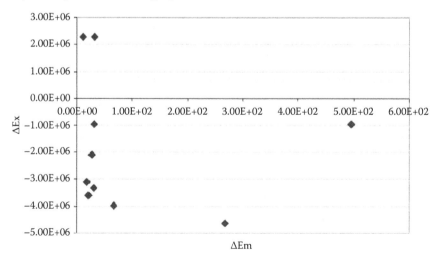

FIGURE 5.4
$\Delta Em/\Delta Ex$ ratio of 10 sampled points.

change in ecosystem autorganization (eco-exergy) and the emergy flow of the atmospheric mercury. This may be due to the low concentration of mercury in the atmosphere, below the effective concentration for vegetal ecosystems composition, or to a different management of the areas. Further studies will be conducted considering a wider number of sampling plots, and also different pollutants such as H_2S.

This study has been elaborated to test the analysis of orientors with a case of ecosystem response to a pollutant inflow. In general, the $\Delta Ex/\Delta Em$ ratio is able to identify the effect of an input, a nutrient or a pollutant, evaluating the ecosystem health variation.

References

Bacci, E., C. Gaggi, E. Lanzillotti, S. Ferrozzi, and L. Valli. 2000. Geothermal power plants at Mt. Amiata (Tuscany–Italy): Mercury and hydrogen sulphide deposition revealed by vegetation. *Chemosphere* 40 (8): 907–11.

Bastianoni S. 1998. A definition of "pollution" based on thermodynamic goal functions. *Ecological Modelling* 113 (1–3,2): 163–66.

———. 2008. Eco-exergy to emergy flow ratio. In *Encyclopedia of ecology,* 979–83. Amsterdam: Elsevier.

Bastianoni, S., and N. Marchettini. 1997. Emergy/exergy ratio as a measure of the level of organization of systems. *Ecological Modelling* 99:33–40.

Bastianoni, S., F. M. Pulselli, and M. Rustici. 2006. Exergy versus emergy flow in ecosystems: Is there an order in maximization? *Ecological Indicator* 6:58–62.

Bendoricchio, G., and S. E. Jørgensen. 1997. Exergy as goal function of ecosystems dynamic. *Ecological Modelling.* 102 (1): 5–15.

Bosco, S., V. Nicolardi, S. Focardi, F. Coppola, C. Gaggi, and S. Bastianoni. 2008. Integration of analytical and systems approaches for the description of the influence of mercury emissions on Mount Amiata ecosystem. *Proceedings of the XI Meeting of the Italian Chemistry Society (Environmental Chemistry),* 152.

Bosco, S., R. Ridolfi, G. Moro, N. Marchettini, and S. Bastianoni. 2007. Emergy values of hydrothermal genesis elements. *Proceedings of ECEM'07.*

Brown, M. T., and R. A. Herendeen. 1996. Embodied energy analysis and EMERGY analysis: A comparative view. *Ecological Economics* 19 (3): 219–35.

Fath, B. D., B. C. Patten, and J. S. Choi. 2001. Complementarity of ecological goal functions. *J Theor Biol* 208:493–506.

Jørgensen, S. E. 2008. Exergy. In *Encyclopedia of ecology,* 1498–1509. Amsterdam: Elsevier.

Jørgensen, S. E., B. D. Fath, S. Bastianoni, J. C. Marquez, F. Müller, S. N. Nielsen, B. C. Patten, E. Tiezzi, and R. E. Ulanowicz. 2007. *A new ecology systems perspective.* Amsterdam: Elsevier.

Jørgensen, S. E., N. Ladegaard, M. Debeljak, and J. C. Marques. 2005. Calculations of exergy for organisms. *Ecological Modelling* 185:165–75.

Jørgensen, S. E., and H. Mejer. 1977. Ecological buffer capacity. *Ecological Modelling* 3:39–61.

Jørgensen, S. E., S. N. Nielsen, and H. Mejer. 1995. Emergy, environ, exergy and eco-logical modelling. *Ecological Modelling* 77:99–109.

Jørgensen, S. E., and Y. M. Svirezhev. 2004. *Towards a thermodynamic theory for ecological systems*. Oxford, UK: Elsevier.

Keitt, T. H. 1991. Hierarchical organization of energy and information in a tropical rain forest ecosystem. M.S. thesis, University of Florida, USA.

Lotka, A. J. 1922. A contribution to the energetics of evolution. *Proceedings of National Academic of Sciences* 8:147–55.

Marques, J. C., and S. N. Nielsen. 1998. Applying thermodynamic orientors: The use of exergy as an indicator in environmental management. In *Eco targets, goal functions, and orientors*, eds. F. Müller and M. Leupelt. Berlin: Springer-Verlag.

Müller, F., and M. Leupelt. 1998. *Eco targets, goal functions, and orientors*. New York: Springer-Verlag.

Odum, E. P. 1977. The emergence of ecology as a new integrative discipline. *Science* 195:1289–93.

Odum, H. T. 1988. Self organization, transformity and information. *Science* 242:1132–39.

———. 1996. *Environmental accounting, emergy and decision making*. New York: Wiley.

Susani, L., F. M. Pulselli, S. E. Jørgensen, and S. Bastianoni. 2006. Comparison between technological and ecological exergy. *Ecological Modelling* 193:447–56.

Tiezzi, E. 2006. *Steps towards an evolutionary physics*. Southampton, UK: WIT Press.

6

Natural Capital Security/Vulnerability Related to Disturbance in a Panarchy of Social-Ecological Landscapes

Nicola Zaccarelli, Irene Petrosillo, and Giovanni Zurlini

CONTENTS

6.1 Ecosystem Health, Integrity, and Security: Diverse Notions for One Possible Single Frame

The need for a measurable and clear definition of ecosystem level properties has become a central issue in evaluating healthy ecosystems (Costanza 1992; Mageau et al. 1995), to provide insights into system dynamics at multiple scales (Zaccarelli, Petrosillo, and Zurlini 2008), and to possibly foresee systems' responses to future shocks or changes in disturbance regimes.

Ecosystem health is defined as the condition of normality in the linked processes and functions that constitute ecosystems (Rapport 1995), and is defined in terms of vigor, resilience, and organization (Mageau et al. 1995). *Ecosystem integrity* refers to "the unimpaired condition in which ecosystems show little or no impact from human actions" (Angermeier and Karr 1994). The notions of ecosystem health and integrity represent different but related intellectual constructs (Karr and Chu 1999), so that an intact ecosystem is also healthy, but a healthy ecosystem may not necessarily be characterized by integrity. Such concepts are considered both a matter of social values and requirements for persistence or resilience of ecosystems to supply man with valuable services (Rapport 1995). However, they are mainly focused on the ecological state of ecosystems, which includes humans only in terms of the possible impacts they may have, and do not consider human perception or human role in supporting such properties.

Over the last decade, scholars have produced new theories and concepts to help identify and value the ways in which environmental change, both natural and human-induced, can affect human well-being and security. The notions of ecosystem health and ecosystem integrity are important ones, but in order to better integrate human and socioeconomic processes into a more system-oriented framework to address environmental change and sustainable development, concepts such as an ecosystem's services and security in provisioning and quality of natural capital (i.e., environmental security) may be useful (Costanza et al. 1997; Müller et al. 2008).

The relationship between the environmental system and the security of human and natural capital has been the object of much research and the subject of many publications in recent decades, but only lately is it becoming an important focus of international environmental policy. The reason, according to Müller et al. (2008), is that the observation of environmental security demonstrates that the environment is a transnational issue, and its security is an important dimension of peace, national security, and human well-being. Thus, for a particular country, security depends on the interplay of the level of environmental pressures, mainly due to human activities, with environmental sensitivity, deemed as the intrinsic propensity of ecosystem goods and services to be affected by human impacts.

The definition of *environmental security* by Müller et al. (2008) builds on the concepts of ecosystem health and integrity and links the risk to the ecological state of a system, whose level of health and integrity has to be defined, together with the human perception of that risk, and represents an added value, as it draws directly on the notions of social-ecological systems (SESs; Gunderson and Holling 2002), adaptive systems (Berkes and Folke 1998; Levin 1998), and natural capital (Chiesura and de Groot 2003; de Groot 2006; Haines-Young et al. 2006). The idea of natural capital is not only a useful framework in which one can consider as a whole the output of goods and services associated with an entire landscape, viewed as a mosaic of different land cover elements (Haines-Young 2000), but helps to address and bridge values and uses society places on the environment with the health of the system or its integrity (as in the case of preserved areas or natural parks; Petrosillo et al. 2006).

Processes and patterns within SESs have been and are so interlinked that it is often very hard to distinguish what is natural from what is not because they have been interacting and co-evolving historically, and society has always influenced and shaped the ecological components and, partially, processes of SESs.

In the real geographic world, SESs materialize as social-ecological landscapes (SELs), and we believe that it is time to focus on the landscape system as a whole instead of components, surrogates, or proxies. Such systems are the most appropriate subjects to study for planning and management as they are open systems, hierarchically structured, and self-organizing, with historical trajectories, memory, and learning capabilities, and with different anthropogenic and natural processes dominating and interacting across different scales (Kay 2000; Gunderson and Holling 2002). SELs are organized in a panarchy of nested levels of organization (Gunderson and Holling 2002) where each system follows an adaptive cycle and interacts with other levels through top-down or bottom-up connections. One of the essential features of the panarchy is that it turns hierarchies into dynamic structures. Individual levels have nonlinear multi-stable properties that can be stabilized or destabilized through critical connections between levels.

6.2 Environmental Security in SESs

The notion of environmental security has been historically linked to international conflicts caused by environmental degradation, e.g., through overuse of renewable resources, pollution, or impoverishment in the space of living (Tuchel 2004; Herrero 2006; Liotta 2006). The concept of environmental security has been developed mainly by international policy researchers and has focused on the role of the scarcity of renewable resources such as cropland, forests, water, and fish stocks. Statistical data demonstrate that agriculture and natural resource availability play an important role in many events of acute violence, which often occur in rural areas (De Soysa et al. 1999). The decrease in quantity and quality of resources, rapid global population growth, and unequal access to resources are the basic drivers behind increasing environment-related security risks. Notably, renewable resources like water and land are crucial factors in security issues, especially with respect to instability and migration between and within countries or regions. Moreover, environmental degradation often results in changes in important ecological and landscape processes that can have irreversible impacts on critical renewable resources such as water, fiber, food, and clean air.

In order to introduce environmental security in the context of landscape sciences, a new definition has been provided by Müller et al. (2008): Environmental security, in an objective sense, aims to evaluate the level of threats to acquire and sustain landscape values in terms of ecosystem goods and services at multiple scales and, in a subjective sense, represents the level of fear that such values will be attacked and possibly lost.

So, environmental security aims at providing expected services and safety, and protecting valuable assets from harm, even during times of increased threat or risk (i.e., the risk of compromising a possible useful state of health of a specific system). Security is achieved through both prospective (preventative) and retrospective (mitigation) actions on the part of governments, agencies, and people. Perceptions of security by individuals, communities, and societies are strongly linked to human well-being and to the satisfaction of the population. The major challenge of environmental security concerns the global environmental change, focusing on the interactions between ecosystems and mankind, the effects of global environmental change on environment degradation, the effects of increasing social request for resources, and the erosion of ecosystem services and environmental goods. Because land use change by humans is one of the major factors affecting global environmental change (Millennium Ecosystem Assessment [MEA] 2005), the question then arises as to how such environmental stresses and the associated risks might vary geographically or evolve over time. Environmental security addresses the risks to, or vulnerability (fragility) of, ecosystem goods and services, as well as the subjective perception of those risks (Petrosillo et al. 2006; Zurlini and Müller 2008).

Environmental security, as the opposite of environmental vulnerability/fragility, is a multilayered, multiscale, and complex notion, existing in both the objective biophysical and social realms, and the subjective realm (Morel and Linkov 2006; Zaccarelli, Petrosillo, and Zurlini 2008). Security is value laden, and related to our normative systems that today recognize concepts like ecosystem functions and services, ecosystem health, integrity, and sustainability as fundamental values for the survival and well-being of mankind. The relevant objects of environmental security are complex, adaptive systems that, in the real geographic world, are SELs. Therefore, we can address environmental security more appropriately in terms of SEL security. The subjective perception of security is fundamental at all levels of human organization, from the individual to government entities, and a "threat" is an abstract concept existing in the domains of feelings and cognition.

6.3 Exercising Environmental Security from a Landscape Perspective

In this paper environmental security is described from the viewpoint of landscape sciences. This approach seems to be particularly suitable because landscapes are comprised of the abiotic and biotic ecological structures and processes of an area and their interrelations with processes and structures of the human society (*cf.* article number one of the "European Landscape Convention"; European Union [EU] 2004).

The rapid progress made in the conceptual, technical, and organizational requirements for generating synoptic multiscale views and explanations of the earth's surface and landscapes provides an outstanding potential support (1) to quantitatively describe real landscapes, habitat mosaics, or land use types; (2) to evaluate and monitor ecological processes by remotely sensed response variables; and (3) to relate response variables to ecological targets by observing at different times ecological changes in targets pattern as well as in their scales (Simmons et al. 1992). Since different processes appear to dominate at different scales (Allen and Starr 1982; O'Neill et al. 1986), multiscale studies have been increasingly conducted (Wu and Qi 2000), giving emphasis to the identification of scale domains (Li 2000; Brown et al. 2002). Such domains are self-similarity intervals of the scale spectrum over which, for a particular phenomenon, patterns do not change or change monotonically with scale. The likelihood of sharp shifts appears linked to an ecosystem's resilience, which is the capacity of a system to undergo disturbance and maintain its functions and controls (Gunderson and Holling 2002).

The development of new, integrated, system-specific evaluation and prediction models for environmental security at multiple scales, framed in terms of both subjective and objective observable quantities in the geographical real-world domain, is necessary to formulate and evaluate ideas relevant to environmental security in SELs. Toward this goal, we exercise an evaluation framework with real landscape disturbances and demonstrate its interpretive power by examining actual disturbance maps relative to land use for a panarchy of SELs in Apulia, an administrative region in southern Italy (Figure 6.1). We exemplify concepts and methods with reference to the recent works of Zurlini et al. (2006, 2007) and Zaccarelli, Petrosillo, Zurlini, and Riitters (2008) in the framework of potential environmental security evaluation with a view toward understanding how disturbances might impact biodiversity and ecosystem service providers through land use and habitat modification. Even though we exercise the framework only based on the objectively observed dynamics of land use and land cover, we believe this framework can represent a common basis for assessing security of SELs both

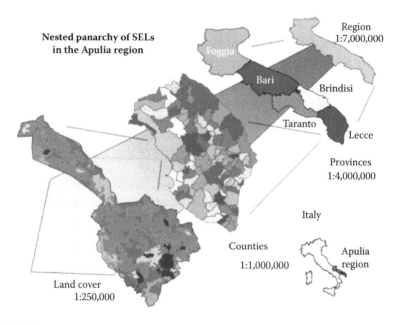

FIGURE 6.1
An example of a panarchy of nested SELs in Apulia, an administrative region in southern Italy. Three main levels of governance hierarchy can be identified (one region, five provinces, and 258 counties) embodying different social, economic, and cultural constraints. The entire region and each sub-region can be described in terms of their unique social-ecological landscapes based on land use/land cover composition supporting ecosystem service providers. Source/ sink patterns of disturbance and cross-scale effects in a panarchy of social-ecological land-scapes. Zaccarelli, N., I. Petrosillo, G. Zurlini and H. Rutters. 2008. *Ecology and Society* 13:26.

objectively and subjectively, at all levels of human organization, by replacing the traditional interpretation of results strictly in ecological terms with an alternate interpretation in terms of environmental security of SELs.

6.4 Disturbance of What, and to What

A fundamental difficulty with SELs is that their complexity makes it hard to forecast the future with any sense of reliability. One way of dealing with this problem is to look retrospectively at the observed trends of effects caused by past exposure to stressors or events and, on this basis, to create future scenarios, taking into account the anticipated changes of the driving forces at work and of their consequent disturbances (Zaccarelli, Petrosillo, and Zurlini 2008). An approach to investigate the interactions between patterns and processes at the landscape level is to look at temporal changes detected by remote sensing, and to ask whether they are significantly associated with different scales (*cf.* Zurlini et al. 2004). If such processes change in type and intensity across scales, the ability of ecosystems to resist lasting change caused by disturbances—their resilience (Gunderson et al. 1997; Gunderson and Holling 2002)—will change accordingly, since habitat/ecosystem resilience and scaling are expected to be intertwined (Peterson 2000). We strongly believe that estimating retrospective resilience at multiple scales is paramount to defining the historical profile of systems as it reveals how they reacted to pressures in the past. In addition, it can also tell us a great deal about current system dynamics, and prospectively how the system might respond to future external shocks (Walker et al. 2002). The future system trajectories can at least be compared with each other to assess whether management scenarios have more or less effect on trajectories, that is, whether proposed actions will move the system in expected directions at expected rates (Antrop 2005).

Since landscape mosaic is mostly defined by vegetation cover, we use land cover change as a measure of disturbance and historical stress. Disturbances have been defined as "any relatively discrete event in space and time that disrupts ecosystem, community, or population structure and changes resources, substrates, or the physical environment" (Pickett and White 1985). Land cover change is a disturbance because converting forest to agriculture land, or vice versa, alters soil biophysical and chemical properties and associated animal and microbial communities, and agricultural practices such as crop rotation or fire alter the frequency of these disturbances. New land cover types can be juxtaposed and shifted within increasingly fragmented remnant native land cover types, and changes in the structure of the landscape can disturb nutrient transport and transformation (Peterjohn and Correll 1984), species persistence and biodiversity (Fahrig and Merriam 1994; With and Crist 1995), and invasive species (With 2004).

To detect change, we applied a standardized differencing change detection technique based on the use of the Normalized Difference Vegetation Index or "greenness" index (NDVI; Pettorelli et al. 2005; Zurlini et al. 2006). From a set of Landsat TM 5 images for June 1997 and June 2001, after registration, calibration, and atmospheric correction, we derived NDVI values for each pixel and calculated the standardized difference NDVI image. A pixel is considered to be "changed" or "disturbed" whenever it falls within a predefined upper or lower percentile of the empirical distribution of the standardized difference values (see Zurlini et al. 2006, or Zaccarelli, Petrosillo, Zurlini, and Riitters 2008 for technical details). We used the percentile of 10% on the distribution of standardized differences for the study area. In other words, we define disturbance as any detectable alteration of land cover reflecting significant and relatively frequent vegetation changes that are mainly assignable to fast human-driven processes.

In this study, a change in a farming practice is like the use of a prescribed fire, which most ecologists would agree is a disturbance even if it does not change the land cover. In the context of environmental security, the justification is that observed changes in NDVI can clearly demonstrate that not only could agricultural fields be more dynamic than other types of land-cover systems, but also that, for instance, agricultural fields could spread distur-bance agents in the landscape to other neighboring land uses like natural areas or permanent cultivations where most of the ecosystem's services pro-viders reside (Zaccarelli, Petrosillo, Zurlini, and Riitters 2008).

In summary, taking into account the different sources of error, we believe that it is possible by this procedure to capture most of the significant, real human-driven disturbances detectable at the resolution of Landsat imagery. However, there might be cases where occasionally NDVI does not capture disturbance when it is, in fact, there, for example, in agricultural fields that went from one crop to another crop of a different type but retained exactly the same or similar NDVI values. On the contrary, this standardized differ-encing change detection technique based on NDVI is far more robust against false positives, so it is very unlikely that an undisturbed location could be marked as disturbed (*cf.* Pettorelli et al. 2005; Zaccarelli, Petrosillo, Zurlini, and Riitters 2008).

In Apulia, typical contagious disturbances are related to land use or land cover and reflect changes associated with urban sprawl, conversion of grasslands to cultivation fields, new olive grove tillage, and farming prac-tices such as fire; the use of herbicides, pesticides, and fertilizers; and crop rotation. Unlike other disturbances such as storms and hurricanes, or clear cutting, the extent and duration of contagious disturbance events in Apulia are dynamically determined by the interaction of the disturbance with the landscape mosaic.

6.5 Ecosystem Service Providers

The services provided by ecosystems are community- or ecosystem-wide, and even have landscape-wide attributes. Nonetheless these services can often be characterized by the component populations, species, functional groups (guilds), food webs, habitat types or mosaics of habitats, and land uses that collectively produce them, i.e., the ecosystem service providers (ESPs). Because disturbances are inflicted at multiple scales, various species and habitats could be differentially affected by disturbances in the same place, and a potentially useful way to appreciate these differences is to look at how disturbances are patterned in space at multiple scales (Zurlini et al. 2006, 2007).

Ecosystem services, classified according to the Millennium Ecosystem Assessment (2005), and their direct and indirect ecosystem service providers are given in Table 6.1. Functional units refer to the unit of study for assess-ing functional contributions of ecosystem service providers (Kremen 2005); spatial scale indicates the scale(s) of operation of the service. The appropriate ecological level for defining the components is service dependent and scale

TABLE 6.1

Ecosystem Services, Classified According to the Millennium Ecosystem Assessment (2005) and Their Direct and Indirect Ecosystem Service Providers (ESPs)

Service	Direct and Indirect ESPs/ Organization Level	Functional Units	Spatial Scale
Aesthetic and cultural	All biodiversity, landscape land use/cover	Species, populations, communities, habitats, landscapes	Local – global
Ecosystem goods	Diverse species, supporting landscape land use/cover	Species, populations, communities, habitats, landscapes	Local – global
UV protection	Biogeochemical cycles, microorganisms, supporting landscape land use/cover	Biogeochemical cycles, functional groups, landscape	Global
Purification of air	Microorganisms, plants, landscape land use/cover	Biogeochemical cycles, populations, species, functional groups	Regional – global
Flood mitigation	Landscape land use/cover	Communities, habitats, landscape	Local – regional
Drought mitigation	Landscape land use/cover	Communities, habitats, landscape	Local – regional
Climate stability	Landscape land use/cover	Communities, habitats, landscape	Local – global
Pollination	Insects, birds, mammals, and supporting landscape land use/cover	Species, populations, functional groups, communities, habitats, landscapes	Local
Pest control	Invertebrate parasitoids and predators and vertebrate predators and supporting landscape land use/cover	Species, populations, functional groups, communities, habitats, landscapes	Local – regional
Purification of water	Landscape land use/cover, soil microorganisms, aquatic microorganisms, aquatic invertebrates, and supporting landscape land use/cover	Species, populations, functional groups, communities, habitats, landscapes	Local – regional
Detoxification and decomposition of wastes	Leaf litter and soil invertebrates, soil microorganisms, aquatic microorganisms, and supporting landscape land use/cover	Species, populations, functional groups, communities, habitats, landscapes	Local – regional
Soil generation and soil fertility	Leaf litter and soil invertebrates, soil microorganisms, nitrogen-fixing plants, plant and animal production of waste products, and supporting landscape land use/cover	Species, populations, functional groups, communities, habitats, landscapes	Local

(continued)

TABLE 6.1

Ecosystem Services, Classified According to the Millennium Ecosystem Assessment
(2005) and Their Direct and Indirect Ecosystem Service Providers (ESPs) (continued)

Service	Direct and Indirect ESPs/ Organization Level	Functional Units	Spatial Scale
Seed dispersal	Ants, birds, mammals, and supporting landscape land use/cover	Species, populations, functional groups, communities, habitats, landscapes	Local
Disturbance regulation	Landscape land use/cover, supported parasitoids, and vertebrate predators	Species, populations, functional groups, communities, habitats, landscapes	Local – regional

Note: Functional units refer to the unit of study for assessing functional contributions of ecosystem service providers; spatial scale indicates the scale(s) of operation of the service.

Source: Managing ecosystem services: What do we need to know about their ecology? Claire Kremen. *Ecology Letters.* (8)5:468-479. Published Online 18 April 2005. Blackwell Science.

dependent; nonetheless, most services in Table 6.1 are directly dependent on landscape land use/land cover. Different providers of the same ecosystem service may operate across a range of spatial and temporal scales and that demands a multiscale approach. Most ecosystem services can be broadly classified as operating on local, regional, global, or multiple scales. For example, native pollinators that provide pollination on crops generally operate at a local scale, while forests and permanent cultivations contribute to climate regulation at local, regional, and global scales. Understanding the spatial scales at which ecosystem services operate will be essential to developing landscape-level conservation and land management plans. How much and with which configuration pattern must a forest in a watershed area be maintained to provide clean water for downstream communities? How many patches of natural habitat should there be, and how should patches be distributed within an agricultural landscape, to provide pollination and pest control services for crops? The answers to these questions will determine how much and how set-asides should be distributed, and areas zoned for different land uses and land covers, in order to protect and manage the service.

We can characterize ecosystem services locally by conducting a functional inventory to identify the component ESPs and measuring or estimating the importance of each ESP's contribution (Kremen 2005). In general, the functional importance of each ESP in a certain environment will depend on both its effectiveness at performing the service and its abundance (Balvanera et al. 2005). Both efficiencies and abundances may respond to eroded or disturbed amount and configuration of habitats, predators, and competitors, as well as to changing physical or biophysical parameters. Functional contributions of ESPs have been measured or estimated for disparate processes including pollination, bioturbation, dung burial, water-flow regulation, carbon sequestration, leaf decomposition, disease dilution, and disturbance regulation.

One of the few functional contributions of ESPs measured across multiple spatial scales is, to our knowledge, the disturbance regulation provided by natural areas and permanent cultivations in Apulia as shown by Zaccarelli, Petrosillo, Zurlini, and Riitters (2008).

6.6 Scales and Patterns of Coupled Disturbance and ESPs in a Panarchy of SELs

SELs are organized in a panarchy of nested levels of organization, which draws on the notion of hierarchies of influences between embedded scales (Gunderson and Holling 2002). Understanding environmental security in SELs requires understanding how the actions of humans as a keystone species (*sensu* O'Neill and Kahn 2000) shape the environment across a range of scales by taking into account the scales and patterns of human land use as ecosystem disturbances. Anthropogenic disturbances such as changes in land use are determined by the social components of SELs, which consist of groups of people organized in a hierarchy at different levels (e.g., household, village, county, province, region, and nation). Decision hierarchies of social systems are intertwined with the hierarchies found at the ecosystem or landscape level (Gunderson and Holling 2002). Within this panarchy (Gunderson and Holling 2002), the participants have differing views as to which system states are desirable at each level. Any given land use system in the panarchy is likely to overlap multiple ownership and jurisdictional boundaries, and fall under different levels of administrative decision and control, so that for the Apulia region study area are at least three (Figure 6.1).

Social-ecological systems may have different dynamics when compared with the ecological component alone because the social domain contains the element of human intent. Thus, management actions can deliberately avoid or seek the crossing of actual and perceived thresholds (Walker et al. 2006). It is not yet clearly shareable whether a common framework of system dynamics could be used to examine and explain both social and ecological systems. Europe is a good place to test models because European landscapes are the result of consecutive reorganizations of the land for a long time to adapt uses and spatial structures to meet changing societal demands (Antrop 2005). Human influence dominates landscape dynamics in space and time (O'Neill and Kahn 2000), thus defining limiting constraints at "higher scales" and altering the detailed functioning of ecological processes at "lower scales." So, land use decisions affect both ecological and social structures and processes, and vice versa.

We hypothesize that the characteristic scales of particular phenomena like anthropogenic changes should entrain and constrain ecological processes, and be related to the scales of human interactions with the biophysical environment (Holling 1992). If the patterns or scales of human land use change, then the structure and dynamics of SEL as a whole can change accordingly, leading to transitions between alternative phases, when the integral structure of the systems is changed (Kay 2000; Li 2002).

In human-driven landscapes, evaluating the disturbance patterns of land use at multiple scales clearly has potential for quantifying and assessing environmental condition, processes of land degradation, subsequent impacts on ESPs and human resources in SELs, and their consequences on environmental security. Land uses and covers within SEL mosaics not only might be disturbed by various agents, but also might act as a "source" or a "sink" as to the potential spread of disturbance to neighbor areas, which may occur because of disturbance agents like, for instance, fire, pesticides, herbicides, pests, disease, alien species, and urban sprawl. Any landscape element (land use/land cover) in SELs contributes to the overall proportion of disturbance in the region, through its composition of disturbed locations, and to the overall disturbance connectivity through its configuration. Such landscape elements represent, in turn, functional units for assessing functional contributions of ESPs at different scale(s) of operation of the services (Kremen 2005). Accordingly, such landscape elements might also act positively as a "source" or a "sink" as to the potential spread of ecosystem services by providers.

In this paper we advance the measure of the functional importance of ESPs provided by natural areas and permanent cultivations based on their effectiveness at performing the services. We assume that such efficiency will result directly by both how much disturbance surrounds ESPs' locations at different neighborhoods and how disturbance is spatially arranged. Thus, the vulnerability (security) of ESPs at multiple scales can be interestingly explored through the analysis of both the scales and patterns of disturbance in the surroundings of locations supporting ESPs.

6.7 Disturbance Patterns at Multiple Scales

Many authors (Li and Reynolds 1994; Riitters et al. 1995) have suggested focusing on a few key measures of pattern and, particularly, on the two most fundamental measures of pattern, which are composition and configuration. Therefore, we characterize landscape patterns of disturbance in terms of the amount (composition) and spatial arrangement of disturbance (configuration or connectivity).

We make use of moving windows to measure composition (Pd, the proportion of disturbed pixels within a window) and configuration (Pdd, contagion as the proportion of shared edges between disturbed pixels on changed pixels' edges within a window) of disturbance patterns at multiple scales (i.e., window sizes), as detected on satellite imagery (*cf.* Zurlini et al. 2006). The measurements were made for each pixel at multiple scales by using 10 square arbitrarily chosen window sizes in pixel units of 3, 5, 9, 15, 25, 45, 75, 115, 165, and 225, thus the window area ranges from 0.81 ha to 5852.25 ha. For each pixel a profile of Pd or Pdd is defined by the set of values measured at different window sizes. Profiles were aggregated (i.e., averaged) and a mean profile derived applying a broad land use type classification spanning the whole SEL mosaic except for urban areas. We considered four classes roughly coincident to the second level of the European CORINE land-cover classification (Heymann et al. 1994), and in particular arable lands covering 41.7% of the region (CORINE code 2.1), permanent cultivations covering 30.3% (CORINE code 2.2), heterogeneous agricultural area covering 13.8% (CORINE codes 2.3 and 2.4), and natural areas with a 14.2% (CORINE codes 3.1, 3.2, 3.3, and 4.1).

The [Pd, Pdd] phase space (Figure 6.2) and the use of a convergence point (CP, an asymptotic point for a window exactly equal to the entire study region) can be very useful to provide the appropriate dynamic representation of different SELs in the panarchy, as traced by their recent disturbance history (Zurlini et al. 2006, 2007; Zaccarelli, Petrosillo, Zurlini, and Riitters 2008). For any given location (i.e., pixel on the Landsat-derived disturbance map) in

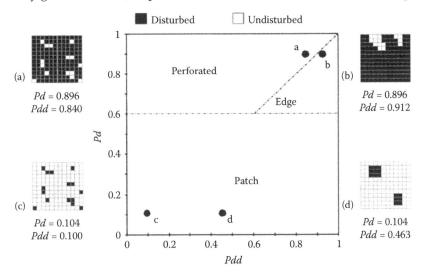

FIGURE 6.2
The graphical model used to identify disturbance categories from local measurements of Pd and Pdd in a fixed-area window. Pd is the proportion of disturbed and Pdd is disturbance connectivity (modified from Riitters et al. 2000). Four simple examples of binary landscapes (a, b, c, d) are presented by the side of the [Pd, Pdd] space for different combinations of composition and configuration: (a) highly disturbed but perforated by undisturbed areas (perforated disturbance), (b) highly disturbed but with clumped undisturbed areas (edge disturbance), (c) low level and highly fragmented disturbance (spread disturbance), and (d) low level and clumped disturbance (patchy disturbance). Disturbance patterns in socio-ecological system and multiple scales. Zurlini G., H. K. Rutters, N. Zaccarelli, G. Petrosillo , K. B. Jones, I. Rossi. June 2006. *Ecological Complexity* (3)2: 119-128. Elsevier Limited.

each land use, the trajectory converging to the CP in [Pd, Pdd] space describes the accumulation profile of disturbance pattern at increasing neighborhoods surrounding that location. If trends in [Pd, Pdd] space were similar for two different locations, then both locations have experienced in their surrounding landscapes the same "disturbance profiles" in terms of amount and configuration. For example, at a given geographic location, the trend in Pd with increasing window size can be interpreted with respect to the disturbances experienced by that location at different spatial lags. A small window (local scale) with high Pd combined with a large window (large scale) with lower Pd implies a local heavy disturbance embedded in a larger region of lighter disturbance. Locations characterized by constant Pd over window size experience equal amounts of disturbance across spatial scales.

Figure 6.3 presents the mean accumulation profiles of disturbance pattern for the four major land-use classes derived for the upper hierarchical level (i.e., regional level) of the panarchy of SELs for the Apulia region, while Figure 6.4 shows means profiles for three of the second levels of SELs (i.e., province level) of the study area. All land-use disturbance trajectories in the Apulia panarchy are located near the lower left corner in the [Pd, Pdd] pattern space, with a certain invariance of disturbance composition (Pd) at increasing disturbance clumping (Pdd). Land uses have distinct disturbance profiles at multiple scales with paths fairly parallel to the Pdd axis almost up to the CP value of the entire region, and with increasing disturbance composition (Pd) usually ranging from natural areas to arable land.

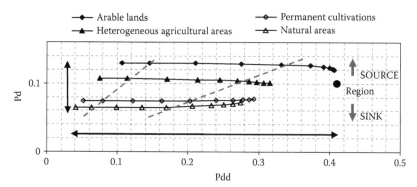

FIGURE 6.3
Convergence point (CP) and disturbance trajectories for the four broad land-use classes of Apulia at multiple scales (10 window sizes in increasing size order from left to right) at regional Pd = 0.10 are represented in the same state space [Pd, Pdd] of Figure 6.2. Dashed lines attempt to connect identical window sizes among different land uses to exemplify cross-scale disturbance mismatches, e.g., between arable lands and natural areas. Source and sink trajectories (vertical arrows on the right) are identified in respect to the regional CP. Black arrows at bottom and left indicate the vulnerability components of ESPs, one due to disturbance composition, the other to disturbance configuration (see text). Source/sink patterns of disturbance and cross-scale effects in a panarchy of social-ecological landscapes. Zaccarelli, N., I. Petrosillo, G. Zurlini and H. Rutters. 2008. *Ecology and Society* 13:26.

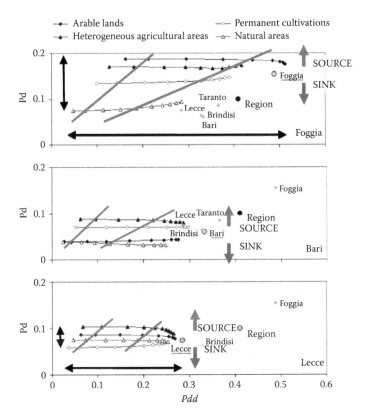

FIGURE 6.4
Trends of disturbance profiles at multiple scales (10 window sizes in increasing order from left to right) of the four land uses for three different provinces of Apulia are presented to show their reciprocal source-sink role. Convergence points for the five provinces and for the Apulia region are shown for comparison. Vulnerability estimates for the five different provinces spanning Apulia as indicated by convergence points, and comparison of disturbance accumulation profiles at multiple scales (10 window sizes in increasing order from left to right) of the same land use within the same province. The CP for the entire Apulia region is shown for comparison. The arrow indicates the direction of vulnerability (see text). Source/sink patterns of disturbance and cross-scale effects in a panarchy of social-ecological landscapes. Zaccarelli, N., I. Petrosillo, G. Zurlini and H. Rutters. 2008. *Ecology and Society* 13:26.

Interestingly, for the assessment of ESP vulnerability, the trajectories of disturbance accumulation profiles at multiple scales on the [Pd, Pdd] state space also indicate whether and where land-use disturbances might act as a "source" or a "sink" across scales in respect to their potential spread to neighbor areas. If a mean profile is always larger than the CP of reference and has a convex trend downward to the CP (e.g., arable lands in Figure 6.3), land use acts as a potential disturbance source to the neighbor mosaic because of local heavy disturbance embedded in a larger region of fewer disturbances. Conversely, if a mean profile of a land use is below the CP with a concave trend upward to the CP (e.g., natural areas, Figure 6.3), land-use locations can be potentially affected by neighbor disturbances (sink) because of local low disturbance embedded in a larger region of heavy disturbances. Disturbance profiles at multiple scales for the four land uses in three different provinces of the Apulia region, and province CPs are shown in Figure 6.4.

Theoretically, spatial "mismatches" are expected when the spatial scales of management and the spatial scales of ecosystem processes are not aligned, possibly leading to disruptions of the SEL, inefficiencies, and/or loss of important components of the ecological system (Cumming et al. 2006). In practice, within SEL mosaics, each land use and land cover has its own disturbance due to human management, even in the case of natural areas, because of the presence of fields and human settlements. Thus, spatial scale mismatches in [Pd, Pdd] space can occur for differences in both disturbance accumulation profiles related to the management of different land uses and accumulation rate of disturbance clumping at different spatial lags. Any two geographic locations with the same accumulation trajectory in [Pd, Pdd] space experience the same multiscale disturbance profile with no spatial scale mismatches, which might occur in some cases for permanent cultivations and natural areas (Figures 6.3 and 6.4). Conversely, dissimilar trends imply differences in spatial profiles of disturbance with consequent scale mismatches of disturbance. Social processes that can lead to mismatches are primarily inherent in land occupancy, which constitutes the hierarchy of social institutions that run the allocation, use, and management of land resources.

The differences in Pdd values between window points tell an interesting story about the cross-scale spatial accumulation rate of disturbance clumping of each land use. Such differences are more pronounced and range from natural areas to arable land (Figure 6.3), meaning that fields have been merged and enlarged to enhance farming efficiency, resulting in almost homogeneously farmed landscapes (e.g., Foggia, Figure 6.4).

Arable lands and heterogeneous areas (source) generally show at the same scales not only higher disturbance composition (Pd), but also cross-scale contagion accumulation increments in disturbance higher than those for permanent cultivations and natural areas (sink).

Distances in the [Pd, Pdd] state space between two land use profiles at the same window size (scale; Figure 6.4) directly draw the attention to spatial scale mismatches of disturbance among land use that can lead to their reciprocal potential role as disturbance source or sink at the same and cross-scales, with possible consequent changes in the structure and dynamics of SELs.

6.8 Discussion: ESP Vulnerability/ Security across Multiple Scales

Ecosystem services can often be characterized by the component populations, species, functional groups (guilds), food webs, habitat types or mosaics of habitats, and land uses that collectively produce them, i.e., the ecosystem service providers (ESPs). Because disturbances are inflicted at multiple scales, species, functional groups, food webs, habitats, and landscapes could be differentially affected by disturbances in the same place, and a potentially useful way to appreciate these differences is to look at how disturbances are patterned in space at multiple scales (Zurlini et al. 2006, 2007).

For an environmental security interpretation of the [Pd, Pdd] space, we have to look not only at the disturbance accumulation profiles at multiple scales (context) of various land-use and land-cover locations, but also at the role those profiles might play as "source" or "sink" across scales within SEL land-use mosaics with respect to the potential spread of disturbance agents to neighbor areas.

The [Pd, Pdd] pattern space has already been interpreted in terms of vulnerability/fragility independently of single location membership to a definite land use (Zurlini et al. 2006); in this case, vulnerability could be highest for scale domains where disturbance is most likely and clumped for trajectories of location clusters.

In our case, we reasonably assume that ESPs reside in natural areas and permanent cultivations, which support most of the component populations, species, functional groups (guilds), food webs, habitat types or mosaics of habitats, and land uses that collectively produce ecosystem services. In Figures 6.3 and 6.4, black arrows indicate the extent of vulnerability components for ESPs: one due to disturbance composition, the other to disturbance configuration. While the first component of vulnerability (disturbance composition) is rather straightforward to evaluate, the second deserves more attention in regard to the specific traits of ESPs. For the first component, the higher the contrast and the difference between arable land disturbance profile (source) and natural area disturbance profile (sink), the higher will be the vulnerability of ESPs residing in natural areas, because there will be the chance of a location with ESPs surrounded by high disturbances.

The reading of [Pd, Pdd] space in terms of vulnerability gradients (or its reverse, environmental security), where vulnerability is highest anywhere disturbance regime is most likely and clumped, is justified by evidence coming, for instance, from metapopulation simulations that show that increasing spatial aggregation of the disturbance regime always decreases habitat occupancy of species, increases extinction risk, and expands the threshold amount of habitat required for persistence, with more marked effects on species with short dispersal distances (Kallimanis et al. 2005). This will help interpret the vulnerability component due to disturbance configuration. This is also particularly central to the dispersal of alien species and therefore to the spatial distribution of risk of competition from alien species. Poor dispersers spread more in landscapes in which disturbances are concentrated in space ("contagious" disturbance), whereas good dispersers spread more in landscapes where disturbances are small and dispersed ("fragmented" disturbance) (With 2004).

The same interpretive framework can be used to compare portions of the SEL, such as provinces in the [Pd, Pdd] space (Figure 6.4), as to their CP, given by its overall Pd and Pdd values. In this way, provinces can be ranked according to the relative vulnerability of ESPs. Thus, the ESPs of the province of Foggia turn out to be the most vulnerable. We can also compare the vulnerability of each single land use at multiple scales among different provinces by looking at its disturbance profiles. In this case, differences in disturbance due to traditional, low-intensity, local land-use practices of agriculture and forestry can be revealed, which have greatly promoted habitat diversity in the European human-dominated landscapes during the last centuries.

Natural areas and permanent cultivations are shown to have higher natural capital value, and higher potential for regulating landscape dynamics and compensating for disturbances in the SELs of Apulia (Zaccarelli, Petrosillo, Zurlini, and Riitters 2008). Consequently, in an environmental security framework, natural areas and permanent cultivations must be considered intrinsically more vulnerable for two different reasons: (1) because they can be affected by internal disturbances, and (2) as a whole because they act as a sink in respect to arable lands, which generally act as a potential source of disturbance agents that could affect neighboring land uses.

6.9 Conclusions

The need to identify, quantify, and evaluate natural and human-induced ecological and social processes, and their corresponding spatial patterns, in order to support an informed planning and management of socio-ecological landscapes (SELs) as well as to face the implications in the context of environmental security has become an urgent issue (Tischendorf 2001; Müller et al. 2008). Different approaches have been proposed to explicitly address such a complex topic. Walker et al. (2002) captured the current state of understanding on how to measure and manage for resilience in social-ecological systems with a set of scenarios and simple models to guide in the identification and manipulation of the system's resilience on an ongoing basis and during times of crisis. Zurlini et al. (2004) proposed a set of tools by coupling a consolidated patch-based multiscale analysis of habitats with ecological remote sensed indices at high spatial and temporal resolution, providing outstanding potential for high-frequency remote monitoring in ecosystem features related to ecosystem resilience and health.

Linking multiscale spatial pattern analysis to remote sensed change detection seems a promising approach for addressing ESPs' security and ecosystem health. Such an approach is an effort to develop an operational and measurable indication of ecosystem-level properties and attributes deeply grounded on insights from empirical analyses and system-based theory

(Carpenter et al. 2001; Zaccarelli, Petrosillo, and Zurlini 2008). Any region in the panarchy of SELs is characterized by the spatial composition (what and how much there is) and configuration (how it is spatially arranged) of landscape elements like land use/land cover. Any landscape element contributes to the overall proportion of disturbance in the region through its composition of disturbed locations, and to the overall disturbance connectivity through its configuration. Such landscape elements represent, in turn, functional units for assessing functional contributions of ESPs at different scales of operation of the service. In rural landscapes, elements with higher land-cover dynamics (disturbance) might act as a source of the potential spread of disturbance to neighboring nonagricultural areas (sink) where most of the ESPs reside.

A landscape perspective of disturbance source-sink patterns at multiple scales is ultimately required to evaluate how changes in landscape structure, e.g., habitat fragmentation, may affect the potential spread of disturbance agents and invasive species. That is essential to assess the vulnerability of ESPs according to their ecological characteristics.

This study points out that management of disturbance in the study region will primarily depend more on broader-scale than local-scale patterns of the drivers of disturbance (Figure 6.3), and clarifies how natural areas and permanent cultivations (i.e., olive groves and vineyards) will act in the interplay of disturbance patterns within SELs, regulating landscape mosaic dynamics and compensating for disturbances across scales. The roles of natural areas and permanent cultivations in providing disturbance regulation across scales in Apulia have consequences for regional SELs since they may govern if and how disturbances associated with land-use intensification will affect the functional contribution of ESPs as well as the ecosystem health and integrity status of different ecosystems.

The [Pd, Pdd] space helps to draw attention to spatial scale mismatches among land uses for disturbance accumulation profiles, which can determine their reciprocal role as disturbance source or sink at cross-scales because of their potential spread to neighbor areas with possible consequent changes in the structure and dynamics of SELs. Such a perspective enriches and improves the accepted way of looking at habitat conservation and evaluation of ecosystem services (Zaccarelli, Riitters et al. 2008) by explicitly integrating scales and space into the analysis, and possibly overcoming some of the limitations connected to a single-scale economic measure, like the value estimate of ecosystem services produced by conserving another unit of habitat (Dasgupta et al. 2000). To support a better-informed decision-making process Armsworth and Roughgarden (2003) recommend basing land-use decisions not simply on the marginal value of ecosystem services added by an additional unit of habitat, but also on the additional stability to ecosystem services that another unit of habitat would add. We argue that the spatial configuration at multiple scales of conservation efforts (which unit and where to conserve) is essential to providing a more effective additional value and stability to ecosystem services. Therefore, it is important to understand not only the relationship between ecosystem service and habitat area, but also the relationship between the distribution of ecosystem service providers and landscape configuration of disturbance at multiple scales, another relationship that may demonstrate nonlinearity.

Even though we attribute to arable land most of the disturbance observed, we acknowledge that agricultural land-use intensification might not only mean a decrease in habitat occupancy with consequent higher extinction, but it could also occasionally make more resources available to enhance populations of some species, since the higher productivity of land use compared with generally less-productive natural systems may provide more resources, such as vegetation biomass and fruits for birds, mammals, and butterflies (Tscharntke et al. 2005).

Understanding the spatial and temporal scales at which ecosystem services and human disturbance operate is essential to developing landscape-level conservation and land-management plans, thus addressing environmental security and ecosystem health issues. So, for instance, how should patches of natural habitat be distributed within an agricultural landscape to provide pollination and pest control services for crops? Or, how should patches of natural habitat and disturbances be distributed within an agricultural landscape to reduce the spread of invasive species? Tentative answers to these questions can be formulated based on observed and simulated disturbance patterns in the Apulia region (Zurlini et al. 2007) to establish the extent and how set-asides should be arranged, and the land zoned for different land uses and land covers, in order to protect and manage the service.

Current approaches to conserving biodiversity may benefit by incorporating greater understanding of how people and nature interact within complex adaptive systems (Gunderson and Holling 2002) like SELs, so that scale mismatches of different land uses in land tenure and thresholds of potential concern for environmental security can be identified and managed for a key set of ecological response variables. That could be the basis for intentionally planning and managing the adaptability of the SEL, which is arguably the key to human management of environmental security.

References

Allen, T. F. H., and T. B. Starr, eds. 1982. *Hierarchy: Perspectives for ecological complexity.* Chicago: University of Chicago Press.

Angermeir, P. L., and J. R. Karr. 1994. Biological integrity versus biological diversity as policy directives: Protecting biotic resources. *BioScience* 44:690–97.

Antrop, M. 2005. Why landscapes of the past are important for the future. *Landscape and Urban Planning* 70:21–34.

Armsworth, P. R., and J. E. Roughgarden. 2003. The economic value of ecological stability. *Proceedings of the National Academy of Sciences of the United States of America* 100:7147–51.

Balvanera, P., C. Kremen, and M. Martinez. 2005. Applying community structure analysis to ecosystem function: Examples from pollination and carbon storage. *Ecological Application* 15:360–75.

Berkes, F., and C. Folke, eds. 1998. *Linking social and ecological systems: Management practices and social mechanisms for building resilience.* Cambridge: Cambridge University Press.

Brown, J. H., V. K. Gupta, B.-L. Li, B. T. Milne, C. Restrepo, and G. B. West. 2002. The fractal nature of nature: Power laws, ecological complexity and biodiversity. *Phil Trans R Soc Lond B* 357:619–26.

Carpenter, S. R., and M. G. Turner. 2001. Editorial. Panarchy 101. *Ecosystems* 4:389.

Chiesura, A., and R. de Groot. 2003. Critical natural capital: A socio-cultural perspective. *Ecological Economics* 44:219–31.

Costanza, R. 1992. Toward an operational definition of ecosystem health. In *Ecosystem health. New goals for environmental management*, eds. R. Costanza, B. G. Norton, B. D. Haskell, 239–56. Washington, DC: Island Press.

Costanza, R., R. d'Arge, R. S. de Groot, S. Farber, M. Grasso, B. Hannon, K. Limburg, S. Naeem, R. V. O'Neill, J. Paruelo, R. G. Raskin, P. Sutton, and M. van den Belt. 1997. The value of the world's ecosystem services and natural capital. *Nature* 387:253–60.

Cumming, G. S., D. H. M. Cumming, and C. L. Redman. 2006. Scale mismatches in social-ecological systems: Causes, consequences, and solutions. *Ecology and Society* 11, no. 14. http://www.ecologyandsociety.org/vol11/iss1/art14/

Dasgupta, P., S. Levin, and J. Lubchenco. 2000. Economic pathways to ecological sustainability. *Bioscience* 50:339–45.

de Groot, R. S. 2006. Function-analysis and valuation as a tool to assess land use conflicts in planning for sustainable, multifunctional landscapes. *Landscape and Urban Planning* 75:175–86.

De Soysa, I., N. P. Gleditsch, M. Gibson, M. Sollenberg, and H. Westing. 1999. *To cultivate peace: Agriculture in a world of conflict.* PRIO Report 1/99. PRIO, Oslo. http://www.futureharvest.org/peace/report.pdf

European Union [EU]. 2004. The European Landscape Convention, Council of Europe Treaty Series (CETS) no. 176. http://www.coe.int/t/dg4/cultureheritage/Conventions/Landscape/florence_en.asp

Fahrig, L., and G. Merriam. 1994. Conservation of fragmented populations, *Conservation Biology* 8:50–59.

Gunderson, L. H., and C. S. Holling, eds. 2002. *Panarchy: Understanding transformations in human and natural systems.* Washington, DC: Island Press.

Gunderson, L. H., C. S. Holling, L. Pritchard, G. D. Peterson. 1997. *Resilience in ecosystems, institutions, and societies.* Beijer Discussion Paper Series, Number 95. Stockholm, Sweden: Beijer International Institute of Ecological Economics, Royal Swedish Academy of Sciences.

Haines-Young, R. H. 2000. Sustainable development and sustainable landscapes: Defining a new paradigm for landscape ecology. *Fennia* 178:7–14.

Haines-Young, R. H., C. Watkins, C. Wale, and A. Murdock. 2006. Modelling natural capital: The case of landscape restoration on the South Downs, England, *Landscape and Urban Planning* 75:244–64.

Herrero, S. T. 2006. Desertification and environmental security: The case of conflicts between farmers and herders in the arid environments of the Sahel. In *Desertification in the Mediterranean region: A security issue,* eds. W. G. Kepner, J. L. Rubio, D. A. Mouat, and F. Pedrazzini, 109–32. Dordrecht, The Netherlands: Springer.

Heymann, Y., C. Steenmans, G. Croissille, and M. Bossard. 1994. *Corine land cover. Technical guide.* Luxembourg: Office for Official Publications of the European Communities.

Holling, C. S. 1992. Cross-scale morphology, geometry, and dynamics of ecosystems. *Ecological Monographs* 62:447–502.

Kallimanis, A. S., W. E. Kunin, J. M. Halley, and S. P. Sgardelis. 2005. Metapopulation extinction risk under spatially autocorrelated disturbance. *Conservation Biology* 19:534–46.

Karr, J. R., and E. W. Chu, eds. 1999. *Restoring life in running waters: Better biological monitoring.* Washington, DC: Island Press.

Kay, J. J. 2000. *Ecosystems as self-organizing holarchic open systems: Narratives and the second law of thermodynamics.* In *Handbook of ecosystem theories and management,* eds. S. E. Jørgensen and F. Müller, 135–60. Boca Raton, FL: CRC Press–Lewis Publishers.

Kremen, C. 2005. Managing ecosystem services: What do we need to know about their ecology. *Ecology Letters* 8:468–79.

Levin, S. A. 1998. Ecosystems and the biosphere as complex adaptive systems. *Ecosystems* 1:431–36.

Li, B. L. 2000. Fractal geometry applications in description and analysis of patch patterns and patch dynamics. *Ecological Modeling* 132:33–50.

———. 2002. A theoretical framework of ecological phase transitions for characterizing tree-grass dynamics. *Acta Biotheoretica* 50:141–54.

Li, H., and J. F. Reynolds. 1994. A simulation experiment to quantify spatial heterogeneity in categorical maps. *Ecology* 75:2446–55.

———. 1995. On definition and quantification of heterogeneity. *Oikos* 73:280–84.

Liotta, P. H. 2006. The Poseidon prairie. In *Desertification in the Mediterranean region: A security issue,* eds. W. G. Kepner, J. L. Rubio, D. A. Mouat, and F. Pedrazzini, 87–108. Dordrecht, The Netherlands: Springer.

Mageau, M. T, R. Costanza, and R. E. Ulanowicz. 1995. The development and initial testing of a quantitative assessment of ecosystem health. *Ecosystem Health* 1:201–13.

Millennium Ecosystem Assessment [MEA]. 2005. *Ecosystems and human well-being: Synthesis.* Washington, DC: Island Press.

Morel, B. and I. Linkov. 2006. *Environmental security and environmental management— The role of risk assessment.* Dordrecht, The Netherlands: Springer, supported by NATO Security through Science.

Müller, F., K. B. Jones, K. Krauze, B.-L. Li, S. Victorov, I. Petrosillo, G. Zurlini, and W. G. Kepner. 2008. Contributions of landscape sciences to the development of environmental security. In *Use of landscape sciences for the assessment of environmental security,* eds. I. Petrosillo, F. Müller, K. B. Jones, G. Zurlini, K. Krauze, S. Victorov, B.-L. Li, and W. G. Kepner, 1–17. Dordrecht, The Netherlands: Springer.

O'Neill, R. V., D. L. DeAngelis, J. B. Waide, and T. F. H. Allen, eds. 1986. *A hierarchical concept of ecosystems.* Princeton: Princeton University Press.

O'Neill, R. V., and J. R. Kahn. 2000. Homo economicus as a keystone species. *Bioscience* 50:333–37.

Peterjohn, W. T., and D. L. Correll. 1984. Nutrient dynamics in an agricultural water-shed. Observations on the roles of a riparian forest. *Ecology* 65:1466–75.

Peterson, G. D. 2000. Scaling ecological dynamics: Self-organization, hierarchical structure, and ecological resilience. *Climatic Change* 44:291–309.

Petrosillo, I., G. Zurlini, M. E. Corlianò, N. Zaccarelli, and M. Dadamo. 2006. Tourist perception of recreational environment and management in a marine protected area. *Landscape and Urban Planning* 79:29–37.

Pettorelli, N., J. O. Vik, A. Mysterud, J.-M. Gaillard, C. J. Tucker, and N. C. Stenseth. 2005. Using the satellite-derived NDVI to assess ecological responses to environmental change. *Trends in Ecology and Evolution* 20:503–10.

Pickett, S. T. A., and P. S. White, eds. 1985. *The ecology of natural disturbance and patch dynamics.* Orlando, FL: Academic Press.

Rapport, D. J. 1995. Ecosystem health: An emerging integrative science. In *Evaluating and monitoring the health of large-scale ecosystems,* eds. D. J. Rapport, C. Gaudet, and P. Calow, 5–31. Heidelberg: Springer-Verlag.

Riitters, K. H., R. V. O'Neill, C. T. Hunsaker, J. D. Wickham, D. H. Yankee, S. P. Timmins, K. B. Jones, and B. L. Jackson. 1995. A factor analysis of landscape pattern and structure metrics. *Landscape Ecology* 10:23–29.

Riitters, K. H., J. D. Wickham, R. V. O'Neill, K. B. Jones, and E. R. Smith. 2000. Global-scale patterns of forest fragmentation. *Conservation Ecology* 4. http://www.consecol.org/Joumal/vol4/iss2/art3/

Simmons, M. A., V. I. Cullinan, and J. M. Thomas 1992. Satellite imagery as a tool to evaluate ecological scale. *Landscape Ecology* 7:77–85.

Tischendorf, L. 2001. Can landscape indices predict ecological processes consistently? *Landscape Ecology* 16:235–54.

Tscharntke, T., A. M. Klein, A. Kruess, I. Steffan-Dewenter, and C. Thies. 2005. Landscape perspectives on agricultural intensification and biodiversity—Ecosystem service management. *Ecology Letters* 8:857–74.

Tuchel, D. 2004. *Romanians warn of Danube Delta ecological disaster.* Environmental News Service, 28 September.

Walker, B. H, S. R. Carpenter, J. Anderies, N. Abel, G. S. Cumming, M. Janssen, L. Lebel, J. Norberg, G. D. Peterson, R. Pritchard. 2002. Resilience management in social-ecological systems: A working hypothesis for a participatory approach. *Conservation Ecology* 6 (1): 14. http://www.consecol.org/vol6/iss1/art14

Walker, B. H., L. H. Gunderson, A. P. Kinzig, C. Folke, S. R. Carpenter, and L. Schultz. 2006. A handful of heuristics and some propositions for understanding resilience in social-ecological systems. *Ecology and Society* 11, no. 13. http://www.ecologyandsociety.org/vol11/iss1/art13/

With, K. A. 2004. Assessing the risk of invasive spread in fragmented landscapes. *Risk Analysis* 24:803–15.

With, K. A., and T. O. Crist. 1995. Critical thresholds in species' responses to landscape structure. *Ecology* 76:2446–59.

Wu, J., and Y. Qi. 2000. Dealing with scale in landscape analysis: An overview. *Geogr Info Sci* 6:1–5.

Zaccarelli, N., I. Petrosillo, and G. Zurlini. 2008. Retrospective analysis. In *Encyclopedia of ecology, vol. 4: Systems ecology,* eds. S. E. Jørgensen and B. D. Fath, 3020–29. Oxford: Elsevier.

Zaccarelli, N., I. Petrosillo, G. Zurlini, and H. Riitters. 2008. Source/sink patterns of disturbance and cross-scale effects in a panarchy of social-ecological landscapes. *Ecology and Society* 13, 26. http://www.ecologyandsociety.org/vol13/iss1/art26

Zaccarelli, N., K. H. Riitters, I. Petrosillo, and G. Zurlini. 2008. Indicating disturbance content and context for preserved areas. *Ecological Indicators* 8:841–53.

Zurlini, G., and F. Müller. 2008. Environmental security. In *Encyclopedia of ecology, vol. 2: Systems ecology*, eds. S. E. Jørgensen and B. D. Fath, 1350–56. Oxford: Elsevier.

Zurlini, G., K. H. Riitters, N. Zaccarelli, and I. Petrosillo. 2007. Patterns of disturbance at multiple scales in real and simulated landscapes. *Landscape Ecology* 22:705–21.

Zurlini, G., K. H. Riitters, N. Zaccarelli, I. Petrosillo, K. B. Jones, and L. Rossi. 2006. Disturbance patterns in a social-ecological system at multiple scales. *Ecological Complexity* 3:119–28.

Zurlini, G., N. Zaccarelli, and I. Petrosillo. 2004. Multi-scale resilience estimates for health assessment of real habitats in a landscape. In *Ecological indicators for assessment of ecosystem health*, eds. R. Costanza, F. Xu, and S. E. Jørgensen, 305–32. Boca Raton, FL: CRC Press–Lewis Publishers.

7

Species Richness in Space and Time as an Indicator of Human Activity and Ecological Change

---

Erich Tasser, Georg Niedrist, Patrick Zimmermann,
and Ulrike Tappeiner

CONTENTS

7.1 Introduction

Biodiversity is vital to the health of the planet and its people. It secures our food supply, provides a source of medicine and new technologies, and helps regulate our climate. According to the Secretariat of the Convention on Biological Diversity, "A major challenge for the twenty-first century will be making the conservation and sustainable use of biodiversity a compelling basis for development policies, business decisions, and consumer desires" (2000). But biodiversity is particularly affected by global change (*cf.* Li and Reynolds 1994; Dramstad et al. 1996; Forman 1997; Gustafson and Gardner 1996). For this reason, it is essential that indicators of biodiversity be integrated in sustainability monitoring systems. Data derived from such systems should then be made available to local administrations, policy and decision makers, as well as stakeholders to aid them in decision-making processes. However, it is difficult to determine which indicators are suitable to assess the effects of global changes on spatiotemporal biodiversity, as only limited quantifiable data on biological diversity are available (Honnay et al. 2003; Dierssen 2006). This makes it nearly impossible to meaningfully compare different space levels as well as different time periods. Numerous attempts have been made to develop and establish monitoring systems at ecosystem and regional scales (Olsen et al. 1999; Hoffmann-Kroll et al. 2003). Some systems have been up and running for a few decades, such as the British Countryside Survey (Haines-Young et al. 2003). However, available data still do not stretch back over a sufficiently long time span. In addition, there is still no formal definition of what is an adequate measure of biodiversity (Yoccoz et al. 2001; Büchs 2003; Dudley et al. 2005).

Biodiversity is multifaceted and hierarchical and cannot be measured, per se (Noss 1990). Indicators should ideally cover different aspects of the three levels of genetics, species, and ecosystem; however, this is a huge challenge (Hermy and Cornelis 2000). Single indicators that cover the whole range of biodiversity do not exist, and there is no consensus on how to use or design indicators for different aspects of biodiversity (Purvis and Hector 2000; Waldhardt 2003) nor an agreement on how to implement them (Büchs et al. 2003). In principle, vascular plants have been suggested as being among the best indicator groups for biodiversity evaluation as a whole as they play a decisive role in terrestrial ecosystems, being structural, autotrophic, and stationary organisms. Other species indicators that are frequently used, such as the number of red list or stenotopic species, are, according to Duelli and Obrist (2003), less effective because species-rich areas (hot spots) and areas harboring rare or red list species seldom coincide. In the studies presented in this paper, we therefore focused on the species richness of vascular plants, where the well-grounded knowledge about their diversity at the ecosystem level should serve as the basis for statements about space and time borders.

Our intention was to apply this knowledge to developing new landscape indicators and to compare the results with those of well-known and commonly used indicators at the different spatial levels. Due to their heterogenic political, socioeconomic, biogeographical, and natural site conditions we used the Alps as our test region. The Alps are not only the highest inner-European mountain chain, crossing eight different countries; they also contain a large variety of landscapes, species, and land-use types. From the valleys to the mountaintops, a variety of flora and fauna patterns can be observed. Slope inclination and altitudinal climate gradients influence the natural dynamics of soil and topography, and thus the typology of land use and habitat varies. This includes glaciers as well as viniculture sites and steppe-like vegetation islands. Furthermore, in recent decades agriculture in the Alps has lost much of its significance (Tappeiner et al. 2003; Streifeneder et al. 2007). The main reasons for this are unfavorable conditions such as a shortened vegetation period, and difficult terrain with steep slopes and small arable plots, which incur high production costs. Mountain agriculture, therefore, cannot compete in national and international markets. Thus, from the 1950s onward, marginal land with low yields has successively been taken out of agricultural use. However, this development varies greatly in intensity between regions (*cf.* Tappeiner, Borsdorf, et al. 2008): while in the South Tyrolean region Unterland-Überetsch, one of the most productive regions of the Alps, only about 6% of farming land has been abandoned within the last 150 years, the figure stands at 33% for the Tyrolean uphill areas, 37% in the region around Innsbruck, and reaches 67% in the Carnia region. The peak of decline in agricultural use occurred in the 1950s and 1960s.

To summarize, using the Alps as our test region, this chapter addresses three main questions: (a) Can indicators be developed that illustrate vascular plant diversity over spatial borders?; (b) Are there general trends in biodiversity? Can general statements be made?; and (c) What have been the prevailing trends in land use since the nineteenth century and how have they affected biodiversity? What changes are to be expected in the future?

7.2 Driving Forces of Change

Biodiversity of the Alps today is threatened by global changes, as it is in many regions all over the world. In recent decades the Alps have experienced massive changes in agriculture and forestry (Pan et al. 1999; Nusser 2001; Lütz and Bastian 2002; Tappeiner, Borsdorf, et al. 2008). This is mainly due to the increased use of machinery, new production techniques, improved breeds, different national and regional subsidizing instruments, or even just the public stance on the respective sectors. The most decisive break occurred in the second half of the last century, when production for self-sufficiency changed toward production for the market. Suddenly farmers were faced with a problem that they had not previously encountered with such severity, i.e., global competition (Lambin et al. 1999; MacDonald et al. 2000; Veldkamp et al. 2001). This necessitated a rethink, a reorientation, which meant in many cases abandoning farming altogether, and it particularly affected farmers in the Alps. As recently as 50 years ago they were using the few favored valley areas to grow cereal and field crops. Hay was made on steep slopes and high-altitude mountain meadows, and animals were driven up to the mountain pastures for the summer. Today all this has changed: in favored areas many fields are farmed more intensively; marginal areas, however, are farmed less intensively or have been abandoned altogether. As a result of such developments, nearly 36% of farmers in the Alps have given up farming within the last 20 years (1980–2000), with a decidedly heterogeneous picture emerging across the Alps (Streifeneder et al. 2007). Around a quarter of all farms in Germany and Austria closed down within the last 20 years, while in Slovenia the figure is more than half of all farms. Of the remaining farms, around 40% are run as a sideline. Today on average about 20% of the former agricultural land lies fallow, with figures in some regions approaching 70% (Tappeiner et al. 2006; Tasser 2007). At the same time many favored areas are now farmed more intensively and have been adapted to modern forms of land use, which has caused far-reaching ecological changes, both positive and negative. Depending on the land use, completely different habitats with typical phytosociological communities develop. The distribution of habitats in the cultural landscape remains constant as long as there are no changes to land use. But if areas are, for example, abandoned, then they are subject to a natural succession. After a certain time the first tree seedlings germinate, which thicken slowly to a young forest and then to a closed timber forest. The speed of this succession depends considerably on the altitude and the climate conditions. A forest can establish itself up to the potential timber line; above that a dwarf shrub belt develops, followed by alpine grass mats.

Closely linked with this development is climate change. Since the end of the last "Little Ice Age" in the mid-nineteenth century it has already become ostensibly warmer in the Alps (approximately +1.8°C). This increase in temperature is, according to Kromp-Kolb and Formayer (2005), on the one hand, of natural origin (quasi a return to the "normal condition"). On the other hand, a substantial part (approximately 0.6°C–0.9°C) might also be attributed to anthropogenic causes. If the climate becomes warmer, then the limits described above are pushed upward. The forest will thereby encroach on areas that had been unsuitable until now. At the same time glaciers retreat. On the scree slopes first pioneer plants germinate, which thicken to alpine mats in the course of time. At lower altitudes it is mainly deciduous woods that will suppress coniferous woods (see Kräuchi and Kienast 1993). These changes have a dramatic effect on the structure and function of ecosystems, e.g., negative and positive effects on biodiversity (Li and Reynolds 1994; Dramstad et al. 1996; Forman 1997; Gustafson and Gardner 1996), on geomorphological processes and soil erosion (Thomas and Allison 1993; Lütz and Bastian 2002), on biochemical cycles and hydrological processes (Nagasaka and Nakamura 1999; Weber et al. 2001), on the contamination of the soil surface and the groundwater, on the carbon cycle (Piussi and Farrell 2000; Bahn et al. 2008), and they also affect human health (Lütz and Bastian 2002). In addition, changes in land use lead to alterations in the interaction between ecosystems and near-ground air layers and may affect the transport of delicate and latent warmth, of CO_2, nutrients, and pollutants (Tenhunen et al. 2008; Wohlfahrt et al. 2008). Agricultural land use produces a cultural landscape that supports many aspects of conservation and contributes to preserving an attractive area in which to live, work, and recuperate, for both indigenous people and tourists alike (Kienast 1993; Usher 1999).

7.3 Spatial Aspects of Biodiversity and Species Richness: From Ecosystem to the Landscape

Several of our publications deal with both the direct and indirect effects of agricultural land use on the composition of grassland plant communities, starting with the publication of Tasser et al. (1999). On the basis of three study areas in a north-south transect of the Eastern Alps, a close correlation between current land use and emerging phytosociological plant communities was established. Depending on the type and intensity of land use, different communities with distinct compositions establish themselves. On fertilized, intensively used hay meadows mainly hemicryptophytes (life forms defined according to Raunkiaer [1934]) grow. With declining intensity of land use, not only the number but also the cover of lignified chameophytes increases. We also demonstrated that even the diversity of vascular plants decreases significantly both with intensified use and with time elapsed since abandonment. To extrapolate these findings to a broader basis, the isolated investigations were extended spatially. The results of 936 vegetation relevées of agriculturally used grasslands, distributed within the Central Alps, containing detailed information on land use, form the data base to analyze the relationships between land use, site parameters, and biodiversity (Niedrist et al. 2008). In the course of these studies it emerged that lightly used areas were not only the habitat with the highest number of vascular plants, but also showed an above-average ecosystem diversity (number of different plant communities, Figure 7.1). Intensively used hay meadows, on the other hand, presented the lowest diversity of species and ecosystems as well as a great homogeneity in their composition, as supported by the high Evenness Index.

Regarding biodiversity, these results highlight that all grassland communities are not created equal. Moreover, the site and land-use conditions must also be considered. This is particularly meaningful with a large spatial view of biodiversity. In order to obtain the desired results, it is a more useful approach to calculate biodiversity indicators and/or species-richness indicators at the level of habitats rather than at that of single phytosociological communities. Within the grassland areas themselves, one must differentiate between intensively and lightly used agricultural grassland habitats, as well as wet and xeric grassland habitats and unused alpine grassland habitats on calcareous and acid soils (see Figure 7.2). The grassland habitats represent important but certainly not unique habitats in the Alps. They cover approximately 30% of the total Alpine region (Tappeiner, Borsdorf, et al. 2008). For a complete evaluation of the landscape biodiversity, all other habitats must also be considered. Using the method of Braun-Blanquet (1964), we collected about 5,240 vegetation relevées (145 syntaxa) that were distributed across the central Eastern Alps. These relevées include both our own relevées and

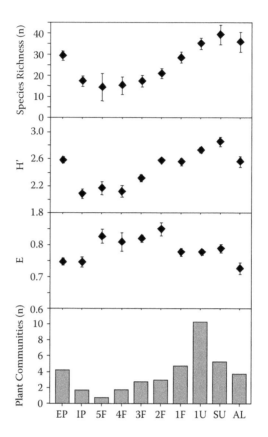

FIGURE 7.1
Comparison of mean species richness (n), Shannon–Wiener index (H'), and evenness-index (E), as well as frequency-weighted incidence of plant communities (n) in differently used grassland in the Central Alps. Mean ± s.e. EP, extensively used pastures; IP, intensively used pastures; 5F, fertilized hay meadows, mown five times; 4F, fertilized hay meadows, mown four times; 3F, fertilized hay meadows, mown three times; 2F, fertilized hay meadows, mown two times; 1F, fertilized alpine meadow, mown once; 1U, unfertilized alpine meadows, mown once; SU, unfertilized alpine meadows, sporadically mown every 2 to 5 years; AL, abandoned land. *Source*: Modified from Niedrist, G., E. Tasser, C. Lüth, J. Dalla Via, and U. Tappeiner. 2008. Plant diversity declines with recent land use changes in European Alps. *Plant Ecology* 202 (2):195–210.

relevées from the literature of the years 1994–2007, which was examined using rigorous quality assurance processes. Even though the number of vegetation relevées in the various habitats differed greatly (*cf.* Figure 7.2), we were able to show that from a minimum of 30 relevées upward, the mean species number for a given habitat did not change significantly (see also Tasser et al. [2008]). Subsequently, potential mean and potential absolute species richness was calculated for all habitats. The potential mean species richness quantifies the mean number of species from all vegetation relevées of a single habitat type, whereas the potential absolute species richness indicates

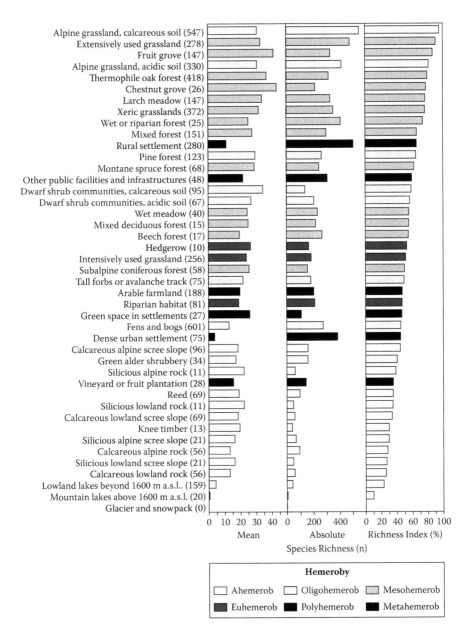

FIGURE 7.2
Mean species richness of vascular plant species, mean number of species from all vegetation relevées of each habitat type, absolute number of vascular plant species, sum of species reported within all vegetations relevées of each habitat, and richness index average between potential mean and potential absolute species richness, related to the respective maximum value of the main habitats in the Eastern Alps (Peer 1991). The number of vegetation relevées is reported in brackets; their hemeroby (see Steinhardt et al. 1999) is indicated by the shade of gray.

the sum of species reported within all vegetations relevées of each habitat. From the outset, a first comparison between the individual habitats shows substantial differences between the mean and absolute richness. On average, oligohemerobe (hemeroby index after Steinhardt et al. [1999] = 2) and meso-hemerobe (hemeroby index = 3) grassland habitats hold the highest vascular plant richness. In the center span predominantly mesohemerobe wood habitats and eu- to polyhemerobe grassland habitats (hemeroby index = 4–5) as well as ahemerobe rock and scree slope habitats are to be found. Generally, artificial habitats (hemeroby index = 6 and 7), lakes, and glaciers exhibit the smallest mean species richness. The absolute species richness clearly deviates from this distribution. Near natural and extensively used grasslands do indeed shelter high mean and absolute species richness; nevertheless, there are also some habitats with low mean species richness, but with a high absolute richness. Examples of this are settlements, fens, and bogs. On the other hand, only a few different species grow in habitats with mean or even high species richness (e.g., scree slopes, skirt habitats, dwarf shrub communities, and Knee timber habitats). When standardizing the individual richness values (percentage of the maximum value) and averaging both values to a mean richness index, then oligo- and mesohemerobe habitats are to be found in the upper range. Habitats that are strongly affected by humans appear in the middle to lower range. Natural alpine and nival habitats as well as lakes have the lowest richness.

Studies at the habitat level provide detailed findings about the changes taking place in a specific environmental site but do not lend themselves to generalizations (Meentemeyer 1978; O'Neill et al. 1991). Society today expects researchers to provide answers to regional and global issues, such as the future impact of changes in land use or climate on biodiversity (Secretariat of the Convention on Biological Diversity 2000). Research findings must be valid on the spatial and temporal scale of emotional, social, and economic life, the level where decisions are often made. Statements should therefore relate to agriculture as a whole if not to an entire region. At the same time they have to map trends for assessing and evaluating future developments. Applied ecological research is increasingly moving in this direction (Ludwig et al. 2003; Del Barrio et al. 2006; Seidl et al. 2007). This results in a need for more methods and tools to investigate larger spatiotemporal relationships (Dodson and Marks 1997, Olseth and Skartveit 1997, Thornton et al. 1997). At the landscape level there is already a large number of indicators that quantify landscape structure, composition, and spatial configuration, thus enabling an analysis of the arrangement, number, and size of patches using areal statistics (e.g., O'Neill et al. 1991; Turner, Gardner, and O'Neill 2001). As a result, we can draw conclusions about the landscape's complexity, diversity, homogeneity, fragmentation, and anthropogenic effects (Jaeger 2000; Papadimitrou 2000). The indicators, however, permit no direct statement regarding the species diversity at landscape level. Increasing the number of samples to cover spatially expanded landscape (e.g., relevées using the

Braun-Blanquet method) is not feasible because of the immense logistic and financial investment required for process-oriented ecosystem surveys. Nor is it always possible to transfer the understanding gained from one scale directly to another; in fact it is often not (O'Neill et al. 1991). New approaches need to be found to combine findings from ecosystems research with those from landscape research.

Our study of biodiversity in South Tyrol is representative of such an approach (Tasser et al. 2008). South Tyrol is the northernmost province of Italy, bordering Austria, a region amid the Alps. We defined for South Tyrol a set of biodiversity indicators to measure changes in biodiversity at the municipal level as comprehensively as possible, to present them in a readily understandable way and to assess them in terms of their effect on sustainability. The set consists of several landscape indicators, such as landscape diversity, compartmentation, degree of anthropogenic influence (hemeroby), degree of naturalness of watercourses, and the intensity of agricultural use. These indicators tell us a lot about the state and development of the landscape, taking into consideration the established knowledge that human land use, in the form of compartmentation (Moser et al. 2006), urban sprawl, and agricultural and forest use (Tasser et al. 2007), is a major cause. In addition, we developed two more indicators that take into account the quality of individual habitats in terms of the occurrence of vascular plants. The comprehensive collection of vegetation relevées of the habitats (Figure 7.3) is used as a basis for these

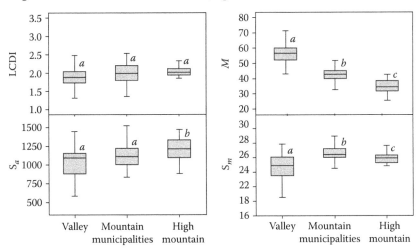

FIGURE 7.3
Biodiversity indicators of the valley, mountain, and high mountain municipalities in South Tyrol. S_m, area-weighted mean species richness of vascular plants; S_a, frequency weighted absolute species richness; LCDI, land cover diversity index; M, hemeroby index. Differences in biodiversity indicators among municipality types were tested with the Bonferroni post-hoc test adjusted for multiple contrasts. Letters indicate significant differences among municipality types at $P < 0.05$.

indicators. The area-weighted mean species richness of vascular plants (S_m) takes into account how many species are found on average in a vegetation relevée collected for the habitat, and their spatial dominance is considered by the area weighting. S_m is calculated with the formula

$$S_m = \sum_{i=1}^{n} A_i m_i$$

where A_i is the area proportion of habitat i and m_i is the mean species number of habitat i. The frequency weighted absolute species richness (S_a) is calculated as

$$S_a = \sum_{i=1}^{s} \frac{F_i}{f_i}$$

It accounts for the occurrence of individual species and down weights frequently occurring species by the ratio F_i/f_i, where F_i is the frequency of species i in the occurring habitats and f_i is the frequency of species i in all habitats. Unlike S_m, the spatial extent of the habitats has no influence on this indicator. This means that now two more indicators are available for including α-diversity (species diversity within a bioceonosis) and γ-diversity (total diversity within an area of investigation) into analyses.

Our results depicted that landscape structure (land cover diversity index, LCDI) turns out to be highest in mountain municipalities where mountain farming is common (Figure 7.3). Because of the natural conditions, mountain farming is never as intensive as farming in the agriculturally favorable areas. The dramatic land-use intensification in the agriculturally favorable areas includes the conversion of complex natural ecosystems to simplified managed ecosystems and the intensification of resource use, resulting in a greatly reduced biodiversity during the last few decades (see also Tscharntke et al. 2005). This is why valley municipalities (mean altitude of 583.9 ± 41.0 m a.s.l.) in South Tyrol have lower potential mean and absolute species richness than mountain (1463.5 ± 29.9 m a.s.l.) and high mountain (1990.0 ± 36.1 m a.s.l.) municipalities. Nickel and Hildebrandt (2003) made similar observations: they identified species numbers as well as the percentage of specialists and pioneer species as suitable indicators of biotic conditions due to management intensity in agricultural ecosystems. In addition, the results of Büchs et al. (2003) for spiders as well as Döring and Kromp (2003) for carabid beetles led to similar conclusions. The influence of site factors is also evident: with an increase in non-usable areas (such as rocks, glaciers, and alpine grasslands), the hemeroby (M) decreases. By the same token, a rise in the number of ahemerobe habitats, which are often species-poor (see Figure 7.2), means a significant decrease of S_m. Steeper areas are also less suited for settlements and intensive agriculture. Both S_m and S_a thus increase with the slope angle, while M decreases. Another unequivocal result was the correlation between S_a and habitat richness (LCDI). It is well known that the more habitats there are, the higher the absolute species richness is (e.g., Hermy and Cornelis 2000; Dauber et al. 2003; Moser et al. 2006).

7.4 Temporal Aspects of Biodiversity and Species Richness: From the Past to the Future

The knowledge gained from the study in South Tyrol formed the starting point for the evaluation of the consequences of changes in agrarian land use on the biodiversity in the entire Alpine region between 1865 and 2000 (Zimmermann et al. under review). Representative Alpine study areas were chosen in eight different agrarian structure regions all over the Alps (see Tappeiner et al. 2003), altogether covering 35 municipalities (in sum 1565 km²). Regions were subdivided into 10 ecoregions to allow altitudinal factors to be examined. Changes in land use and habitat type were mapped on the basis of historical maps and airborne images (for the methodical approach see Tasser et al. 2009). As airborne images are not continuously available for every single year, the used photographs were taken within certain time-spans (1950–1961, 1979–1990, and 1998–2003). Since the spatial and temporal coverage of historical land-use maps is even less dense, they cover the range between 1800 and 1879. The four time-steps are hereafter referred to as nineteenth century, 1950s, 1980s, and 2000. Collections of vegetation relevées from the literature provided information on plant biodiversity for each habitat. Considering the spatial and the floristic data, landscape biodiversity was expressed as frequency weighted absolute species richness S_a, area weighted mean species richness S_m, and by Rao's quadratic entropy Q as a measure of dissimilarity (Anderson et al. 2006). Q gives the portion of unshared species out of the total number of species present in the habitats.

Six land-use trends dominated the landscapes development of the Alps in the past 150 to 200 years: abandonment of grassland, continuous grassland farming, change from mixed agriculture to grassland farming, specialization in vine and fruit farming, continuous mixed agriculture, and strong urban sprawl.

Specialization in vine and fruit farming had the most dramatic effects on biodiversity (Figure 7.4). In this region, fields and meadows were nearly completely transformed into intensive permanent cultures in the mid-twentieth century. The spread of permanent cultures has led to strong landscape homogenization, the loss of habitats, and a decrease of species richness. This trend is typical for the agriculturally favorable valley bottoms, e.g., in

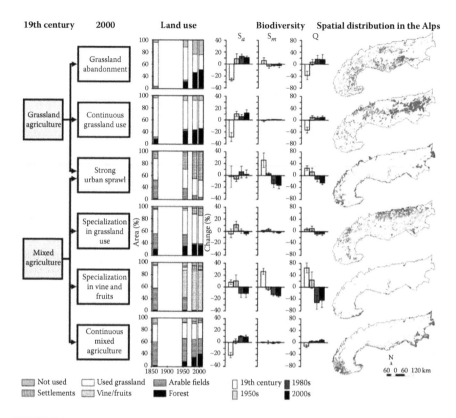

FIGURE 7.4

Main land-use trends since the nineteenth century and biodiversity change. For the indicators the relative deviation from the average of all years is shown. S_m, area-weighted mean species richness of vascular plants; S_a, frequency weighted absolute species richness; Q, Rao's quadratic entropy.

the Adige Valley in Italy or the French pre-alpine regions. Similar negative effects are produced also by the urban sprawl, even if S_a increases slightly thanks to the many cultivated plants and synanthropic species. S_m and Q significantly decreased, on the other hand, because of the substantial urban sprawl. A moderate abandonment, recognizable in the presence of a continuous grassland use and continuous mixed use (mean: 20% abandoned), resulted in rising biodiversity. The continuous grassland use is the most frequent trend within the agriculturally used ecoregions and is rare in the colline zone and in the Southern Alps. The maintenance of mixed agriculture occurs mostly at the southern margins of the Alps. In France, this trend is characterized by a higher portion of arable land than in Italy, where grassland is more dominant. The specialization in grassland use for dairy and cattle farming shows a strong decrease in the area of fields, while the grassland area itself remains stable or increases. The trend can mainly be observed in the montane valleys, especially in the Northern Alps. As the total agricultural area was not always maintained, this trend is partly coupled with reforestation or, as in some favorable areas, with growing settlements. Therefore a light decrease of the biodiversity is connected with this development.

Where abandonment is the dominant trend (mean: 63% abandoned), land-use change had different effects according to the indicator. Significant change occurred if the use of species-rich extensive meadows was significantly reduced (especially in the subalpine zone), which led to a continuous drop of S_m. However, the invasion of dwarf shrubs and trees increased habitat heterogeneity and the potential species pool, expressed through rising Q and S_a. Due to landscape homogenization as a result of a strong reforestation, these two measures peaked in the 1980s, whereupon a decline set in. A strong abandonment of grassland can be found everywhere in the subalpine zone and in the slope regions of the lower altitudes, especially in the Southern Alps. Through this study we could thus link the historical land-use change with some aspects of biodiversity change.

Similar trends to those in the Alps with probably comparable effects on biodiversity are also found in other regions of the world: in many places, e.g., in Northwestern Europe or the mountain regions of the United States, agriculture has been marginalized as an economic activity, often with land abandonment and urban sprawl as the result (Romero-Calcerrada and Perry 2004; EEA 2006). Where agriculture is maintained, management tends to intensify as a response to market pressures. In other regions of the world (e.g., in Asia or Central and South America) agriculture has gained importance as the population quickly increases. An enlargement of the agriculturally used areas is the result (Farrow and Winograd 2001; Nusser 2001; Gautam et al. 2003). Land abandonment and intensification are thus going hand in hand. Species-rich extensive farming systems in Southern and Eastern Europe suffer particularly from this combined trend (Bruns et al. 2000).

Following on these findings, the question arises how developments in the future will affect biodiversity. Before being able to answer this question, however, future scenarios must be generated, which we did for the project area Stubaital as an example (North Tyrol, Austria) (see Tappeiner, Tasser, et al. 2008). Future patterns of change were projected through two different methodological approaches: (a) a Markovian model and (b) participatory involvement. Markovian models allow for the estimation of rates of change between two dates, and project changes in type for a third date, assuming that rates of change are constant. A transition matrix from 1988 to 2003 was generated to estimate functional land-use type changes, using the area-changed values. Subsequently, the values of this matrix were transformed to probability values according to the eigenvector methods described in

Gomez-Mendoza et al. (2006). Hence, a status quo scenario in a 15-year time span was available, but no additional conditions could be added. To minimize the impact of this methodological limitation, we used in addition a scenario technique based on the judgment of experts in stakeholder workshops. Such workshops allow for a comprehensive approach to predicting the effects of contrasting policy scenarios over an imagined period, where the personal experience of the stakeholders plays a significant role. The approach used (Bayfield et al. 2008) examined three likely future scenarios of different European agricultural policy strategies (Midgley et al. 2005):

1. Status quo—gradual reduction of farm income support; continuation of restrictive planning policies
2. Reduced area-based support—rapid reduction of area-based direct payments in favor of environmental or cultural landscape payments linked to labor; continuation of restrictive planning
3. Rural diversification—enhanced rural development policy with positive planning

Since the EU's 2003 Common Agricultural Policy (CAP) reform and the Single Farm Payment (SFP) scheme are expected to change the allocation of resources on- and off-farm, the payments become decoupled from production and will increasingly go more in the direction of supporting rural development (Gorton et al. 2008). The scenario of a rural diversification especially takes these possible policy changes into account. Also Tappeiner, Tasser, et al. (2008) could establish that in comparison to the Markovian model this scenario provokes the biggest changes in the landscape. The diversification scenario suggests stronger changes for the valley floor, which would be almost dominated by a built-up area and intensively used grassland. Hence, we regard the results of the Markovian model and the scenario of rural diversification of the stakeholder workshop as a good basis for biodiversity projections and we generated scenario maps identifying the locations of the inferred changes (see Tappeiner, Tasser, et al. 2008).

The results now show the following trends (Figure 7.5): (a) Regarding all considered indicators the intensively used ecoregion (valley bottom) is characterized by the smallest diversity. Traditionally or extensively used ecoregions (valley slopes and subalpine zones), however, clearly possess more variety. (b) The historical and future changes of land use significantly affect the development of species richness in the different ecoregions:

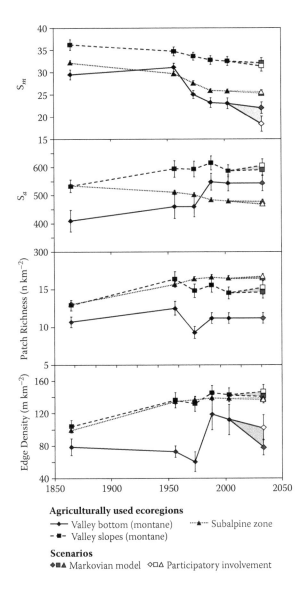

FIGURE 7.5

Biodiversity trends $\bar{x}$ ± s.e. since the nineteenth century and future scenarios through the example of the agriculturally used ecoregions of the Stubai Valley, Tyrol, Austria.

1. Starting from 1950 in the valley bottom, the suppression of traditional land-use forms (fields, open orchard meadows) and the increase of the residential area led to a decrease of mean species richness and ecosystem diversity. This trend will continue in the future and/or

will even strengthen considerably under the scenario of a rural diversification (stakeholder scenario). On the other hand, these developments also result in an increase in absolute species richness, since settlements in absolute terms count among the habitats with highest species richness (Figure 7.2). The landscape connectivity (edge density), however, will decrease in the future in any case (using the Markovian model even more clearly).

2. Until 1950, the valley slopes were characterized by an increase of biodiversity. Since then only slight changes have occurred. The scenarios vary likewise between constant measures and/or slightly rising values within the stakeholder approach. This increase results particularly from small-area abandonments of marginal areas, which increase the diversity in the grassland-dominated region. The only exception to this development is S_m. The mean species richness decreases slightly but constantly and this is attributable predominantly to the abandonment of extensively used, species-rich habitats and to the spreading of species-poor coniferous forests.

3. This development, which is due to the abandonment of agrarian areas, becomes even more apparent in the subalpine belt. There the species richness (S_m, S_a) decreases constantly by the succession of the abandoned, formerly extensively used areas. On the other hand, this development, which will intensify in the future, leads to an increase of the ecosystem diversity. Formerly wide agriculturally used areas break down into a mosaic of smaller succession stages, which leads to an increase of patch richness and edge density.

7.5 Conclusion

At present there are a great number of biodiversity indicators. They differ more or less according to the question of spatial scale and thus represent different aspects of biodiversity. There are, however, hardly any indicators of the same aspects of biodiversity at different scale levels. Our indicators of the mean and absolute species richness of vascular plants represent an exception to this. They are based on individual measurements at the ecosystem level, and are then aggregated to habitats in order to extrapolate the habitats at the landscape level. We thus linked data from single ecosystem measurements (vegetation relevées) with results at landscape level (shifts in areas), and derived statements about species richness on a larger scale. These statements have more relevance to the public without losing their relationship with free land measurements. Our findings therefore underline the importance of agricultural use, particularly site-adapted extensive use, in the conservation of biodiversity. Both intensification measures and abandonment cause a decrease of the species richness at all spatial levels.

Acknowledgments

We thank Dr. Ian Wells for correcting and enhancing the English text. This work was partly funded within the framework of the INTERREG-IIIA Italy/Austria projects "MASTA" and "DNA-Chip-Entwicklung zur Charakterisierung und Valorisierung von Bergheu."

References

Anderson, M. J., K. E. Ellingsen, and B. H. McArdle 2006. Multivariate dispersion as a measure of beta diversity. *Ecology Letters* 9:683–93.

Bahn, M., M. Rodeghiero, M. Anderson-Dunn, S. Dore, S. Gimeno, M. Drösler, M. Williams, C. Ammann, F. Berninger, C. Flechard, S. Jones, S. Kumar, C. Newesely, T. Priwitzer, A. Raschi, R. Siegwolf, S. Susiluoto, J. Tenhunen, G. Wohlfahrt, A. Cernusca. 2008. Soil respiration in European grasslands in relation to climate and assimilate supply. *Ecosystems* 11 (8): 1352–67.

Bayfield, N., P. Baranock, M. Furger, M. T. Sebastia, G. Domınguez, M. Lapka, E. Cudlinova, L. Vescova, D. Gianelle, A. Cernusca, U. Tappeiner, and M. Drössler. 2008. Stakeholder perceptions of the impacts of rural funding scenarios on mountain landscapes across Europe. *Ecosystems* 11 (8): 1368–82.

Braun-Blanquet, J. 1964. *Pflanzensoziologie*. Vienna: Springer-Verlag.

Bruns, D., D. Ipsen, and I. Bohnet. 2000. Landscape dynamics in Germany. Landscape and urban planning 47:143–58.

Büchs, W. 2003. Biodiversity and agri-environmental indicators—General scopes and skills with special reference to the habitat level. *Agriculture, Ecosystems and Environment* 98:35–78.

Büchs, W., A. Harenberg J. Zimmermann, and B. Weib. 2003. Biodiversity, the ultimate agri-environmental indicator? Potential and limits for the application of faunistic elements as gradual indicators in agroecosystems. *Agriculture, Ecosystems and Environment* 98:99–123.

Dauber, J., M. Hirsch, D. Simmering, R. Waldhardt, A. Otte, and V. Wolters. 2003. Landscape structure as an indicator of biodiversity: Matrix effects on species richness. *Agriculture, Ecosystems and Environment* 98:321–29.

Del Barrio, G., P. A. Harrison, P. M. Berry, N. Butt, M. E. Sanjuan, R. G. Pearson, and T. Dawson. 2006. Integrating multiple modelling approaches to predict the potential impacts of climate change on species' distributions in contrasting regions: Comparison and implications for policy. *Environmental Science and Policy* 9 (2): 129–47.

Dierssen, K. 2006. Indicating botanical diversity—Structural and functional aspects based on case studies from Northern Germany. *Ecological Indicators* 6:94–103.

Dodson, R., and D. Marks. 1997. Daily air temperature interpolated at high spatial resolution over a large mountainous region. *Climate Research* 8:1–20.

Döring, T. F., and B. Kromp. 2003. Which carabid species benefit from organic agriculture? A review of comparative studies in winter cereals from Germany and Switzerland. *Agriculture, Ecosystems and Environment* 98:153–61.

Dramstad, W. E., J. D. Olson, and R. T. T. Forman. 1996. *Landscape ecology principles in landscape architecture and land use planning*. Cambridge, MA: Harvard University Graduate School of Design.

Dudley, N., D. Baldock, R. Nasi, and S. Stolton. 2005. Measuring biodiversity and sustainable management in forests and agricultural landscapes. *Philosophical Transactions of the Royal Society* 360:457–70.

Duelli, P., and M. K. Obrist. 2003. Biodiversity indicators: The choice of values and measures. *Agriculture, Ecosystems and Environment* 98:87–98.

EEA. 2006. *Integration of environment into EU agriculture policy—The IRENA indicator-based assessment report. EEA report no. 2/2006*. Copenhagen: European Environment Agency.

Farrow, A., and M. Winograd. 2001. Land use modelling at the regional scale: An input to rural sustainability indicators for Central America. *Agriculture, Ecosystems and Environment* 85 (1–3): 249–68.

Forman, R. T. T. 1997. *Land mosaics. The ecology of landscapes and regions*. Cambridge: Cambridge University Press.

Gautam, A. P., E. L. Webb, G. P. Shivakoti, M. A. Zoebisch. 2003. Land use dynamics and landscape change pattern in a mountain watershed in Nepal. *Agriculture, Ecosystems and Environment* 99 (1–3): 83–96.

Gomez-Mendoza, L., E. Vega-Pena, M. I. Ramirez, J. L. Palacio-Prieto, and L. Galicia. 2006. Projecting land-use change processes in the Sierra Norte of Oaxaca, Mexico. *Applied Geography* 26 (3–4): 276–90.

Gorton, M., E. Douarin, S. Davidova, and L. Latruffe. 2008. Attitudes to agricultural policy and farming futures in the context of the 2003 CAP reform: A comparison of farmers in selected established and new Member States. *Journal of Rural Studies* 24 (3): 322–36.

Gustafson, E. J., and R. H. Gardner. 1996. The effect of landscape heterogeneity on the probability of patch colonisation. *Ecology* 77 (1): 94–107.

Haines-Young, R., C. J. Barr, L. G. Firbank, M. Furse, D. C. Howard, G. McGowan, S. Petit, S. M. Smart, and J. W. Watkins. 2003. Changing landscapes, habitats and vegetation diversity across Great Britain. *Journal of Environmental Management* 67:267–81.

Hermy, M., and J. Cornelis. 2000. Towards a monitoring and a number of multifaceted and hierarchical biodiversity indicators for urban and suburban parks. *Landscape and Urban Planning* 49:149–62.

Hoffmann-Kroll, R., D. Schafer, and S. Seibel. 2003. Landscape indicators from ecological area sampling in Germany. *Agriculture, Ecosystems and Environment* 98:363–70.

Honnay, O., K. Piessens, W. Van Landuyt, M. Hermy, and H. Gulinck. 2003. Satellite based land use and landscape complexity indices as predictors for regional plant species diversity. *Landscape and Urban Planning* 63:241–50.

Jaeger, J. 2000. Landscape division, splitting index, and effective mesh size: New measures of landscape fragmentation. *Landscape Ecology* 15:115–30.

Kienast, F. 1993. Analysis of historic landscape patterns with a Geographical Information System—A methodological outline. *Landscape Ecology* 8:103–18.

Kräuchi, N., and F. Kienast. 1993. Modelling subalpine forest dynamics as influenced by a changing environment. *Water, Air, Soil Pollution* 68:185–97.

Kromp-Kolp, H., and H. Formayer. 2005. *Schwarzbuch klimawandel*. Salzburg: Ecowin.

Lambin, E. F., X. Baulies, N. Bockstael, G. Fischer, T. Krug, R. Leemans, E. F. Moran, R. R. Rindfuss, Y. Sato, D. Skole, B. L. Turner, and C. Vogel. 1999. *Land-use and land-cover change LUCC: Implementation strategy. A core project of the International Geosphere-Biosphere Programme and the International Human Dimensions Programme on Global Environmental Change. IGBP Report 48, IHDP Report 10*. Stockholm: IGBP.

Li, H., and J. F. Reynolds. 1994. A simulation experiment to quantify spatial heterogeneity in categorical maps. *Ecology* 75 (8): 2446–55.

Ludwig, R., W. Mauser, S. Niemeyer, A. Colgan, R. Stolz, H. Escher-Vetter, M. Kuhn, M. Reichstein, J. Tenhunen, A. Kraus, M. Ludwig, M. Barth, and R. Hennicker. 2003. Web-based modelling of energy, water and matter fluxes to support decision making in mesoscale catchments—The integrative perspective of GLOWA-Danube. *Physics and Chemistry of the Earth* 28 (14–15): 621–34.

Lütz, M., and O. Bastian. 2002. Implementation of landscape planning and nature conservation in the agricultural landscape: A case study from Saxony. *Agriculture, Ecosystems and Environment* 92:159–70.

MacDonald, D., J. R. Crabtree, G. Wiesinger, T. Dax, N. Stamou, P. Fleury, J. Gutierrez Lazpita, and A. Gibon. 2000. Agricultural abandonment in mountain areas of Europe: Environmental consequences and policy response. *Journal of Environmental Management* 59: 47–69.

Meentemeyer, V. 1978. Macroclimate and lignin control of litter decomposition rates. *Ecology* 59 (3): 465–72.

Midgley, J. L., D. M. Shucksmith, R. V. Birnie, A. Geddes, N. Bayfield, and D. Elston. 2005. Rural development policy and community data needs in Scotland. *Land Use Policy* 22:163–74.

Moser, B., J. A. G. Jäger, U. Tappeiner, E. Tasser, and B. Eiselt. 2006. Modification of the effective mesh size for measuring landscape fragmentation to solve the boundary problem. *Landscape Ecology* 22 (3): 447–59.

Nagasaka, A, and F. Nakamura. 1999. The influences of land-use changes on hydrology and riparian environment in a northern Japanese landscape. *Landscape Ecology* 14 (6): 543–56.

Nickel, H., and J. Hildebrandt. 2003. Auchenorrhyncha communities as indicators of disturbance in grasslands Insecta, Hemiptera—A case study from the Elbe flood plains northern Germany. *Agriculture, Ecosystems and Environment* 98:183–99.

Niedrist, G., E. Tasser, C. Lüth, J. Dalla Via, and U. Tappeiner. 2008. Plant diversity declines with recent land use changes in European Alps. *Plant Ecology* 10.1007/s11258-008-9487-x.

Noss, R. F. 1990. Indicators for monitoring biodiversity: A hierarchical approach. *Conservation Biology* 4:355–64.

Nusser, M. 2001. Understanding cultural landscape transformation: A re-photographic survey in Chitral, eastern Hindukush, Pakistan. *Landscape and Urban Planning* 57 (3–4): 241–55.

Olsen, A. R., J. Sedransk, D. Edwards, C. A. Gotway, W. Liggett, S. Rathbun, K. H. Reckhow, and L. J. Young. 1999. Statistical issues for monitoring ecological and natural resources in the United States. *Environmental Monitoring and Assessment* 54:1–45.

Olseth, J. A., and A. Skartveit. 1997. Spatial distribution of photosynthetically active radiation over complex topography. *Agricultural and Forest Meteorology* 86:205–14.

O'Neill, R. V., S. J. Turner, V. I. Cullinan, D. P. Coffin, T. Cook, W. Conley, J. Brunt, J. M. Thomas, M. R. Conley, and J. Gosz. 1991. Multiple landscape scales: An intersite comparison. *Landscape Ecology* 5: 137–43.

Pan, D., G. Domon, S. de Blois, and A. Bouchard. 1999. Temporal 1958–1993 and spatial patterns of land use changes in Haut-Saint-Laurent Quebec, Canada, and their relation to landscape physical attributes. *Landscape Ecology* 14 (1): 35–52.

Papadimitriou, F. 2000. Modelling indicators and indices of landscape complexity: An approach using G.I.S. *Ecol Indic* 2:17–25.

Peer, T. 1991. *Karte der aktuellen vegetation südtirols 1:200.000.* Vienna: Autonome Provinz Bozen-Südtirol, Lithopress.

Piussi, P., and E. P. Farrell. 2000. Interactions between society and forest ecosystems: Challenges for the near future. *Forest Ecology and Management* 132 (1): 21–28.

Purvis, A., and A. Hector. 2000. Getting the measure of biodiversity. *Nature* 405:212–19.

Raunkiaer, C. 1934. *The life forms of plants and statistical plant geography.* Oxford: Oxford University Press.

Romero-Calcerrada, R., and G. L. W. Perry. 2004. The role of land abandonment in landscape dynamics in the SPA "Encinares del rio Alberche y Cofio," Central Spain, 1984–1999. *Landscape and Urban Planning* 66 (4): 217–32.

Secretariat of the Convention on Biological Diversity. 2000. *Sustaining life on earth. How the Convention on Biological Diversity promotes nature and human well-being.* Secretariat of the Convention on Biological Diversity.

Seidl, R., W. Rammer, D. Jäger, W. S. Currie, and M. J. Lexer. 2007. Assessing trade-offs between carbon sequestration and timber production within a framework of multi-purpose forestry in Austria. *Forest Ecology and Management* 248 (1–2): 64–79.

Steinhardt, U., F. Herzog, A. Lausch, E. Müller, and S. Lehmann. 1999. Hemeroby index for landscape monitoring and evaluation. In *Environmental indices—System analysis approach*, eds. Y. A. Pykh, D. E., Hyatt, and R. J. Lenz. Oxford, UK: EOLSS Publishers.

Streifeneder, T., U. Tappeiner, F. Ruffini, G. Tappeiner, and C. Hoffmann. 2007. Selected aspects of agro-structural change within the Alps. A comparison of harmonised agro-structural indicators on a municipal level. *Journal of Alpine Research* 3:41–52.

Tappeiner, U., A. Borsdorf, and E. Tasser, eds. 2008. *Mapping the Alps.* Heidelberg: Spektrum.

Tappeiner, U., G. Tappeiner, A. Hilbert, and E. Mattanovich. 2003. SUSTALP. *Evaluation of EU-instruments: Their contribution to a sustainable agriculture and environment in the Alps.* Berlin: Blackwell.

Tappeiner, U., E. Tasser, G. Leitinger, A. Cernusca, and G. Tappeiner. 2008. Effects of historical and likely future scenarios of land use on above- and belowground vegetation carbon stocks of an alpine valley. *Ecosystems* 10.1007/s10021-008-9195-3.

Tappeiner, U., E. Tasser, G. Leitinger, and G. Tappeiner. 2006. Landnutzung in den Alpen: Historische Entwicklung und zukünftige Szenarien. In *Die Alpen im Jahr 2020. Alpine space—Man and environment, vol. 1*, eds. R. Psenner and R. Lackner, 23–39. Innsbruck: University Press.

Tasser, E. 2007. Vom wandel der landschaft. In *Bergwelt im wandel merlin*, eds. S. Hellebart and M. Machatschek, 48–59. Klagenfurt: Verlag des Kärntner Landesarchivs.

Tasser, E., S. Prock, and J. Mulser. 1999. The impact of land-use on the vegetation in the mountain region. In ECOMONT: Ecological effects of land-use changes in mountain areas of Europe, eds. A. Tappeiner and U. Cernusca, 235–46. Vienna and Berlin: Blackwell Wiss.-Ver.

Tasser, E., F. V. Ruffini, and U. Tappeiner. 2009. An integrative approach for analysing landscape dynamics in diverse cultivated and natural mountain areas. *Landscape Ecology* 24 (5): 611–28.

Tasser, E., E. Sternbach, and U. Tappeiner. 2008. Biodiversity indicators for sustainability monitoring at municipality level: An example of implementation in an alpine region. *Ecological Indicators* 8:204–23.

Tasser, E., A. Teutsch, W. Noggler, and U. Tappeiner. 2007. Land-use changes and natural reforestation in the Eastern Central Alps. *Agriculture, Ecosystems and Environment* 118:115–29.

Tenhunen, J., R. Geyer, S. Adiku, U. Tappeiner, M. Bahn, N. Q. Dinh, O. Kolcun, A. Lohila, K. Owen, M. Reichstein, M. Schmidt, Q. Wang, M. Wartinger, G. Wohlfahrt, and A. Cernusca. 2008. Influences of land use change on ecosystem and landscape level carbon and water balances in mountainous terrain of the Stubai Valley, Austria. *Global and Planetary Change* (in press).

Thomas, D. S. G., and R. J. Allison, eds. 1993. *Landscape sensitivity*. New York: Wiley.

Thornton, P. E., S. W. Running, and M. A. White. 1997. Generating surfaces of daily meteorological variables over large regions of complex terrain. *Journal of Hydrology* 190:214–51.

Tscharntke, T., A. M. Klein, A. Kruess, I. Steffan-Dewenter, and C. Thies. 2005. Landscape perspectives on agricultural intensification and biodiversity—Ecosystem service management. *Ecol Lett* 8:857–74.

Turner, M. G., R. H. Gardner, and R. V. O'Neill. 2001. *Landscape ecology in theory and practice—Pattern and process*. New York: Springer.

Usher, M., ed. 1999. *Landscape character. Perspectives on management and change*. Edinburgh, UK: The Stationary Office.

Veldkamp, A., P. H. Verburg, K. Kok, G. H. J. de Koning, J. Priess, and A. R. Bergsma. 2001. The need for scale sensitive approaches in spatially explicit land use change modelling. *Environmental Modeling and Assessment* 6:111–21.

Waldhardt, R. 2003. Biodiversity and landscape—Summary, conclusions and perspectives. *Agriculture, Ecosystems and Environment* 98:305–309.

Weber, A., N. Fohrer, and D. Möller. 2001. Long-term land use changes in a mesoscale watershed due to socio-economic factors—Effects on landscape structures and functions. *Ecological Modelling* 140:125–40.

Wohlfahrt, G., A. Hammerle, A. Haslwanter, M. Bahn, U. Tappeiner, and A. Cernusca. 2008. Seasonal and inter-annual variability of the net ecosystem CO_2 exchange of a temperate mountain grassland: Effects of weather and management. *Journal of Geophysical Research* 113, D08110, doi:10.1029/2007JD009286.

Yoccoz, N. G., J. D. Nichols, and T. Boulinier. 2001. Monitoring of biological diversity in space and time. *Trends in Ecology and Evolution* 16:446–53.

Zimmermann, P. D., E. Tasser, and U. Tappeiner. (under review). Historical land-use trends and biodiversity change in the European Alps. *Agriculture, Ecosystem and Environment*.

8

Landscape Development Intensity and Pollutant Emergy/Empower Density Indices as Indicators of Ecosystem Health

Mark T. Brown and Kelly Chinners Reiss

CONTENTS

8.1 Introduction

In this chapter, an index of Landscape Development Intensity (LDI) is presented as a tool to evaluate ecosystem health. The LDI was recently proposed by Brown and Vivas (2005) following earlier work of Brown (1980) and an evaluation of the relationship of development intensity to water quality in the St. Marks River watershed in Florida (Brown et al. 1998; Parker 1998). The LDI is an index based on nonrenewable areal empower density of land uses. The LDI has been used as a human disturbance gradient in developing wetland bio-indicators of ecosystem health (Lane and Brown 2006; Reiss 2006) and in developing a Stream Condition Index (Fore et al. 2007). Recently the LDI was tested as an indicator of human disturbance against a large wetland data set in Ohio (Mack 2006).

Here we propose a new method for calculating the LDI of a landscape unit based on a $\log_{10}$ scale of the ratio of the nonrenewable areal empower density of the landscape unit to an areal empower density of the environmental baseline of the landscape unit. The environmental baseline is the average renewable areal empower density. In addition, we propose a spatial averaged LDI for point source pollutants, especially those associated with pollutants such as nutrients, metals, and other toxins. In general, metals, nutrients, and toxins have high Unit Emergy Values (UEVs) and as a result, when excess concentrations occur, they are capable of instigating significant changes in ecosystem processes, which often result in declines in ecosystem health.

8.2 Emergy, Time, and Area

Emergy is defined as the amount of energy of one type (usually solar) that is directly or indirectly required to provide a given flow or storage of energy or matter. The units of emergy are emjoules (abbreviated eJ) to distinguish them from energy joules (abbreviated J). We propose that the Greek letter epsilon (ε) be used for emergy in equations. Solar emergy is expressed in solar emergy joules (seJ, or solar emjoules). Emergy per unit time is *empower*, in units of emjoules per time. Solar empower is solar emjoules per time (e.g., seJ/time). We propose that the Greek letter omega (ω) be used for empower in equations.

When the emergy required to make something is expressed as a ratio to the available energy of the product, the resulting ratio is called a *transformity*. The solar emergy required to produce a unit flow or storage of available energy is called *solar transformity* and is expressed in solar emergy joules per joule of output flow (seJ/J). The transformity of solar radiation is assumed equal to one (1.0 seJ/J). We propose that the Greek letter tau (τ) be used for transformity in equations.

Specific emergy is the unit emergy value of matter defined as the emergy per mass, usually expressed as solar emergy per gram (seJ/g). Solids may be evaluated best with data on emergy per unit mass for its concentration. Because energy is required to concentrate materials, the unit emergy value of any substance increases with concentration. Elements and compounds not abundant in nature therefore have higher emergy/mass ratios when found in concentrated form since more work was required to concentrate them, both spatially and chemically. We propose that the Greek letter sigma (σ) be used for specific emergy in equations.

8.2.1 Area Empower Intensity

The following paragraphs provide background on our choice of the terminology for emergy per unit time per unit area, *areal empower intensity*. In the past we have proposed the terms areal empower density to describe emergy per unit time per unit area; however, in light of our need to define a new concept of emergy per unit time per unit volume, we suggest differentiating between *intensity* and *density* following the lead of physics.

In physics, especially related to sound, *intensity* is a measure of the time-averaged energy flux, or in other words, the amount of energy that is transported past a given area of a medium per unit of time. Intensity is the energy per time per area (energy*time^{-1}*area^{-1}), and since the energy per time ratio is equivalent to the quantity *power*, intensity is simply the power per area. Emergy intensity, then, is the emergy per time per area, and since emergy per time is *empower*, emergy intensity is empower per area. It should be noted that *energy intensity* as used in economics is defined as the measure of the energy efficiency of a nation's economy. It is calculated as units of energy per unit of gross domestic product (GDP). We suggest that the term *areal empower intensity* be used to describe emergy per unit time per unit area and that we use the Greek letters alpha, omega, iota ($\alpha\omega\iota$) to denote it in equations.

8.2.2 Environmental Emergy Density

In physics, *density* is defined as the ratio of the mass of any substance to the volume occupied by it (usually expressed in kg/m^3). *Energy density* is usually defined as the amount of energy stored per unit volume, or per unit mass, depending on the context (usually expressed in J/g or J/L). When considering concentrations of pollutants in environmental systems, it is often appropriate to express them as concentrations (i.e., mg/L, μg/L, ppm, ppb). Since pollutants can be expressed as emergy using their specific emergy (seJ/g) then concentrations of pollutants in the environment, especially in aqueous environments, can be expressed as emergy density (i.e., seJ/m^3 or seJ/L). We propose that the terminology *emergy density* be used to describe emergy per unit volume (seJ/volume) in environmental systems and that the Greek letters epsilon delta ($\varepsilon\delta$) be used to denote it in equations.

8.2.3 Environmental Empower Density

In engineering, the term *power density* refers to power per unit volume. It is often used to describe the amount of power delivered by an energy source,

divided by some measure of the size or mass of the source. In the environment, when pollutants are released over time their emergy per unit time
per unit volume can be calculated from the pollutant's specific emergy and
the quantities released. We have used the term *empower density* and, more
recently, *areal empower density* to describe emergy per unit time per unit area.
However, in keeping with engineering and physics definitions of density, we
suggest the term *empower density* be used to describe emergy per unit time
per unit volume (seJ*time^{-1}*volume^{-1}) and that the Greek letters phi delta ($\phi\delta$)
be used to denote it in equations.

8.3 Landscape Development Intensity

In a previous paper, Brown and Vivas (2005) suggested that the ecological
health of landscapes and the ecosystems within them is strongly related
to levels of human activity that can affect adjacent ecological communities
through direct, secondary, and cumulative impacts. In that paper, land use
data were used in a development intensity measure derived from energy use
per unit area and calculated as an index of Landscape Development Intensity
(LDI), relating the index to direct measures of water quality in watersheds
and the biological condition of hydrologically isolated wetlands measured
through a field-based rapid assessment method.

8.3.1 Area Weighted LDI Calculation

As the method matured in its use, several limitations were realized that
resulted in redefining LDI based on a background areal empower intensity.
In previous publications (Brown and Vivas 2005; Lane and Brown 2006; Reiss
2006), LDI was calculated using a simple area weighted relationship between
nonrenewable areal empower intensity of land uses as follows:

$$LDI_{Total} = \sum \%LU_i * LDI_i \qquad (8.1)$$

where
  LDI_{total} = LDI ranking for landscape unit
    $\%LU_i$ = percent of the total area of influence in land use *i*
    LDI_I = landscape development intensity coefficient for land use *i*

The area weighted LDI calculation served as a human disturbance gradient for development of indices of wetland condition for marshes, cypress
domes, and riparian forested wetlands in Florida (Figure 8.1). Using community condition indices for three separate communities, significant

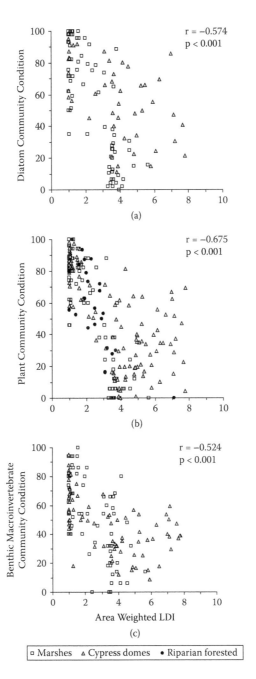

FIGURE 8.1
Area weighted LDI calculation as a human disturbance gradient with indices of wetland condition for (a) diatoms, (b) plants, and (c) benthic macroinvertebrate communities for marshes, cypress domes, and riparian forested wetlands in Florida. Values represent Pearson r correlation coefficients.

($p < 0.001$) correlations were found for the area weighted LDI with diatom community condition (Pearson r = −0.574), plant community condition (Pearson r = −0.675), and the benthic macroinvertebrate community condition (Pearson r = −0.524) using Minitab, version 15.1.

LDIs for larger areas calculated in this manner involved the averaging of logs (since individual land use LDIs are natural logs of their areal empower intensity). This method of calculating average LDIs for landscape areas composed of several land use types inserted significant bias in favor of the land uses with lower areal empower intensity. Further, it was apparent that impact of human disturbance intensity should in some way be related to the background renewable areal empower intensity of the landscape. That is to say, the effect of the nonrenewable emergy intensity is proportionally smaller if the background renewable emergy intensity is greater. This led to redefining the LDI in relation to the renewable background areal empower intensity. Finally, a limitation resulted from defining strict classes of land use types and limiting the calculation of LDIs to known land uses and their areal empower intensity. By redefining the LDI based on the nonrenewable empower intensity of land uses rather than a predetermined LDI for each land use, more flexibility in application of the method may result.

8.3.2 Renewable Background Areal Empower Intensity LDI Calculation

The calculation of a landscape, basin, or watershed LDI requires a land use/land cover map of the landscape unit of interest, aerial empower density multipliers for land use types (Table 8.1 is an example for Florida land uses), and the ability to calculate areas of land use within the landscape unit. The step-by-step procedure is as follows.

First, areas of each land use type within the landscape unit are summed and expressed as percent of total area. Second, percent of land use types are multiplied by the nonrenewable areal empower intensity of each type and summed. Then the following equations are applied:

$$\text{LDI} = 10 * \log\left(\alpha\varpi\iota_{\text{Total}}/\alpha\varpi\iota_{\text{Ref}}\right) \tag{8.2}$$

where

LDI = Landscape Development Intensity index for a given landscape unit

$\alpha\varpi\iota_{\text{Total}}$ = Total areal empower intensity (sum of renewable background areal empower intensity and nonrenewable areal empower density of land uses)

$\alpha\varpi\iota_{\text{Ref}}$ = Renewable areal empower intensity of the background environment (Florida = 1.97 E15 seJ*ha^{-1}*yr^{-1}; Vivas 2007).

The total areal empower intensity ($\alpha\varpi\iota_{\text{Total}}$) is calculated as follows:

$$\alpha\varpi\iota_{\text{total}} = \alpha\varpi\iota_{\text{Ref}} + \Sigma(\%\text{LU}_i * \alpha\varpi\iota_i) \tag{8.3}$$

where

%LU$_i$ = Percent of the total area in land use i

$\alpha \varpi \iota_i$ = The nonrenewable empower intensity for land use i

Table 8.1 lists common land use types found in the Florida landscape. The second column lists typical nonrenewable areal empower intensities for land uses. The third column lists LDIs for 1 ha of the various land use types calculated using Equations (8.1) and (8.2) and the Florida renewable areal empower intensity of the background environment (1.97 E15 seJ*ha^{-1}*yr^{-1}).

This method facilitates the calculation of LDIs for any area, using the areal empower intensity of land uses. The LDI scale begins with zero (i.e., equal to average renewable empower of the landscape unit), and there is

TABLE 8.1

Landscape Development Intensity (LDI) Coefficients for Typical Land Uses

Notes	Land Use	Nonrenewable Areal Empower Intensity (E15 seJ^{-1}*ha^{-1}*yr^{-1})	LDI$_{FI}$[a]
1	Natural land/open water	0.0	0.00
2	Pine plantation	0.5	1.00
3	Low intensity open space/recreational	0.5	1.02
4	Unimproved pastureland (with livestock)	0.5	1.04
5	Improved pasture (no livestock)	2.0	3.07
6	Low intensity pasture (with livestock)	3.4	4.34
7	High intensity pasture (with livestock)	5.9	6.03
8	Medium intensity open space/recreational	6.1	6.10
9	Citrus	7.8	6.94
10	General agriculture	15.1	9.38
11	Row crops	20.3	10.53
12	High intensity agriculture (dairy farm)	50.4	14.25
13	Recreational/open space (high intensity)	123.0	18.02
14	Single-family residential (low density)	197.5	20.05
15	Transportation—2-lane highway	308.0	21.97
16	Single-family residential (med. density)	658.3	25.25
17	Single-family residential (high density)	921.7	26.71
18	Transportation—4-lane highway, low intensity	2533.7	31.10
19	Multifamily residential (low density)	4213.3	33.30
20	Institutional	4042.2	33.12
21	Transportation—4-lane highway, high intensity	5020.0	34.06
22	Low intensity commercial (comm. strip)	5173.4	34.19
23	Industrial	5210.6	34.23
24	High intensity commercial (mall)	8372.4	36.28
25	Multifamily residential (high rise)	12771.7	38.12

(continued)

TABLE 8.1 (continued)

Landscape Development Intensity (LDI) Coefficients for Typical Land Uses

Notes	Land Use	Nonrenewable Areal Empower Intensity (E15 seJ^{-1}*ha^{-1}*yr^{-1})	LDI$_{Fl}$[a]
26	Central business district (avg. 2 stories)	16150.3	39.14
27	Central business district (avg. 4 stories)	29401.3	41.74

[a] $\text{LDI} = 10 * \log\left[(\alpha\varpi\iota_i + \alpha\varpi\iota_{ref})/\alpha\varpi\iota_{ref}\right]$

where

$\alpha\varpi\iota_i$ = nonrenewable areal empower density of Land Use *i*

$\alpha\varpi\iota_{ref}$ = areal empower density of background environment;

Florida = 1.97E+15 seJ^{-1}*ha^{-1}*yr^{-1}

Notes:

1. Nonrenewable empower density for natural systems = 0.

2. Doherty (1995).

3. Average of empower densities of 2 and 4.

4. Based on 0.09 cows/ha/yr (27 acres/animal) (Kalmbacher and Ezenwa 2006). Empower density to support 0.09 cows: 0.53 E15 sej/ha/yr (Brandt-Williams 2002).

5. Brandt-Williams (2002).

6. Based on 0.57 steer/ha/yr (1.76 ha/animal) (Arthington et al. 2007). Empower density to support 0.57 steer: 3.38 E15 sej/ha/yr = Improved Pasture (5) + 1.61 E15 sej/ha/yr

7. Based on 2 steer/ha/yr (Brandt-Williams 2002). Empower density to support two steer: 5.93 E15 sej/ha/yr

8. Assume three times intensity of improved pasture. In an urban landscape applies generally to grassy lawns (Falk 1976).

9. Brandt-Williams (2002).

10. Average of all crops (Brandt-Williams 2002).

11. Average of empower densities for 6 row crops (Brandt-Williams 2002).

12. Brandt-Williams (2002).

13. Based on the emergy evaluation for a golf course (Behrend 2000).

14. Parker (1998) and Brown (1980). Assumes 1.5 units per hectare.

15. Parker (1998).

16. Parker (1998) and Brown (1980). Assumes 5 units per hectare.

17. Based on Brown (1980). Assumes 7 units per hectare.

18. Brown and Vivas (2005).

19. Parker (1998) and Brown (1980). Assumes 32 units per hectare.

20. Brown (1980).

21. Vivas and Brown (2007).

22. Vivas and Brown (2007).

23. Parker (1998) and Brown (1980).

24. Vivas and Brown (2007).

25. Parker (1998) and Brown (1980). Assumes 97 units per hectare.

26. Brown (1980).

27. Brown (1980).

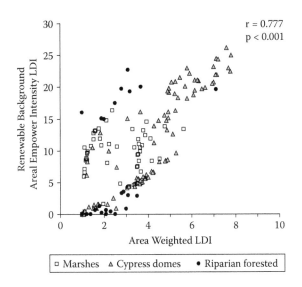

FIGURE 8.2
Correlation of renewable background areal empower intensity LDI calculation with the previous area weighted LDI calculation for marshes, cypress domes, and riparian forested wetlands in Florida. Values represent Pearson r correlation coefficients.

no upper limit. The LDI is calculated for the entire area of interest, without averaging logs, but instead calculating the weighted average nonrenewable areal empower intensity. Figure 8.2 shows the significant correlation between the renewable background areal empower intensity LDI calculation with the area weighted LDI calculation (Pearson $r = 0.777$, $p < 0.001$). While calculating an LDI based on land use/land cover appears to capture a great deal of the human disturbance gradient, a modifier that accounts for point source pollutants directly discharged into the ecosystem of interest is proposed below.

8.4 Pollutant Emergy and Empower Density

While the LDI can be used to estimate the impacts of general development intensity on ecological systems, it is well known that toxics and other pollutants have deleterious effects far from their initial source of introduction into the environment. Often they are released as dispersed materials in air and water and later become concentrated in aquatic systems or terrestrial food chains. Under these circumstances, the LDI calculated for land uses may not capture the production or subsequent concentration of the pollutant. If pollutants of known concentration are present in the environment, a Pollutant Density Index (PDI) can be calculated that relates the intensity of pollutants to the average intensity of the reference environment. Where pollutants are discharged into the aquatic environments, their flux in the environment can be deleterious to productive processes. Using the flux of the pollutant and the productivity of the background environment (measured as empower of the environment), an index of Pollutant Empower Density (PED) can be calculated.

8.4.1 Pollutant Density Index

The actions of chemical stressors including metals, toxins, and nutrients may be explained by their emergy and empower density relative to background environments. Table 8.2 lists the Unit Emergy Values (UEVs) for several metals and other pollutants. As UEVs increase, their potential effect within ecosystems increases. Effects can be both positive and negative. The ultimate impact of a pollutant or toxin is not only related to its UEV, but, more important, to its concentration, measured as emergy density in the ecosystem. Genoni et al. (2003) measured concentrations of 25 different elements in trophic compartments and in the physical environment of the Steina River in Germany. They calculated transformities of each element based on global emergy supporting the river ecosystem, which cycles the elements and their Gibbs energy. They suggested that the tendency to bioaccumulate was related to transformity of the elements and the transformity of accumulating compartments (i.e., metals and heavy elements accumulated in high transformity compartments).

The PDI is calculated in much the same manner as the LDI, however, using the standing stock of pollutant (concentration * volume) in the environment

$$PDI = 10 * \log (\varepsilon\delta_{Total} / \varepsilon\delta_{Ref}) \tag{8.4}$$

where
    PDI = Pollutant Density Index for a given environmental volume
    $\varepsilon\delta_{Total}$ = Total emergy density of the volume (sum of reference emergy density and pollutant emergy density [$\varepsilon\delta_i$])
    $\varepsilon\delta_{Ref}$ = Emergy density of the background environment (freshwater = 1.45 E8 seJ/L; Odum et al. 2000)

The total emergy density ($\varepsilon\delta_{Total}$) is calculated as follows:

$$\varepsilon\delta_{Total} = \varepsilon\delta_{Ref} + \Sigma \ \varepsilon\delta_i \tag{8.5}$$

where
    $\varepsilon\delta_i$ = Emergy density of pollutant *i*

Where emergy density of a stressor is significantly higher than the average of the ecosystem components it is released into, one might expect significant changes in ecosystem function. For instance, because of the very high

TABLE 8.2

Unit Emergy Values (UEVs) of Selected Metals, Nutrients, and Pesticides

Item	Specific Emergy (seJ/g)	Source
Elements		
Silicon	5.07E+08	See Appendix 8.1
Aluminum	1.74E+09	"
Iron	2.78E+09	"
Calcium	3.84E+09	"
Sodium	5.10E+09	"
Potassium	5.44E+09	"
Magnesium	6.75E+09	"
Titanium	2.26E+10	"
Hydrogen	1.00E+11	"
Phosphorus	1.08E+11	"
Carbon	1.49E+11	"
Manganese	1.56E+11	"
Sulfur	2.70E+11	"
Barium	2.81E+11	"
Chlorine	3.12E+11	"
Chromium	4.01E+11	"
Fluorine	4.84E+11	"
Zirconium	5.61E+11	"
Nickel	7.39E+11	"
Copper	2.06E+12	"
Nitrogen	7.02E+12	"
Lead	1.40E+13	"
Arsenic	6.68E+13	"
Uranium	7.80E+13	"
Cadmium	9.36E+14	"
Silver	1.75E+15	"
Mercury	2.09E+15	"
Gold	4.53E+16	"
Pesticides		
Herbicides	1.7E+10	From Pimentel (1980)
Insecticides	2.7E+10	From Pimentel (1980)

specific emergy of most metals (Table 8.2), their concentrations need only be in the parts per billion range to still have emergy densities greater than most natural ecosystems. Table 8.3 lists several metals and other pollutants and their EPA water quality criteria. Most of the metals have acute and chronic concentrations in the parts per billion range, while the criteria for nutrients are recommendations only. The sixth and seventh columns list the emergy

TABLE 8.3

US EPA Water Quality Criteria, Resulting Emergy Density, and Calculated PDI

Parameter	Units	Acute[a]	Chronic[a]	EPA Recommended[b]	Emergy Density[c] (seJ/L)	PDI[d]
Aluminum	µg/L	750	87		1.31E+06	0.09
Chromium	µg/L	16	11		6.42E+06	0.42
Copper	µg/L	13	9		2.68E+07	1.64
Lead	µg/L	65	2.5		9.10E+08	19.55
Arsenic	µg/L	340	150		2.27E+10	50.27
Cadmium	µg/L	2	0.25		1.87E+09	26.01
Mercury	µg/L	1.4	0.77		2.93E+09	30.21
Pesticide (Chlordane)	µg/L	2.4	0.0043		6.48E+04	0.00
Phosphorus (total)	µg/L			10	1.50E+05	0.01
Nitrogen (total)	mg/L			0.52	8.32E+06	0.54

[a] From the US EPA National Recommended Water Quality Criteria from May 2005. http://www.epa.gov/waterscience/criteria/wqcriteria.html

[b] EPA document EPA-822-B-00-013 from December 2000. http://www.epa.gov/waterscience/criteria/nutrient/ecoregions/lakes/

[c] Calculated as the product of UEV in Table 8.2 and acute quantity of constituent.

[d] PDI = 10 * log ($\varepsilon\delta_{Total}/\varepsilon\delta_{Ref}$) where $\varepsilon\delta_{Ref}$ = 1.45 E8 seJ/L for freshwater (Odum et al. 2000).

density (seJ/L) for the acute concentrations and the PDI, assuming the pollutant is in a freshwater environment.

For instance, using the specific emergy of materials in Table 8.2 and the average concentrations of pollutants in landfill leachate in Table 8.4 the emergy density of the pollutants in the leachate totals 1.56 E10 seJ*L^{-1} and the PDI when calculated using the emergy density of freshwater (1.45E8 seJ*L^{-1}; Odum et al. 2000) is 33.6. When compared with the LDI indices in Table 8.1 it is apparent that the emergy density of leachate is extremely high. This emergy density of all the constituents is about two orders of magnitude greater than the emergy density of freshwater.

8.4.2 Pollutant Empower Density

Where pollutants are discharged into aquatic environments, their flux in the environment can be deleterious to productive processes. Concentrations at any one point can be calculated, but more important is the flux of the pollutant measured as the empower per unit volume (seJ*time^{-1}*volume^{-1}) as compared to the empower of the background environment. An example is the point discharge of pollutants into a stream. The stream ecosystem has characteristic productivity measured by its total empower (without the pollutant

TABLE 8.4

Concentrations, Emergy Density, and PDI of Some Landfill Leachate Constituents

Constituent	Units	Range[a]	Median Value[a]	Emergy Density[b] (seJ/L)	PDI[c]
Arsenic	mg/L	0.0002–1.6	0.8	5.34E+10	45.69
Barium	mg/L	0.08–5	2.5	7.03E+08	8.13
Cadmium	mg/L	0.0007–0.15	0.03	2.81E+10	39.35
Copper	mg/L	0.004–9	4.5	9.27E+09	28.65
Lead	mg/L	0.005–1.6	0.8	8.32E+09	27.64
Mercury	mg/L	0.0002–0.05	0.025	5.23E+10	45.47
Nickel	mg/L	0.02–2.227	1.12	8.28E+08	9.07
Phosphorus	mg/l	5–10	7.5	8.10E+08	8.95
			Total	1.54E+11	56.18

[a] Englehardt et al. (2006).
[b] Calculated as the product of UEV in Table 8.2 and quantity of constituent.
[c] $PDI = 10 * \log (\varepsilon\delta_{Total}/\varepsilon\delta_{Ref})$ where $\varepsilon\delta_{Ref} = 1.45$ E8 seJ/L (Odum et al. 2000).

discharge). The index of PED is calculated using the flux of the pollutant and the productivity of the background environment (measured as the empower of the environment) as follows:

$$PED = 10 * \log (\phi\delta_{Total}/\phi\delta_{Ref}) \tag{8.6}$$

where
   PED = Pollutant Empower Density index for a given landscape unit
   $\phi\delta_{Total}$ = Total empower density (sum of background empower density and pollutant empower density)
   $\phi\delta_{Ref}$ = Empower density of the background environment

The total empower density ($\phi\delta_{Total}$) is calculated as follows:

$$\phi\delta_{total} = \phi\delta_{Ref} + \Sigma \phi\delta_i \tag{8.7}$$

where
   $\phi\delta_i$ = Empower density of pollutant i

Table 8.5 lists typical empower densities of aquatic ecosystems as examples of the background productivity (reference environment) to be used in Equation (8.6).

TABLE 8.5

Empower Density of Aquatic Ecosystems

Ecosystem	Empower Intensity (seJ*m²*yr⁻¹)	Empower Density (seJ*m³*yr⁻¹)	Source
Freshwater systems			
Subtropical spring	3.80E+11	1.90E+11	Collins and Odum 2000
Subtropical lake	9.40E+11	4.09E+11	Brown and Bardi 2001
Subtropical herb. wetland	3.69E+11	5.59E+11	Bardi and Brown 2000
Subtropical eutrophic lake	3.30E+12	2.75E+12	Brown and Bardi 2001
Saltwater systems			
Louisiana estuarine sys.	9.60E+09	9.60E+09	Odum and Collins 2002
Oyster reef	7.57E+09	1.51E+10	Odum and Collins 2002
Coral reef	2.60E+11	1.73E+11	McClanahan 1990

8.5 Summary

In this chapter we have:

1. Proposed nomenclature to clearly define concepts of emergy intensity, empower intensity, areal empower intensity, emergy density, and empower density

2. Outlined an approach to calculating a human disturbance gradient using the Landscape Development Intensity (LDI) index

3. Outlined an approach to calculating pollutant emergy and empower density indices that can be used when known discharges of pollutants impact aquatic systems

The indices outlined in this chapter are our attempt to relate impacts for two general sources to potential alteration in ecosystem structure and function, which might collectively be termed ecosystem health. While the LDI has had several rounds of development and evaluation using a reference wetland database in Ohio (Mack 2006), data collected for herbaceous and forested wetlands of Florida (Reiss and Brown 2007), and riverine ecosystems in Arkansas (Vivas and Brown 2007), the PDI and PED indices are new concepts that require thorough vetting. Much research is needed to further develop these concepts and gather empirical evidence required to fully examine the theory.

References

Arthington, J. D., F. M. Roka, J. J. Mullahey, S. W. Coleman, R. M. Muchovej, L. O. Lollis, and D. Hitchcock. 2007. Integrating ranch forage production, cattle performance, and economics in ranch management systems for Southern Florida. *Rangeland Ecology and Management* 60 (1): 12–18.

Bardi, E., and M. T. Brown. 2000. Emergy evaluation of ecosystems: A basis for environmental decision making. In *Emergy synthesis, proceedings of the 1st emergy research conference*, ed. M. T. Brown, 81–100. Gainesville, FL: Center for Environmental Policy, University of Florida.

Behrend, G. 2000. Emergy evaluation of golf course. Class project EES5305. Spring 2000. University of Florida, Gainesville, FL.

Brandt-Williams, S. 2002. *Handbook of emergy evaluation, folio #4: Emergy of Florida agriculture*. Gainesville, FL: Center for Environmental Policy, University of Florida.

Brown, M. T. 1980. Energy basis for hierarchies in urban and regional landscapes. PhD dissertation, University of Florida, Gainesville, FL.

Brown, M. T., and E. Bardi. 2001. *Handbook of emergy evaluation, folio 4: Emergy of ecosystems*. Gainesville, FL: Center for Environmental Policy, University of Florida.

Brown, M. T., N. Parker, and A. Foley. 1998. Spatial modeling of landscape development intensity and water quality in the St. Marks River watershed. Final report to Florida Department of Environmental Protection, Center for Wetlands, University of Florida, Gainesville, FL.

Brown, M. T., and M. B. Vivas. 2005. A landscape development intensity index. *Environmental Monitoring and Assessment* 101:289–309.

Clarke, F. W., and H. S. Washington. 1924. Composition of the earth's crust. U.S. Geological Survey Professional Paper, No. 127.

Collins, D., and H. T. Odum. 2000. Calculating transformities with the eigenvalue method. In *Emergy synthesis, proceedings of the 1st Emergy Research Conference*, ed. M.T. Brown, 265–79. Gainesville, FL: Center for Environmental Policy, University of Florida.

Doherty, S. J. 1995. *Emergy Evaluations of and Limits to Forest Production*. PhD Dissertation, Department of Environmental Engineering. University of Florida, Gainesville FL. 215 pp.

Englehardt, J. D., Y. Deng, J. Polar, D. E. Meeroff, Y. Legrenzi, and J. Mognol. 2006. Options for managing municipal landfill leachate: Year 1 development of iron-mediated treatment processes. Report #0432024-06. Gainesville, FL: State University System of Florida, Florida Center for Solid and Hazardous Waste Management, University of Florida, 137.

Falk, J. H. 1976. Energetics of a suburban lawn ecosystem. *Ecology* 57 (1): 141–50.

Fore, L. S., R. Frydenborg, D. Miller, T. Frick, D. Whiting, J. Espy, and L. Wolfe. 2007. Development and testing of biomonitoring tools for macroinvertebrates in Florida streams (Stream Condition Index and Biorecon). Tallahassee, FL: Florida Department of Environmental Protection.

Genoni, G. P., E. I. Mejer, and A. Ulrich. 2003. Energy flow and elemental concentrations in the Steina River ecosystem (Black Forest, Germany). *Aquatic Sciences* 6:143–57.

Kalmbacher, R. S., and I. V. Ezenwa. 2006. Managing South Florida range for cattle. Publication #SS-AGR-105. Gainesville, FL: Florida Cooperative Extension Service, Institute of Food and Agricultural Sciences, University of Florida.

Lane, C. R., and M. T. Brown. 2006 Energy-based abscissa assessment: Proximal and ultimate relationships between environmental variables and benthic diatoms in marshes of Florida, USA. *Environmental Monitoring and Assessment* 117:433–50.

Mack, J. J. 2006. Landscape as a predictor of wetland condition: An evaluation of the Landscape Development Index (LDI) with a large reference wetland dataset from Ohio. *Environmental Monitoring and Assessment* 120:221–41.

McClanahan, T. 1990. Hierarchical control of coral reef ecosystems. PhD dissertation, University of Florida, Gainesville, FL.

Odum, H. T., M. T. Brown, and S. Brandt-Williams. 2000. *Handbook of emergy evaluation, folio #1: Introduction and global budget.* Gainesville, FL: Center for Environmental Policy, University of Florida.

Odum, H. T., and D. Collins. 2002. Transformities from ecosystem energy webs with the eigenvalue M. In *Emergy synthesis, proceeding of the 2nd Emergy Research Conference*, ed. M. T. Brown, 203–21. Gainesville, FL: Center for Environmental Policy, University of Florida.

Odum, H. T., E. C. Odum, and M. T. Brown. 1998. *Environment and society in Florida.* Boca Raton, FL: CRC Press–Lewis Pub.

Parker, N. M. 1998. Spatial models of total phosphorus loading and landscape development intensity in a North Florida watershed. Master's thesis, University of Florida, Gainesville, FL.

Pimentel, D. 1980. Energy inputs for the production, formulation, packaging, and transport of various pesticides. In *Handbook of energy utilization in agriculture*, ed. D. Pimentel. Boca Raton, FL: CRC Press.

Reiss, K. C. 2006. Florida Wetland Condition Index for depressional forested wetlands. *Ecological Indicators* 6:337–52.

Reiss, K. C., and M. T. Brown, 2007. An evaluation of Florida palustrine wetlands: Application of USEPA levels 1, 2, and 3 assessment methods. *Ecohealth* 4:206–18.

Vivas, M. B. 2007. Development of an index of landscape development intensity for predicting the ecological condition of aquatic and small isolated palustrine wetland systems in Florida. PhD dissertation, Department of Environmental Engineering Sciences, University of Florida, Gainesville, FL.

Vivas, M. B., and M. T. Brown. 2007. Landscape Development Intensity (LDI) coefficients for land use classes of the Little Bayou Meto watershed, Arkansas. Report submitted to the Arkansas Soil and Water Conservation Commission under the Sub-grant Agreement SGA 104. Gainesville, FL: Center for Environmental Policy, University of Florida.

Appendix 8.1

Specific Emergy of Selected Elements

Element	% of Crust by Weight[a]	Weight (g)[b]	Specific Emergy[c] (seJ/g)
Total crust	100	2.82E+25	1.40E+08
Silicon	27.690	7.81E+24	5.07E+08
Aluminum	8.070	2.28E+24	1.74E+09
Iron	5.050	1.42E+24	2.78E+09
Calcium	3.650	1.03E+24	3.84E+09
Sodium	2.750	7.76E+23	5.10E+09
Potassium	2.580	7.28E+23	5.44E+09
Magnesium	2.080	5.87E+23	6.75E+09
Titanium	0.620	1.75E+23	2.26E+10
Hydrogen	0.140	3.95E+22	1.00E+11
Phosphorus	0.130	3.67E+22	1.08E+11
Carbon	0.094	2.65E+22	1.49E+11
Manganese	0.090	2.54E+22	1.56E+11
Sulfur	0.052	1.47E+22	2.70E+11
Barium	0.050	1.41E+22	2.81E+11
Chlorine	0.045	1.27E+22	3.12E+11
Chromium	0.035	9.87E+21	4.01E+11
Fluorine	0.029	8.18E+21	4.84E+11
Zirconium	0.025	7.05E+21	5.61E+11
Nickel	0.019	5.36E+21	7.39E+11
Copper	0.0068	1.92E+21	2.06E+12
Nitrogen	0.0020	5.64E+20	7.02E+12
Lead	0.0010	2.82E+20	1.40E+13
Uranium	0.00018	5.08E+19	7.80E+13
Silver	0.000008	2.26E+18	1.75E+15
Mercury	0.0000067	1.89E+18	2.09E+15
Gold	0.00000031	8.74E+16	4.53E+16

[a] Clarke and Washington (1924).

[b] Calculated as the percent of earth's crust by weight times the weight of the earth's crust. Weight of continental and oceanic crust = 2.82 E25 g.

[c] Annual emergy driving the geobiosphere = 1.58 E25 seJ/yr (Odum et al. 1998). Crust turnover time 2.5 E8 years. Total emergy driving geologic processes = turnover time * annual emergy flow. Emergy total = 1.58 E25 seJ/yr * 2.5 E8 yrs = 3.96 E33 seJ.

9

Ecosystem Services and Ecological Indicators

Robert Costanza

CONTENTS

9.1 What Are Ecosystem Services?

Ecosystem services are defined as "the benefits people obtain from ecosystems" (Costanza et al. 1997; Millennium Ecosystem Assessment 2005). These include provisioning services such as food and water; regulating services such as regulation of floods, drought, and disease; supporting services such as soil formation and nutrient cycling; and cultural services such as recreational, spiritual, and other nonmaterial benefits (Costanza et al. 1997; Daily 1997; de Groot et al. 2002).

This is an appropriately broad and an appropriately vague definition. It includes both the benefits people *perceive*, and those they do not. The conventional economic approach to "benefits" is far too narrow in this regard, and tends to limit benefits only to those that people both perceive and are "willing to pay" for in some real or contingent sense. But the general population's information about the world, especially when it comes to ecosystem services, is extremely limited. We can expect many ecosystem services to go almost unnoticed by the vast majority of people, especially when they are public, nonexcludable services that never enter the private, excludable market. Think of the storm-regulation value of wetlands (Costanza et al. 2008). How can we expect the average citizen to understand the complex linkages between landscape patterns, precipitation patterns, wetlands, and flood attenuation when even the best landscape scientists find this an extremely challenging task? We need to remember the definition of ecosystem services (the benefits provided by ecosystems), and acknowledge that the degree to which the public perceives and understands them is a separate (and very important) question.

In addition, the benefits one receives from functioning ecosystems do not necessarily depend on one's ability to pay for them in monetary units. For example, indigenous populations with no money economy at all derive most of the essentials for life from ecosystem services but have zero ability to pay for them. To understand the value of these ecosystem services we need to understand the trade-offs involved, and these may be best expressed in units of time, energy, land, or other units, not necessarily money, remembering that the local population may or may not understand or be able to quantify these trade-offs. Finally, if one can express the trade-offs (value) in one set of units (numerator) and can express the trade-offs between that numerator and another, then one can convert the trade-offs into the other numerator. For example, if we can express trade-offs in units of time and can estimate the time/money trade-off, we can express the time units in monetary terms.

9.2 Intermediate versus Final Goods and Services

A second issue is that ecosystem services are, by definition, not ends or goals, but means to the end or goal of sustainable human well-being. This does not imply that ecosystems are not also valuable for other reasons, but that *ecosystem services* are defined as the instrumental values of ecosystems as means to the end or goal of human well-being. An important, but different, distinction some authors have made is one between intermediate services and final services (Boyd and Banzhaf 2007). It is certainly true that for the purposes of certain aggregation exercises, adding intermediate and final services would be double counting. But that does not imply that intermediate services are not services. Think of the production of tires in an economy. Some tires are sold directly to consumers and are part of final demand, while others are sold to car companies and are intermediate products, sold to consumers as parts of cars. The tires themselves are indistinguishable from each other, the only difference being who buys them. When calculating gross domestic product (GDP) (which is the aggregate of sales to final demand) it would not be appropriate to count both the tires sold to final demand and the tires sold to car companies, since those tires are already counted as parts of the cars sold to final demand. But tires in both cases, whether intermediate or final products, are means to the end or goal of human well-being and are not ends in themselves. Likewise, ecosystem goods and services, whether intermediate (or "supporting" in the Millennium Assessment typology) services or final services are all contributors to the end of human well-being. Also, ecosystem processes (or functions) and services are not mutually exclusive categories. Some processes or functions are also services; others are not. Some services are intermediate, some are final, and some are partly both.

9.3 Classifying Ecosystem Services

There are several important and useful ways to classify and group ecosystem services. I'll mention just two by way of example: classification according to spatial characteristics and classification according to "excludability/rivalness" status. Table 9.1 groups the 17 ecosystem services listed in Costanza et al. (1997) into five categories according to their spatial characteristics. For example, services like carbon sequestration (an intermediate input to climate regulation) is classified as "Global: non-proximal" since the spatial location of carbon sequestration does not matter. The atmosphere is well mixed, and removing carbon dioxide (or other greenhouse gases) at any location is equivalent to removing it anywhere else. "Local proximal" services, on the other hand, are dependent on the spatial proximity of the ecosystem to the human beneficiaries. For example, "storm protection" requires that the ecosystem doing the protecting be proximal to the human settlements being protected. "Directional flow–related" services are dependent on the flow from upstream to downstream, as is the case for water supply and water regulation. And so on for the other categories listed in Table 9.1.

Another way to classify ecosystem services is according to their "excludability and rivalness" status. Table 9.2 arrays these two characteristics against each other in a matrix that leads to four categories of goods and services. Goods and services are "excludable" to the degree that individuals can be excluded from benefiting from them. Most privately owned, marketed goods and services are relatively easily excludable. I can prevent others from eating the tomatoes I have grown or the timber I have harvested or the fish I have caught unless they pay me. But it is difficult or impossible to exclude others from benefiting from many public goods, like a well-regulated climate, fish in the open ocean, or the aesthetic benefits of a forest. Goods and services are "rival" to the degree that one person's benefiting from them interferes with or is rival with others' benefiting from them. If I eat the tomato or the fish, you cannot also eat it. But if I benefit from a well-regulated climate, you can also do the same. Excludability is largely a function of supply (to what extent

TABLE 9.1

Ecosystem Services Classified According to Spatial Characteristics

1. *Global—non-proximal (does not depend on proximity)*
    1&2. Climate regulation
        Carbon sequestration (NEP)
        Carbon storage
    17. Cultural/existence value

2. *Local proximal (depends on proximity)*
    3. Disturbance regulation/storm protection
    9. Waste treatment
    10. Pollination
    11. Biological control
    12. Habitat/refugia

3. *Directional flow related: flow from point of production to point of use*
    4. Water regulation/flood protection
    5. Water supply
    6. Sediment regulation/erosion control
    8. Nutrient regulation

4. *In situ (point of use)*
    7. Soil formation
    13. Food production/non-timber forest products
    14. Raw materials

5. *User movement related: flow of people to unique natural features*
    15. Genetic resources
    16. Recreation potential
    17. Cultural/aesthetic

Source: From Costanza, R., 2008. Ecosystem Services: Multiple classification
systems are needed. *Biological Conservation* 141:350–52.

TABLE 9.2

Ecosystem Services Classified According to Rivalness and Excludability

	Excludable	Non-excludable
Rival	Market goods and services (most provisioning services)	Open access resources (some provisioning services)
Non-rival	Congestable services (some recreation services)	Public goods and services (most regulatory and cultural services)

Source: From Costanza, R., 2008. Ecosystem Services: Multiple classification systems are needed.
Biological Conservation 141:350–52.

producers can exclude users) and is related to the cultural and institutional mechanisms available to enforce exclusion, while rivalness is a function of demand (how benefits depend on other users) and is more a characteristic of the good or service itself. Table 9.2 places ecosystem services into the four categories that this two-by-two matrix creates.

These two examples should be enough to indicate that there are many useful ways to classify ecosystem goods and services, and our goal is not a single, consistent system, but rather an intelligent pluralism of typologies that will each be useful for different purposes.

9.4 Complex Systems and Ecosystem Services

Ecosystems with embedded humans are complex, dynamic, adaptive systems with nonlinear feedbacks, thresholds, hysteresis effects, etc. (Costanza et al. 1993). Ecosystem services are therefore not the product of a linear chain from production (means) to direct benefits by people (ends) with no feedbacks or any of the other complexities of the real world. All ecosystem services are, by definition, means to the end of human well-being. Ecosystem processes or functions can also be services (they are not mutually exclusive categories), and the same services can be both intermediate and final. The real world is complex and messy and our systems of classification and definition of ecosystem services should recognize that and work with it, not ignore it in a misguided attempt to impose unrealistic order and consistency.

We need to build models of these complex, interconnected systems that can help us better understand the dynamics and patterns of ecosystem services production and benefits. For example, the GUMBO (Global Unified Metamodel of the BiOsphere) model (Boumans et al. 2002) and the more recent MIMES (Multiscale Integrated Models of Ecosystem Services) framework are attempts to do just that. MIMES is a framework to address the magnitude, dynamics, and spatial patterns of ecosystem services at multiple scales (Figure 9.1). The MIMES framework explicitly addresses the linked dynamics of natural, human, built, and social capital, and allows one to integrate site-specific information with regional and global surveys, geographic information system (GIS), and remote sensing data. MIMES is a process-based, spatially explicit, dynamic, nonlinear simulation model (including carbon, water, nitrogen, phosphorous, plants, consumers [including humans] and a range of ecosystem services) under various climate, economic, and policy scenarios. MIMES is spatially scaleable. Each "location" in MIMES includes the percent of the land surface in 11 biomes or ecosystem types: *Open Ocean, Coastal Ocean, Forests, Grasslands, Wetlands, Lakes/Rivers, Deserts, Tundra, Ice/rock, Croplands,* and *Urban.* The relative areas of each biome at each location change in response to urban and rural population growth, economic production, changes in temperature and precipitation, and other variables. Among the biomes, there are exchanges of energy, carbon, nutrients, water, and mineral matter. The model calculates the marginal product of ecosystem services in both an economic production and welfare function as estimates of the prices of each service. The number of

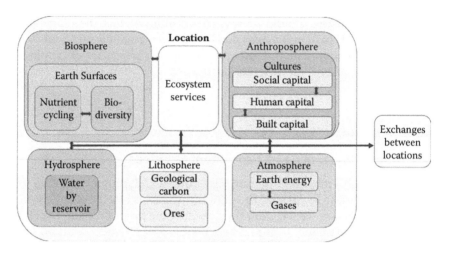

FIGURE 9.1
Basic structure of the MIMES (Multiscale Integrated Models of Ecosystem Services) framework.

"locations" or cells in MIMES is variable and can include either grid or polygon representations of multiple locations. For each of these applications, the basic MIMES structure remains the same, but the parameters must be recalibrated. Conventional economic valuation presumes that people have well-formed preferences and enough information about trade-offs that they can adequately judge their "willingness to pay." These assumptions do not hold for many ecosystem services. Therefore, we must (1) inform people's preferences (for example, by showing them the underlying dynamics of the ecosystems in question using the models like MIMES); (2) allow groups to discuss the issues and "construct" their preferences (again using the MIMES framework to inform the discussions); or (3) use other techniques that do not rely on preferences to estimate the contribution to human welfare of ecosystem services (i.e., using the MIMES to directly infer marginal contributions to welfare). We combine these three methods to develop new and more integrated methods to value ecosystem services.

9.5 Valuation of Ecological Systems and Services

The issue of valuation is inseparable from the choices and decisions we have to make about ecological systems. Some argue that valuation of ecosystems is either impossible or unwise. For example, some argue that we cannot place a value on such "intangibles" as human life, environmental aesthetics, or long-term ecological benefits. But, in fact, we do so every day. When we set construction standards for highways, bridges, and the like, we value human life—acknowledged or not—because spending more money on construction would save lives. Another often-made argument is that we should protect ecosystems for purely moral or aesthetic reasons, and we do not need valuations of ecosystems for this purpose. But there are equally compelling moral arguments that may be in direct conflict with the moral argument to protect ecosystems. For example, the moral argument that no one should go hungry. All we have done is to translate the valuation and decision problem into a new set of dimensions and a new language of discourse. So, while ecosystem valuation is certainly difficult, one choice we do *not* have is whether or not to do it. Rather, the decisions we make, as a society, about ecosystems *imply* trade-offs and therefore valuations. We can choose to make these valuations explicit or not; we can undertake them using the best available ecological science and understanding or not; we can do them with an explicit acknowledgment of the huge uncertainties involved or not, but as long as we are forced to make choices we are doing valuation. The valuations are simply the relative weights we give to the various aspects of the decision problem. Society can make better choices about ecosystems if the valuation issue is made as explicit as possible. This means taking advantage of the best information and models we can muster and making uncertainties about valuations explicit too. It also means developing new and better ways to make good decisions in the face of these uncertainties. Ultimately, it means being explicit about our goals as a society, both in the short term and in the long term.

9.6 Natural Capital and Ecosystem Services

Sustainability has been variously construed (Pezzey 1989; WCED 1987; Costanza 1991), but one useful definition is the amount of consumption that can be sustained indefinitely without degrading capital stocks—including natural capital stocks (Pearce and Turner 1989; Costanza and Daly 1992). Since capital is traditionally defined as produced (manufactured) means of production, the term *natural capital* needs explanation. It is based on a more functional definition of capital as "a stock that yields a flow of valuable goods or services into the future." What is functionally important is the relation of a stock yielding a flow—whether the stock is manufactured or natural is in this view a distinction between kinds of capital and not a defining characteristic of capital itself. For example, a stock or population of trees or fish provides a flow or annual yield of new trees or fish, a flow that can be sustainable year after year. The sustainable flow is *natural income*; the stock that yields the sustainable flow is natural capital. Natural capital may also provide services like recycling waste materials or water catchment and erosion control, which are also counted as natural income. Since the flow of services from ecosystems requires that they function as whole systems, the structure and diversity of the ecosystem is a critical component in natural capital.

To achieve sustainability, we must incorporate natural capital, and the ecosystem goods and services that it provides, into our economic and social accounting and our systems of social choice. In estimating these values we must consider how much of our ecological life support systems we can afford to lose. To what extent can we substitute manufactured for natural capital, and how much of our natural capital is irreplaceable? For example, could we replace the radiation screening services of the ozone layer if it were destroyed? Because natural capital is not captured in existing markets, special methods must be used to estimate its value. These range from attempts to mimic market behavior using surveys and questionnaires to elicit the preferences of current resource users (i.e., willingness-to-pay [WTP]) to methods based on energy analysis (EA) of flows in natural ecosystems (which do not depend on current human preferences at all) (Farber and Costanza 1987; Costanza et al. 1989; Costanza 2004). Because of the inherent difficulties and uncertainties in determining these values, we are better off with an intelligently pluralistic approach that acknowledges and utilizes these different, independent approaches.

The point that must be stressed is that the economic value of ecosystems is connected to their physical, chemical, and biological role in the long-term, global system—whether the present generation of individuals fully recognizes that role or not. If it is accepted that each species, no matter how seemingly uninteresting or lacking in immediate utility, has a role in natural ecosystems (which *do* provide many direct benefits to humans), it is possible to shift the focus away from our imperfect short-term perceptions and toward the goal of developing more accurate values for long-term ecosystem services. Ultimately, this will involve the collaborative construction of dynamic, evolutionary models of linked ecological economic systems that adequately address long-term responses and uncertainties, like those mentioned above.

9.7 Uncertainty and Ecosystem Services

Valuation of ecosystem services and natural capital will always involve many uncertainties. How do we make good decisions in the face of these uncertainties? Current command and control systems of direct environmental regulation are not very efficient at managing environmental resources for sustainability, particularly in the face of uncertainty about long-term values and impacts. They are inherently reactive rather than proactive. They induce legal confrontation, obfuscation, and government intrusion into business in a way that reduces efficiency. Rather than encouraging long-range technical and social innovation, they tend to suppress it. They do not mesh well with the market signals that firms and individuals use to make decisions and do not effectively translate long-term global goals into short-term local incentives.

We need to explore promising alternatives to our current command and control environmental management systems, and modify existing government agencies and other institutions accordingly. The enormous uncertainty about local and transnational environmental impacts needs to be incorporated into decision making. We also need to better understand the sociological, cultural, and political criteria for acceptance or rejection of policy instruments.

For example, one policy instrument designed to incorporate uncertainty into the market system and to induce positive environmental technological innovation is the flexible environmental assurance bonding system (Costanza and Perrings 1990; Costanza and Cornwell 1992). In addition to direct charges for known environmental damages, a company would be required to post an assurance bond equal to the current best estimate of the largest potential future environmental damages; the money would be kept in interest-bearing escrow accounts. The bond (plus a portion of the interest) would be returned if the firm could show that the suspected damages had not occurred or would not occur. If they did, the bond would be used to rehabilitate or repair the environment and to compensate injured parties. Thus, the burden of proof would be shifted from the public to the resource-user and a strong economic incentive would be provided to research the true costs of environmentally innovative activities and to develop cost-effective pollution control technologies. This is a combination of the "polluter pays" principle with the "precautionary" principle to produce the "precautionary polluter pays principle" or 4P approach to environmental management under uncertainty (Costanza and Cornwell 1992).

Instead of being mesmerized into inaction by scientific uncertainty over our future, we should acknowledge uncertainty as a fundamental part of the system. We must develop better methods to model and value ecological goods and services, and devise policies to translate those values into appropriate incentives.

References

Boumans, R., R. Costanza, J. Farley, M. A. Wilson, R. Portela, J. Rotmans, F. Villa, and M. Grasso. 2002. Modeling the dynamics of the integrated earth system and the value of global ecosystem services using the GUMBO model. *Ecological Economics* 41: 529–60.

Boyd, J., S. Banzhaf. 2007. What are ecosystem services? The need for standardized environmental accounting units. *Ecological Economics* 63:616–26.

Costanza, R. 1980. Embodied energy and economic valuation. *Science* 210:1219–24.

———, ed. 1991. *Ecological economics: The science and management of sustainability.* New York: Columbia University Press.

———. 2004. Value theory and energy. In *Encyclopedia of energy, vol. 6,* ed. C. Cleveland, 337–46. Amsterdam: Elsevier.

———. 2008. Ecosystem services: Multiple classification systems are needed. *Biological Conservation* 141:350–52.

Costanza, R., and L. Cornwell. 1992. The 4P approach to dealing with scientific uncertainty. *Environment* 34:12–20, 42.

Costanza, R., and H. E. Daly. 1992. Natural capital and sustainable development *Conservation Biology* 6:37–46.

Costanza, R., R. d'Arge, R. de Groot, S. Farber, M. Grasso, B. Hannon, S. Naeem, K. Limburg, J. Paruelo, R. V. O'Neill, R. Raskin, P. Sutton, M. van den Belt. 1997. The value of the world's ecosystem services and natural capital. *Nature* 387:253–60.

Costanza, R., S. C. Farber, and J. Maxwell. 1989. The valuation and management of wetland ecosystems. *Ecological Economics* 1:335–62

Costanza, R., and C. Perrings. 1990. A flexible assurance bonding system for improved environmental management. *Ecological Economics* 2:57–76.

Costanza, R., L. Wainger, C. Folke, and K.-G. Mäler. 1993. Modeling complex ecological economic systems: Toward an evolutionary, dynamic understanding of people and nature. *BioScience* 43:545–55.

Daily, G., ed. 1997. *Nature's services: Societal dependence on natural ecosystems.* Washington, DC: Island Press.

de Groot, R. S., M. A. Wilson, and R. M. J. Boumans. 2002. A typology for the classification, description and valuation of ecosystem functions, goods and services. *Ecological Economics* 41 (3): 393–408.

Farber, S., and R. Costanza. 1987. The economic value of wetlands systems. *Journal of Environmental Management* 24:41–51.

Millennium Ecosystem Assessment. 2005. Washington, DC: Island Press.

Pearce, D. W., and R. K. Turner. 1989. *Economics of natural resources and the environment.* Brighton: Wheatsheaf.

Pezzey, J. 1989. Economic analysis of sustainable growth and sustainable development. Environment Department working paper no. 15. Washington, DC: The World Bank.

WCED. 1987. *Our common future: Report of the World Commission on Environment and Development.* Oxford: Oxford University Press.

Section II

Assessment of Ecosystem Health

10

Application of Ecological Indicators for the Assessment of Wetland Ecosystem Health

S. E. Jørgensen

CONTENTS

10.1 Introduction: The Importance of Wetlands

Wetlands are a major feature of the landscape in many parts of the world. They are among the most important ecosystems on earth and are sometimes described as the kidney of the landscape. Up to the mid-nineteenth century wetlands were considered storage tanks of diseases and were often given a sinister image. As a consequence of this view and the need for more agriculture land, wetlands have disappeared at alarming rates. They were drained and turned into agricultural land, which has resulted, particularly in industrialized countries, in a massive pollution threat of pesticides and nutrients discharge by agriculture—a pollution that the wetlands and other natural ecosystems (ditches, trees, wind shelterbelts, ponds, forests, and so on) otherwise would eliminate. The lack today of many different small or large ecosystems in the landscape has been a disaster for the abatement of the non-point pollution from agriculture. The various natural ecosystems are crucial for the health of the landscape.

10.2 Ecosystem Services by Wetlands

Today, we have realized the importance of wetlands and their role in landscape. The role of wetlands is manifold, and the following list of ecosystem services offered to the society of wetlands could easily be extended:

1. Production of rice
2. Grazing
3. Production of proteins in general
4. Flood control
5. Enhancement of cycling of nutrients
6. Purification of drainage water and even wastewater
7. Buffer zones
8. Production of sphagnum
9. Peat mining
10. Conservation of high biodiversity
11. Bird reserve
12. Fishery

It is possible to calculate the economic value of wetlands by the services that they are offering the society. Costanza et al. (1997) found that the services offered by wetlands amounted to \$15,000/ha/y. Wetlands produce a biomass corresponding to 18 $MJ/m^2/y$, but if we calculate the exergy (ecological exergy or work capacity; see Chapter 3) that would include information, it would be in the order of 45,000 GJ/ha/y (see Jørgensen [submitted]). With an energy (exergy) price of 1 EURO cent per MJ, the value would be 450,000 EURO/ha or about \$550,000/ha, more than 30 times the value indicated by Costanza et al. (1997). The exergy value considers, however, the total exergy content and therefore indicates the entire value of wetlands included in all possible services, while Costanza et al. only consider the values of the services that we actually use.

10.3 Types of Wetlands and Wetland Processes

There is a wide spectrum of different wetlands with different properties:

1. Salt marshes
2. Mangroves
3. Freshwater marshes

4. Forested wetland
5. Peatlands
6. Swamps
7. Floodplains
8. Tundra
9. Ponds
10. Ditches
11. Riparian wetlands
12. Constructed wetlands, surface wetlands, and subsurface wetlands

All these types of wetlands play a major role in nature and in different landscapes. They have slightly different properties, functions, and roles, but are all very important for the health of the landscape. Let us exemplify the importance of wetlands by focusing on the importance of mangrove wetlands:

1. Physical protection of coast (for instance, against tsunamis)
2. Very high productivity
3. Important nesting areas for mussels, shrimps, crayfish, lobsters, and fish
4. Judicious use in the treatment of wastewater

Because of increasing interest in having wetlands in the landscape to cope with agricultural pollutants, and the possibility of treating wastewater through wetlands, there has been an increase in interest of constructing wetlands, which has resulted in development of models to design constructed wetlands. The model facilitates construction of wetlands. It can also be used to assess the natural wetland area needed for a well-defined water treatment task, whether it be drainage water or wastewater. The design of a constructed or a natural wetland is obtained within 20 minutes by use of the model, named SubWet, developed by the author of this chapter. The model can also be applied as a tool for the everyday management of wetlands.

Figure 10.1 gives an overview of some of the most important processes for which wetlands are being applied, including purification of drainage and wastewater (see point 6 on the list of the services offered by wetlands in Section 10.2):

1. Removal of nitrogen by nitrification and denitrification
2. Decomposition of organic matter including toxic organic compounds such as, for instance, pesticides
3. Adsorption of phosphorus compounds to the wetland soil
4. Uptake of nitrogen and phosphorus by plants
5. Adsorption of heavy metal ions to the soil
6. Uptake of toxic organic compounds and heavy metals by plants

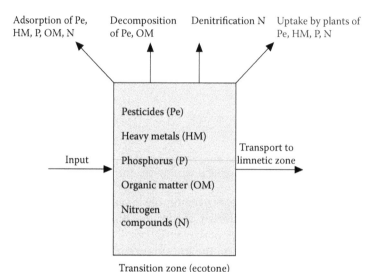

FIGURE 10.1
The processes that wetlands use to reduce the concentration of pesticides, heavy metals, phosphorus, organic matter, toxic organic compounds, and nitrogen compounds. Wetlands may be constructed or they may be ecotones (transition areas between two ecosystems) between terrestrial and aquatic ecosystems, for instance, agriculture land and a lake. Other ecotones than wetlands, for instance, small ponds and ditches, are also able to improve water quality, but wetlands usually offer higher removal efficiency.

By using models, it is possible to design a wetland that is able to treat the wastewater or drainage water to obtain a defined water quality with given concentrations of ammonium-N, nitrate-N, organic-N, total phosphorus, BOD_5, pesticides, and several of the most toxic heavy metals. In other words, the model answers the question, How can we design a wetland (area and flow pattern) to use the six processes listed above to obtain a defined water quality?

10.4 Ecological Indicators Applied to Assess the Ecosystem Health for Wetlands

Different indicators are applied when we consider different applications and types of wetlands.

For constructed wetlands that are used to treat wastewater and drainage water, it is important to assess whether the wetlands are doing the job— meaning, do we obtain the desired water quality? This means that the indicators will be the components that are analyzed according to the defined water

TABLE 10.1

Typical Efficiencies in g/24h*m² of BOD₅ Removal, Nitrogen Removal, and Phosphorus Removal for a Constructed or Natural Wetland for Different Plant Densities

	Removal Efficiencies		
Plant Density (t/ha)	BOD$_5$	Nitrogen	Phosphorus
2	5.0	1.0	0.15
5	7.5	1.6	0.22
10	13	2.1	0.31
15	20	3.0	0.45
20	24	3.6	0.54
25	25	3.8	0.57

quality. If natural wetlands are used to treat wastewater and drainage water, the same indicators—the water quality of the treated water—are of interest. It is of further importance to know the density of plants and the plant species, because a higher density of plants implies higher removal efficiency (see Table 10.1) and plants have different properties and therefore also different ability to participate in the six processes listed in Section 16.3. The difference between the constructed and the natural wetland is, of course, the possibility to adjust the efficiency of the treatment of wastewater and drainage water. It is usually easier to increase the plant density for the constructed wetland and to select the most effective plants. See Figure 10.2, which is a constructed wetland treating wastewater in Tanzania. The selection of the plant density should actually be integrated in the design phase; see, for instance, the wetland model presented in Jørgensen (2009). The flow pattern also has influence on the efficiency of the water treatment. Again, it is possible in the design phase to consider the best flow pattern for a constructed wetland, while it is usually more difficult to force the best flow in a natural wetland, although it is, of course, not impossible. Figure 10.3 shows a selected flow pattern in a constructed wetland in Tanzania. Indicators for wetlands used for water treatment are selected to be able to answer the questions:

1. Is the treatment acceptable?
2. If the treatment is not acceptable and the required water quality is not achieved, it would be natural to ask whether the following could improve the treatment results:
   a. A higher density
   b. Use of other plant species
   c. A change of the flow pattern

For natural wetlands that are not used for treatment of wastewater or drainage water, other issues are of interest for the environmental

FIGURE 10.2
Constructed wetland in Tanzania. A high density of plants are selected in the design phase to ensure high treatment efficiency. The selected plants species are local wetland plants.

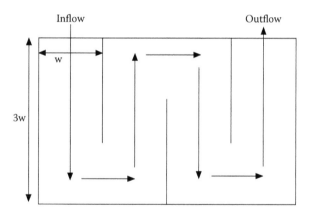

FIGURE 10.3
A flow pattern selected for a constructed wetland in Tanzania is shown. The pattern is selected in the design phase to ensure higher efficiency.

management of the wetlands. All the indicators mentioned in Section I of the book could come into the picture, but we would particularly like to use indicators that could tell us whether a wetland is a healthy natural ecosystem with, for instance, the diversity that could ensure a wide spectrum of buffer capacities that could reduce coming environmental threats, as mentioned in Section I.

The following relevant questions would be significant to answer in this context, and the answers would require assessment of several ecological indicators:

1. Which species are characteristic for the wetland? Are the species normal for wetlands in the region?
2. What is the spatial distribution of the species?
3. What is the diversity, including the biodiversity? Does the diversity give the wetland a wide spectrum of buffer capacities?
4. Is the wetland density of plants and the wetland area able to "absorb" flooding and a tsunami?
5. Are the animals represented in the wetland characteristic for the type of wetland in the region, or are some typical species that would be important for the functioning of the wetland missing?

To answer these five questions, it is recommended that one get information about the following ecological indicators:

1. What is the plant density and area of the wetland?
2. Which are the dominant plant and animal species?
3. Determine the species diversity.
4. Determine the respiration/biomass ratio and the production/biomass ratio—two of E. P. Odum's attributes.
5. Find the most characteristic buffer capacities.
6. What is the distribution of the dominant plant species?
7. Make a conceptual model of the most important energy flows in the wetland and determine the concentration of the species that are included in the flow diagram. Calculate the eco-exergy of the wetland model. Use the ecological indicators mentioned in points 1–6 to answer the five questions and give a general assessment of the ecosystem's health.

10.5 Ecological Indicators and Wetland Development— A Case Study

Mitsch et al. (2005) presented an interesting study of constructed experimental wetlands at the Olentangy River Wetland Research Park. On the campus of Ohio State University, two experimental wetlands were constructed under the leadership of W. J. Mitsch. In wetland number one, 2,400 plants were planted, representing 13 typical species of the region. The second wetland remained unplanted. It was left to "Mother Nature" to develop this wetland.

Since it was a pure research experiment, many indicators were followed during the first 8 to 10 years of these two wetlands. Determinations of the following water-quality indicators were carried out: nutrient concentrations, temperature, pH, dissolved oxygen, turbidity, and redox potential. Simultaneously, the following ecological indicators were followed: productivity and diversity of phytoplankton and of macrophytes, and diversity of macroinvertebrates and birds. Also, the similarity index was calculated for the macrophyte community.

The results of this investigation show that the number of species increased over time for both wetlands. The naturally colonized wetland was behind the planted basin in the early years, but in 1999 the two wetlands had the same number of species. Both wetlands also had increasing primary production of macrophytes, while the natural wetland had a development toward more dominance of *Typha*. It seems, in other words, that the number of species is not very dependent on whether the wetland has been planted or naturally colonized, although the latter will be dominant in the species that are best fitted to the region and to the local conditions.

Since its publication in 2005, the investigations have continued and the results are expected to be published soon. The investigations of the wetland development after 1999 will probably give a clearer picture of the development of the two wetlands, including the possible difference in the development of the planted wetland and the naturally colonized wetland.

10.6 Conclusions

The entire spectrum of indicators presented in Section I may be applied for wetlands as for most other ecosystems. Particularly, the holistic indicators biodiversity and species richness seem to be applied more widely for wetlands than for other ecosystems, because wetlands often have an important role in the landscape as buffer zones, and in the coastal zone for coast protection and reduction of flooding. In this context the thermodynamic indicators could and should also be applied. When wetlands are used for a specific purpose, for instance, reduction of flooding, the choice of indicators gets more evident. For instance, the reduction of flooding is strongly dependent on the plant density and the biomass per area unit. Therefore, they can be applied directly as indicators. When constructed or natural wetlands are applied for treatment of wastewater or drainage water, the indicators that assess the water quality of the treated water are surprisingly not in focus. If the treatment is insufficient several semi-holistic and holistic indicators can be used to improve the quality of the treated water.

References

Costanza, R., R. d'Arge, R. de Groot, S. Farber, M. Grasso, B. Hannon, K. Limburg, S. Naeem, R. V. O'Neill, J. Paruelo, R. G. Raskin, P. Sutton, and M. Van Den Belt. 1997. The value of the world's ecosystem services and natural capital. *Nature* 387:252–60.

Jørgensen, S. E. 2009. *Ecological modelling: An introduction.* Southampton, UK: WIT Press.

———. (in press). Ecosystem services, sustainability and thermodynamic indicators. Ecological Complexity.

Mtisch, W. J., N. Wang, R. Deal , X. Wu, and A. Zuwerink. 2005. Using ecological indicators in a whole-ecosystem wetland experiment. In *Handbook of ecological indicators for the assessment of ecosystem health*, eds. S. E. Jørgensen, R. Costanza, and F.-L. Xu, 213–37. Boca Raton, FL: CRC Press.

11

Application of Ecological Indicators for the Assessment of Ecosystem Health in Estuaries and Coastal Zones

João C. Marques, Fuensanta Salas, J. Patrício,
J. Neto, H. Teixeira, and R. Pinto

CONTENTS

11.1 How Does Environmental Health Relate to Pressures, Human Well-Being, and Ecological Sustainability?

The answer to this question, although it might seem intuitive, is indeed not trivial. In fact, the answer encompasses one of today's major problems for human society, which is to reconcile the concepts of *ecological sustainability* and *sustainable development*.

The idea of *sustainable development* formally emerged about 20 years ago, and since then researchers from different disciplines have tried to understand and define more precisely the meaning of this term. The most widely adopted significance has been "development that satisfies present needs without compromising the possibility of future generations satisfying theirs" (Brundtland 1987). Thus the concept remains obviously elusive and vague, without real sound epistemological, cultural, and scientific foundations.

The bases of sustainability are, of course, biophysical, strictly related to the concept of equilibrium, and must obey the natural laws that also govern human behavior. This implies that the quest for sustainability must necessarily take into account three major issues: (1) time, (2) relationships, and (3) biophysical limits.

Time is a crucial issue in the sense that human society does not evolve in accordance with the environment's capacity to produce the resources required for such development. In fact, in the last centuries, human systems have been driven mostly by nonrenewable resources, which will be exhausted in a limited period of time. This means, of course, that if this trend does not change, the needs of future human generations will be at stake.

Relationships are critical because care of environment and of natural resources might prove impossible to reconcile with the present economic paradigm. In fact, economic instruments often appear to lack the criterion of efficient allocation of resources, since they do not consider things not directly linked to the market. Such relationships imply interdependencies, but there is a need to determine at what scale aspects are interdependent (regional, national, continental, or global).

Finally, biophysical limits are also an unavoidable issue because each local human population can hardly meet its needs for materials, energy, land, waste sinks, and information from its own local resources. Therefore, if possible, the resources of other populations tend to be used, and this increasing trend will be critical in a globalized world.

In other words, sustainable environmental management will only be achieved if we are able to maintain and even increase the economic goods and services required by a developing society while at the same time maintaining and protecting ecological goods and services. Taken together these represent environmental goods and services.

On the other hand, the recognition that humans, with their cultural diversity, are an integral component of ecosystems, and the foreseeable threats represented by a serious world-level environmental degradation put *ecological sustainability* in international agendas. The intimate linkage between the natural and social aspects of ecosystems is reflected as the Ecosystem Approach (e.g., Convention for Biological Diversity). In particular, the need for environmental restoration involves the need to deal with problems such as (a) losses of habitats and species diversity, as well as a decrease in habitats' size and heterogeneity; (b) decrease of population size and changes in dynamics and distribution of many species; (c) habitat fragmentation and inherent increase in the vulnerability of the remaining isolated pockets; (d) decrease of economically relevant services and goods naturally provided by ecosystems.

In this context, the search for ecological sustainability represents a great challenge, namely because although some ecological concepts are well understood, such as the nature of ecosystem structure and functioning, or at least properly defined, others such as carrying capacity, resilience, and ecosystem goods and services are in general still poorly quantified. If the overall carrying capacity, compared with today, as well as the available resources, tends to decrease, people will have to adapt accordingly. But the linking between these ecological concepts and the management framework is also relatively recent, and it is still not clear how to integrate them in a holistic approach to manipulate and manage the environment.

If we look at the restoration, management, or sustainable use of a specific ecosystem, how can we decide the best possible course in an ocean of driving forces that are often contradictory? How can we orient ourselves within such complexity? The application of ecological indicators to assess ecosystem health is, of course, a fundamental tool. But evaluations must nevertheless be interpreted in the scope of a broader conceptual guidance tool, which may act as a kind of compass and might be useful to provide orientation in the process of building management scenarios for dealing with environmental problems. While the ideas here are from the marine and estuarine field, they are applicable to other domains.

11.1.1 Building Management Scenarios to Relate Environmental Pressures, Human Well-Being, and Ecological Sustainability

Environmental problems are intrinsicaly complex and intimately related to the development of human society. Therefore, possible solutions to environmental problems must always be approached by taking into account various viewpoints, and expressing the different perceptions of multiple sectors, which are often contradictory. Nevertheless, we may view those different perspectives and apprehensions in the scope of three major drivers:

1. The search for human well-being and the maintenance of human health and safety
2. The endeavor of ecological sustainability and natural environmental well-being
3. The tolerance to increasing human population pressure and demand for wealth creation

In terms of governance, the definition of human well-being is relatively subjective. In practical terms, the search for human well-being is often taken as synonymous with stakeholders' benefits, and may eventually be expressed by some kind of tentative metrics (e.g., a well-being index) (Figure 11.1). The basic government goal has been to maximize economic goods and services while at the same time protecting ecological goods and services (or at least for business not to be prosecuted for harming the latter).

On the other hand, the endeavor of ecological sustainability is normally associated with the concept of environmental health, which can obviously be approached from different theoretical orientations, all involving inherent uncertainties. A panoply of tools are available to evaluate environmental health, among which are ecological indicators, although probably none of them are entirely suitable. Integrated approaches are proving valuable, especially where the indicators are related to the DPSIR approach (Aubry and Elliott 2006; Borja and Dauer 2008). Additionally, the acknowledgment that much of the nation's economic prosperity depends on ecosystems' functioning, and that many natural ecosystems are threatened, led to the emergence of a new interest in ecosystem goods and services (Figure 11.1). In fact, and

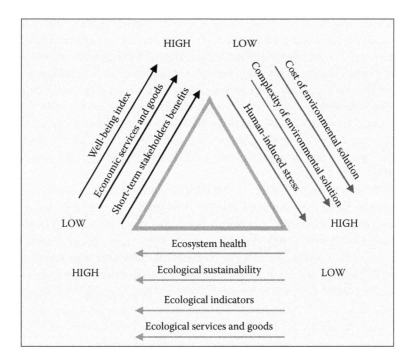

FIGURE 11.1
Ecological Sustainability Trigon illustrating the expected trends and relationships between variables assumed to be correlated with ecological sustainability, human well-being, and population size/pressure.

estuarine and marine areas are an excellent example, economic goods and services at one area depend on their successful functioning elsewhere (e.g., estuarine nursery grounds in one area providing marine commercial stocks in another area).

Finally, human population pressure is increasing as a direct result of population growth, on one hand, but also as a function of increasing resources consumption and pollution related to the pursuit of human needs satisfaction. As a function of such pressure increase, the solution for environmental problems becomes gradually more complex and difficult, as well as the inherent costs (Marques et al. 2009) (Figure 11.1).

Let us assume that ecological sustainability constitutes a major goal for human society. It would be conceptually possible to maximize ecological sustainability and stakeholders' benefits, but only in the condition of having a very low population size and pressure. As well, it would be possible to maximize ecological sustainability and population size, but in such a case we should accept very low stakeholders' benefits. Finally, at least in the short term, we could maximize stakeholders' benefits and the size of the population and its pressure if we give up ecological sustainability. But in the "game of possibles" no conceivable scenario allows maximizing the three simultaneously, although a trade-off between the endeavor of ecological sustainability, the search for human well-being, and the increasing human population pressure would be conceptually possible (Marques et al. 2009) (Figure 11.2).

Nevertheless, this trade-off cannot simply rely on an increasing complexity of the solution to environmental problems. In fact, the benefits from an increasing complexity of the solutions will not increase linearly as a function of that complexity, since the inherent costs (energy and money) will most probably become unsustainable ecologically in the long term (Marques et al. 2009) (Figure 11.3).

Everything said is, to a certain extent, intuitive. But the number of variables that must be taken into account is extremely high and uncertainties regarding their relationships and trends may be confusing. The use of an Ecological Sustainability Trigon (EST) (Marques et al. 2009) may be very useful as a tool to provide orientation, namely in building back-cast management scenarios ("Where do we want to go?" "How do we get there?") instead of the more conservative forecast scenarios ("Where are we going?"). Examples of

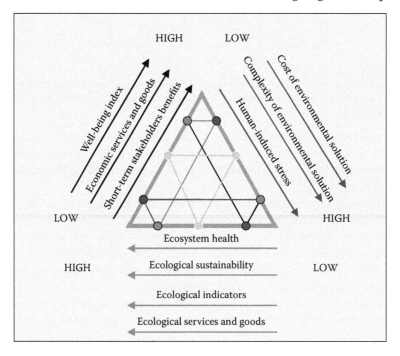

FIGURE 11.2
Proposed use of the Ecological Sustainability Trigon in building management scenarios: analysis of the expected variations and relationships of different variables correlated with ecological sustainability, human well-being, and population size/pressure. Dark gray, good; black, bad; light gray, acceptable.

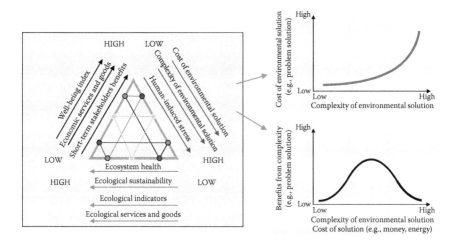

FIGURE 11.3
Proposed use of the Ecological Sustainability Trigon in building management scenarios: expected trend in costs and benefits as a function of an increasing complexity of environmental problem solutions. Dark gray, good; black, bad; light gray, acceptable.

different variables to which the trends are supposed to be correlated may be organized along the three sides of the trigon, with their expected trends varying from low to high (Figure 11.1). Of course, many other variables could be included. Now, using the trigon, one may easily analyze, at least roughly (correlations are far from being linear), the expected variations and relationships between different variables correlated with ecological sustainability, human well-being, and human pressures. This may be extremely useful in building environmental management scenarios and, of course, in interpreting and making operative results obtained from the application of ecological indicators.

This approach has the advantages of addressing and measuring all components with the same species-specific currency, i.e., the human society view, and at the same time describing our behavior, energetics (economy), and dynamics with the same tool used for all other ecosystem components, i.e., ecological theory. Incorporation of our behavior, energetics, and dynamics into the ecosystem health framework poses key challenges for the science of ecology by requiring ecological indicators of ecological status from ecosystem organization and functioning rather than from pressures and vulnerability, but it makes explicit the evaluation criteria for the environmental management scenarios, i.e., scales should match (time), interactions should match (relationships), rates should match (biophysical limit). Additionally, the EST approach appears to be very promising for gap analysis, as well as to address new research questions.

11.1.2 Ecological Indicators and Ecosystems' Health Evaluation

Ecological indicators are commonly used to supply synoptic information about an ecosystem's health. Most often they address an ecosystem's structure and/or functioning accounting for a certain aspect or component, for instance, nutrient concentrations, water flows, macro-invertebrate and/or vertebrate diversity, plant diversity, plant productivity, erosion symptoms and, sometimes, ecological integrity at a systems level.

The main attribute of an ecological indicator is to combine numerous environmental factors in a single value, which might be useful in terms of management and for making ecological concepts compliant with the general public understanding. Moreover, ecological indicators may help in establishing a useful connection between empirical research and modeling since some of them are of use as orientors (also referred to in the literature as goal functions) in ecological models. Such applications proceed from the fact that conventional models of aquatic ecosystems are not effective in predicting the occurrence of qualitative changes in ecosystems, e.g., shifts in species composition, which is because measurements typically carried out, like biomass and production, are not efficient in capturing such modifications (Nielsen 1995). Nevertheless, incorporating these type of changes in structurally dynamic models has been tried (Nielsen 1992, 1994, 1995; Jørgensen et al. 2002) to improve their predictive capability, achieving a better understanding of ecosystems behavior, and consequently better environmental management.

In structurally dynamic models, simulated ecosystem behavior and development (Nielsen 1995; Straskraba 1983) is guided through an optimization process by changing the model parameters in accordance to a given ecological indicator, used as orientor (goal function). In other words, this allows introducing in models parameters that change as a function of changing forcing functions and conditions of state variables, optimizing the model outputs using a stepwise approach. In this case, the orientor is assumed to express a given macroscopic property of the ecosystem, resulting from the emergence of new characteristics arising from self-organization processes.

In general, the application of ecological indicators is not exempt from criticism, the first of which is that aggregation results in oversimplification of the ecosystem under observation. Moreover, problems arise from the fact that indicators account not only for numerous specific system characteristics, but also other kinds of factors, e.g., physical, biological, ecological, socioeconomic, etc. Thus, indicators must be used following the right criteria and in situations that are consistent with their intended use and scope; otherwise, they may lead to confusing data interpretations.

This chapter addresses the application of ecological indicators for assessing biological integrity and environmental quality in coastal ecosystems and transitional waters. What might be the characteristics of a good ecological indicator, or what kind of information, regarding ecosystem responses, can be obtained from the different types of biological data usually taken into account in evaluating the state of coastal areas was already discussed in Chapter 2. Two case studies are used to illustrate whether different types of indicators were satisfactory in describing the state of ecosystems, comparing their relative performances and discussing how can their usage be improved for environment health assessment.

11.2 Brief Review of the Application of Ecological Indicators in Estuarine and Coastal Ecosystems

Almost all estuarine and coastal marine ecosystems around the world have been under severe environmental stress, following the settlement of human activities. Estuaries, for instance, are the transition between marine, freshwater, and land ecosystems, being characterized by distinctive biological communities with specific ecological and physiological adaptations. In fact, we may say that the estuarine habitat does not imply a simple overlap of marine and land factors, constituting instead an individualized whole, with its own biogeochemical factors and cycles, which represents the environment for real estuarine species to evolve. In such ecosystems, besides resources available, fluctuating conditions, namely salinity and type of substrate, are a key issue regarding an organism's ecological distribution and adaptive strategies (see, for instance, McLusky and Elliot 2004; Engle et al. 1994).

The most common types of problems in terms of pollution include illegal sewage discharges associated with nutrient enrichment, pollution due to toxic substances such as pesticides, heavy metals and hydrocarbons, unlimited development, and habitat fragmentation and/or destruction.

In the case of transitional waters, limited water circulation and inappropriate water management tend to concentrate nutrients and pollutants, and to a certain extent we may say that sea pollution begins there (Perillo et al. 2001). Moreover, in estuaries, drainage of harbors and channels modifies geomorphology, water circulation, and other physicochemical features, and consequently the habitat's characteristics. In recent times, perhaps the most important problem is the excessive loading of nutrients, mainly because of fertilizers used in agriculture and untreated sewage waters, which induce eutrophication processes observed all over the world.

Many ecological indicators used and/or tested in evaluating the status of these ecosystems can be found in the literature, resulting nevertheless from just a few distinct theoretical approaches. A number of them focus on the presence/absence of given indicator species, while others take into account the different ecological strategies carried out by organisms, diversity, or the energy variation in the system through changes in the biomass of individuals. A last group of ecological indicators that are thermodynamically oriented or based on network analysis look for capturing the information of the ecosystem from a more holistic perspective (Table 11.1).

TABLE 11.1

Recompilation of the Main Descriptors of *Posidonia oceanica*

Descriptor	Measured Parameters	Information
Upper depth limit	Depth, localization, density, bottom cover, characterization of the substrate	Human impact, hydrodynamism, sedimentary dynamics
Density	Number of shoots on a surface >1600 cm²	Dynamics of the meadow, human impact
Epiphytic coverage	Biomass, diversity	Nutrient concentration, flora and fauna biodiversity
Bottom cover	% of meadow on a given surface (1 to 25 m²)	Dynamics of the meadow, human impact
Leaf biometry	Type, number, size of leaves, leaf surface, coefficient A, biomass, epiphytic coverage, presence of necrosis	Health state of the meadow, human impact, hydrodynamism, herbivore pressure
Lower depth limit	Depth, localization, type, density, bottom coverage, leaf biometry, granulometry, content in organic matter	Water transparency, human impact, hydrodynamism, dynamics of the meadow (regression of colonization)
Population associated to the meadow	Fauna, flora, diversity	Biodiversity, meadow–population interactions
Structure of the matte	Intermattes, "cliff of dead matte," erosive structures, receding, silting up, biodiversity of the endofauna, homogeneity, resistance and compactness, % of plagiotropic rhizomes, width of the matte, physicochemical composition	Dynamics of the meadow, human impact, sedimentary dynamics, study of currents
Biochemical and chemical composition	Elementary composition (C, N, P) phenolic compounds, proteins, carbohydrates, stress enzymes	Dynamics of the meadow, human impact, herbivore pressure
Datation measurement	Lepidochronology, plastochrone interval, paleo-flowering, primary production	Temporal evolution of the production, sedimentation speed, intensity of sexual reproduction, dynamics of the meadow, human impact
Contamination	Metals (Hg, Cu, Cd, Pb, Zn)	Human impact

Source: Pergent-Martini, C., V. Leoni, V. Pasqualini, G. D. Ardizzone, E. Balestri, R. Bedini, A. Belluscio, T. Belsher, J. Borg, C. F. Boudouresque, S. Boumaza, J. M. Bouquegneau, M. C. Buia, S. Calvo, et al. 2005. Descriptors of *Posidonia oceanica* meadows: Use and application. *Ecol Indic* 5:213–30.

11.2.1 Indicators Based on Species Presence versus Absence

Presence or absence of one species or group of species has been one of the most used approaches in detecting pollution effects. For instance, the Bellan (based on polychaetes) or the Bellan-Santini (based on amphipods) indices attempt to characterize environmental conditions by analyzing the dominance of species indicating some type of pollution in relation to the species considered to indicate an optimal environmental situation (Bellan 1980; Bellan-Santini 1980). Several authors consider the use of these indicators inadvisable because often such indicator species may occur naturally in relatively high densities. The point is that there is no reliable methodology to know at which level one of those indicator species can be well represented in a community that is not really affected by any kind of pollution, which leads to a significant exercise of subjectivity (Warwick 1993). Despite these criticisms, even recently, the AMBI index (Borja et al. 2000), based on the Glemarec and Hily (1981) species classification regarding pollution, as well as the BENTIX index (Simboura and Zenetos 2002), have gone back to update such pollution-detecting tools. Roberts et al. (1998) also proposed an index based on macrofauna species that accounts for the ratio of each species abundance in control versus samples proceeding from stressed areas. It is, however, semiquantitative as well as site and pollution type specific.

The AMBI index, for instance, which accounts for the presence of species indicating a type of pollution and of species indicating a reference situation assumed as not polluted, has been considered useful in terms of the application of the European Water Framework Directive in coastal ecosystems and estuaries. In fact, although this index is very much based on the paradigm of Pearson and Rosenberg (1978), which emphasizes the influence of organic matter enrichment on benthic communities, it was shown to be useful in assessing other anthropogenic impacts, such as physical alterations in the habitat, heavy metal inputs, etc., in several European areas of the Atlantic (North Sea, Bay of Biscay, and South of Spain) and Mediterranean coasts (Spain and Greece) (Borja et al. 2000; Borja, Muxika, et al. 2003; Borja, Franco, et al. 2003; Bonne et al. 2003; Gorostiaga et al. 2004; Salas et al. 2004).

Marine benthic macrophytes, in their turn, respond directly to the abiotic and biotic aquatic environments, and thus represent sensitive bioindicators regarding their changes (Orfanidis et al. 2003). On the other hand, a series of algae genera are universally considered to appear when pollution occurs, such as the green algae *Ulva*, *Cladophora*, and *Chaetomorpha*, and the red algae *Gracilaria*, *Porphyra*, and *Corallina*. Additionally, species with high structural complexity, like the Phaeophyta belonging to *Fucus* and *Laminaria* genera, are seen worldwide as the most sensitive to any kind of pollution, even if *Fucus* species may cope with moderate pollution (Niell and Pazó 1978). Finally, marine Spermatophytae are considered indicator species of good water quality.

In the Mediterranean Sea, for instance, the presence of *Cystoseira* and *Sargassum* (Phaeophyta) or *Posidonia oceanica* meadows indicate the water's good quality. Thus, monitoring population density and distribution of such species allows detecting and evaluating the impact of all kinds of activities (Pérez-Ruzafa 2003). *Posidonia oceanica* is possibly the most used indicator of water quality in the Mediterranean because of its sensitivity to disturbances, its wide distribution along the Mediterranean coast, and the good knowledge about the plant and its ecosystem-specific response to a particular impact (e.g., Ruiz et al. 2001; Pergent-Martini et al. 2005). Furthermore, this species is capable of giving information about present and past levels of trace metals in the environment (Pergent-Martini 1998).

Pergent-Martini et al. (2005) identified the descriptors of *Posidonia oceanica*, constituting the first step in opening the way to the use of this species to assess the ecological status of Mediterranean coastal zones (Table 11.1). POMI, the *Posidonia oceanica* Multivariate Index, was developed on these grounds and based on those physiological, morphological, and structural descriptors combined into a variable using a principal component analysis (PCA).

In the same sense, a Conservation Index (Moreno et al. 2001), based on the named marine Spermatophyta, is used in Mediterranean coasts. Along the same lines, Orfanidis et al. (2001) introduced a new Ecological Evaluation Index (EEI) to assess the ecological status of transitional and coastal waters in accordance with the Water Framework Directive (WFD). This index is based on the marine benthic macrophyte classification in two ecological state groups (ESGs I, II), representing alternative ecological states (pristine and degradated). More recently, and also in the scope of the WFD implementation, Scanlan et al. (2007) proposed the Opportunistic Macroalgae Assessment Tool.

11.2.2 Biodiversity as Reflected in Diversity Measures

Biodiversity is a widely accepted concept usually defined as biological variety in nature. This variety can be perceived intuitively, which leads to the assumption that it can be quantified and adequately expressed in any appropriate manner (Marques 2001), although expressing biodiversity as diversity measures had proved to be not an easy challenge. Nevertheless, diversity measures have been perhaps the most commonly used approach, which assumes that the relationship between diversity and disturbances can be seen as a decrease in the former as stress increases.

Looking to a certain systematization, Magurran (1988) classifies diversity measurements into three main categories:

1. Indices that measure the enrichment of the species, such as the Margalef's index, which are, in essence, a measurement of the number of species in a defined sampling unit.

2. Models of the abundance of species, such as the K-dominance curves (Lambshead et al. 1983) or the lognormal model (Gray 1979), which describe the distribution of their abundance, going from situations in which there is a high uniformity to those in which the abundance is very unequal. However, the lognormal model deviation was rejected a long time ago by several authors because of the impossibility of finding any benthic marine sample that clearly responded to such a lognormal distribution model (Shaw et al. 1983; Hughes 1984; Lambshead and Platt 1985).

3. Indices based on the proportional abundance of species aiming to account for species richness and regularity of species distribution in a single expression. Secondarily, these indices can be subdivided into those based on information theory and those accounting for species dominance. Indices derived from the information theory, e.g., Shannon–Wiener, assume that diversity, or information, in a natural system can be measured in a similar way as information contained in a code or message. On the other hand, dominance indices, e.g., Simpson or Berger–Parker, are referred to as measurements that account for the abundance of the most common species.

More recently, a measure called "taxonomic distinctness" has been used in some studies (Warwick and Clarke 1995, 1998; Clarke and Warwick 1999) to assess biodiversity in marine environments, taking into account taxonomic, numeric, ecologic, genetic, and philogenetic aspects of diversity. Nevertheless, it is most often very complicated to meet certain requirements to apply it, such as having a complete list of the species present in the area under study in pristine situations. Moreover, some works have shown that in fact taxonomic distinctness is not more sensitive than other diversity indices usually applied when detecting disturbances (Sommerfield and Clarke 1997), and consequently, since it was proposed, this measure has not been widely used on marine environment quality assessment and management studies.

11.2.3 Indicators Based on Ecological Strategies

The purpose of some indicators is to assess environmental stress effects, taking the ecological strategies followed by different organisms into consideration. That is the case for trophic indices such as the Infaunal Trophic Index (ITI) (Word 1979) and the Feeding Structure Index (FSI), based on organisms' different feeding strategies. Another example is the Nematodes/Copepods Index (Rafaelli and Mason 1981), which accounts for the different behavior of two taxonomic groups under environmental stress situations. Nevertheless, several authors rejected these types of indicators because of their dependence on parameters like depth and sediment particle size, as well as their unpredictable pattern of variation depending on the type of pollution (Gee et al. 1985; Lambshead and Platt 1985).

Other proposals also appeared, such as the Meiobenthic Pollution Index (Losovskaya 1983); the Mollusc Mortality Index (Petrov 1990); the Indice of Trophic Diversity (ITD) (Heip et al. 1985); the Maturity Index (MI) (Bongers et al. 1991); the Polychaeta Amphipoda Ratio (Gómez-Gesteira and Dauvin 2000), revised as Benthic Opportunistic Polychaeta Amphipoda (BOPA) (Dauvin and Ruellet 2007); or the Index of r/K strategies proposed by De Boer et al. (2001).

The Rhodophyceae/Phaeophyceae Index proposed by Feldmann (1937), based on marine vegetation, is frequently used in the Mediterranean Sea. It was established as a biogeographical index and accounts for the fact that the number of Rhodophyceae species decreases from the tropics to the poles. Its application as indicator holds on the higher or lower sensitivity of Phaeophyceae and Rhodophyceae to disturbances. In addition, Belsher (1982) proposed an index based on the qualitative and quantitative dominance of each taxonomic group.

11.2.4 Indicators Based on Species Biomass and Abundance

Other approaches account for the variation of organisms' biomass as a measure of environmental disturbances. Along these lines, we have methods such as SAB (Pearson and Rosenberg 1978), consisting of a comparison between the curves resulting from ranking the species as a function of their representativeness in terms of their abundance and biomass. The use of this method is not advisable because it is purely graphical, which leads to a high degree of subjectivity that impedes relating it quantitatively with the different environmental factors. The ABC method (Warwick 1986) also involves the comparison between the cumulative curves of species biomass and abundance, from which Warwick and Clarke (1994) derived the W-statistic index.

11.2.5 Multimetric Indices

From a more holistic point of view, some authors proposed indices capable of integrating different types of environmental information. A first approach was developed by Satsmadjis (1982) for application in coastal areas, relating sediment particle size to the diversity of benthic organisms. Jeffrey et al. (1985) developed the Pollution Load Index (PLI), Rhoads and Germamo (1986) proposed the Organism Sediment Index (OSI). Vollenweider et al. (1998) developed a Trophic Index (TRIX) integrating chlorophyll a, oxygen saturation, total nitrogen, and phosphorus to characterize the trophic state of coastal waters. In the same way, Fano et al. (2003) proposed the Ecofunctional Quality Index, which considers the macrofaunal and macrophytic abundance/biomass. In a progressively more complex way, other indices such as the Index of Biotic Integrity (IBI) for coastal systems (Nelson 1990), the Benthic Condition Index (Engle et al. 1994), the Benthic Habitat Quality (BHQ) (Nilsson and Rosenberg 1997), the Benthic Response Index (BRI) (Smith et al. 2001), the Biological Quality Index (BQI) (Rosenberg et al. 2004), or the Chesapeake Bay B-IBI (Weisberg et al. 1997), the Carolina Province B-IBI (Van Dolah et al. 1999), and the Virginia Province Benthic Index (VPBI) (Paul et al. 2001) include physicochemical factors, diversity measures, specific richness, taxonomic composition, and the system's trophic structure.

Similarly, a set of specific indices of fish communities have been developed to measure the ecological status of estuarine areas. The Estuarine Biological Health Index (BHI) (McGinty and Linder 1997) combines two separate measures (health and importance) into a single index. The Fish Health Index (FHI) (Cooper et al. 1993) is based on both qualitative and quantitative comparisons with a reference fish community. The Estuarine Ecological Index (EBI) (Deegan et al. 1993) reflects the relationship between anthropogenic alterations in the ecosystem and the status of higher trophic levels, and the Estuarine Fish Importance Rating (FIR) is based on a scoring system of seven criteria that reflect the potential importance of estuaries for the associated fish species. This index is able to provide a ranking, based on the importance of each estuary, and helps to identify the systems of major importance for fish conservation.

Nevertheless, these indices are often not used in a generalized way because they have usually been developed for application in a particular system or area, which makes them dependent on the type of habitat and seasonality.

More recently, Hale and Heltshe (2008) developed the Acadian Province Benthic Index (APBI). With the challenges brought by the WFD implementation, other multimetric indices were proposed, namely the Multivariate-AMBI (M-AMBI) (Muxika et al. 2007) and the Portuguese Benthic Assessment Tool (P-BAT) (Marques et al. 2009; Pinto et al. 2009). A comprehensive revision of the multimetric indices most used in estuarine and coastal marine ecosystems can be found in Marques et al. (2009) and Pinto et al. (2009).

11.2.6 Thermodynamically Oriented and Network-Analysis-Based Indicators

In the last two decades, several functions have been proposed as holistic ecological indicators, intending (a) to express emergent properties of ecosystems arising from self-organization processes in the run of their development, and (b) to act as orientors (goal functions) in models development, as mentioned above. Such proposals resulted from a wider application of theoretical concepts, following the assumption that it is possible to develop a theoretical framework able to explain ecological observations, rules, and correlations on the basis of an accepted pattern of ecosystem theories (Jørgensen and Marques 2001). That is the case of ascendancy (Ulanowicz 1986; Ulanowicz and Norden 1990) and emergy (Odum 1983, 1996), both having originated in the field of network analysis, which appear to constitute suitable system-oriented characteristics for natural tendencies of ecosystems development (Marques et al. 1998). Also exergy (Jørgensen and Mejer 1979, 1981), a concept derived from thermodynamics, which can be seen as energy with a built-in measure of quality, has been tested in several studies (e.g., Nielsen 1990; Jørgensen 1994; Fuliu 1997; Marques et al. 1997, 2003).

11.3 How to Choose the Most Adequate Indicator

The application of a given ecological indicator is always a function of data requirements and data availability. Therefore, in practical terms, the choice of ecological indicators to use in a particular case is a sensible process. Table 11.2 provides a summary of what we consider to be the essential options that have been applied in coastal and transitional waters ecosystems.

Table 11.3 exemplifies the process of selecting the most adequate ecological indicators as a function of data requirements and data availability. In the process of selecting an ecological indicator, data requirements and data availability must be accounted for. Moreover, the complementary use of different indices or methods based on different ecological principles is highly recommended in determining the environmental quality status of an ecosystem.

11.4 Case Studies: Subtidal Benthic Communities in the Mondego Estuary (Atlantic Coast of Portugal) and Mar Menor (Mediterranean Coast of Spain)

11.4.1 Study Areas and Type of Data Used

Different ecological indicators were used in the Mondego estuary, located on the western coast of Portugal, and Mar Menor, a 135 km² Mediterranean coastal lagoon located on the southeast coast of Spain. The lagoon is connected to the Mediterranean at some points by channels through which water exchange takes place with the open sea.

The Mondego estuary, located on the western coast of Portugal, is a typical temperate, small intertidal estuary. As for many other regions, this estuary shows symptoms of eutrophication, which have resulted in an impoverishment of its quality. A more detailed description of the system is reported elsewhere (e.g., Marques, Rodrigues, et al. 1993; Marques, Maranhão, et al. 1993;

TABLE 11.2

Short Review of Environmental Quality Indicators Regarding the Benthic Communities

Type of Indicator	Requirements and Applicability Evaluation	Algorithm
Based on Species Presence vs. Absence	List of species. Subjective in most cases. Only the use of AMBI and BENTIX are likely to be recommended.	**Bellan index** (Bellan 1980): $IP = \sum \dfrac{\text{pollution species indicator}}{\text{no pollution species indicator}}$ Pollution indicator species: *Platenereis dumerilli, Theosthema oerstedi, Cirratulus cirratus,* and *Dodecaria concharum* No pollution indicator species: *Syllis gracillis, Typosyllis prolifera, Typosyllis* spp., and *Amphiglena mediterranea* **Bellan-Santini index** (Bellan-Santini 1980): $IP = \sum \dfrac{\text{pollution species indicator}}{\text{no pollution species indicator}}$ Pollution indicator species: *Caprella acutrifans* and *Podocerus variegatus* No pollution indicator species: *Hyale* sp., *Elasmus pocillamunus,* and *Caprella liparotensis* **AMBI** (Borja et al. 2000): $AMBI = \dfrac{\{(0 \times \%GI) + (1.5 \times \%GII) + (3 \times \%GIII) + (4.5 \times \%GIV) + (6 \times \%GV)\}}{100}$ GI: Species very sensitive to organic enrichment and present under unpolluted conditions GII: Species indifferent to enrichment GIII: Species tolerant to excess of organic matter enrichment GIV: Second-order opportunist species, mainly small-sized polychaetes GV: First-order opportunist species, essentially deposit-feeders

(continued)

TABLE 11.2

Short Review of Environmental Quality Indicators Regarding the Benthic Communities (continued)

Type of Indicator	Requirements and Applicability Evaluation	Algorithm
		BENTIX (Simboura and Zenetos 2002):
		$$BENTIX = \frac{\{(6 \times \%GI) + 2 \times (\%GII + \%GIII)\}}{100}$$
		GI: Species very sensitive to pollution
		GII: Species tolerant to pollution
		GIII: Second-order and first-order opportunist species
Based on Ecological Strategies	List of taxa (species or higher taxonomic groups) and knowledge on their life strategies, which can be in the literature. Subjective. Not recommended.	**Nematodes/copepods ratio** (Rafaelli and Mason 1981):
		$$I = \frac{\text{nematodes abundance}}{\text{copepodes abundance}}$$
		Polychaetes/amphipods ratio (Gómez-Gesteira et al. 2000):
		$$Log10\left(\frac{\text{Polychaetes abundance}}{\text{Amphipodes abundance}} + 1\right)$$
		Infaunal index (Word 1979):
		$$ITI = 100 - 100/3 \times (0n_1 + 1n_2 + 2n_3 + 3n_4)/(n_1 + n_2 + n_3 + n_4)$$
		n_1 = number of individuals of suspensivore feeders
		n_2 = number of individuals of interface feeders
		n_3 = number of individuals of surface deposit feeders
		n_4 = number of individuals of subsurface deposit feeders
Diversity Measures	Quantitative samples; adequate taxa identification; data on species density (number of individuals and/or biomass). In the case of K-dominance curves, time series for the same	**Shannon–Wiener index** (Shannon–Weaver 1963):
		$H' = -\sum p_i \log_2 p_i$
		Where p_i is the proportion of abundance of species i in a community where species proportions are $p_1, p_2, p_3 \cdots p_n$.

locale are desirable. Although not exempt from subjectivity, results might be useful.

Margalef index:

$$D = (S - 1)/\log_e N$$

where S is the number of species found and N is the total number of individuals

Berger–Parker index:

$$D = (n_{max})/N$$

where n_{max} is the number of individuals of the dominant species and N is the total number of individuals

Simpson index:

$$D = \sum n_i(n_i - 1)/N(N - 1)$$

where n_i is the number of individuals of species i and N is the total number of individuals

Average taxonomic diversity index (Warwick and Clarke 1995, 1998):

$$\Delta = [\sum\sum_{i<j} \omega_{ij} x_i x_j]/[N(N - 1)/2]$$

where ω_{ij} is the taxonomic distance between every pair of individuals, the double summation is over all pairs of species i and j $(i,j = 1,2,\ldots, S; i < j)$, and $N = \sum_i x_i$, the total number of individuals in the sample.

When the sample consists simply of a species list the index takes this form:

$$\Delta^+ = [\sum\sum_{i<j} \omega_{ij} x_i x_j]/[S(S - 1)/2]$$

where S is the number of the species in the sample

K-dominance curves (Lambshead et al. 1983):

Cumulative ranked abundance plotted against species rank, or log species rank

(continued)

TABLE 11.2

Short Review of Environmental Quality Indicators Regarding the Benthic Communities (continued)

Type of Indicator	Requirements and Applicability Evaluation	Algorithm
Based on Species Biomass and Abundance	Quantitative benthic samples; taxa identification; species density (number of individuals and/or biomass). Data along gradients in the same system are suitable. Results might be useful.	**ABC curves** (Warwick 1986): K-dominance curves for species abundances and species biomasses on the same graph. The ABC method derived the W-statistic (Warwick and Clarke 1994): $W = \Sigma (B_i - A_i) / 50^*(S - 1)$ where B_i is the biomass of species i, A_i the abundance of species i, and S is the number of species
Indicators Accounting for the Whole Environmental Information	Physical chemical parameters; quantitative benthic samples; taxa identification; species density (number of individuals and/or biomass). Although it is a good idea to integrate the whole environmental information, they are difficult to apply as they need a large amount of data of different nature. B-IBI (Weisberg et al., 1997) is dependent on the type of habitat and seasonality.	**Benthic index of environmental condition** (Engle et al. 1994): Benthic index = (2.3841*Proportion of expected diversity) + (−0,6728*proportion of total abundance as tubifids) + (0.6683*Proportion of total abundance as bivalves) Coefficient of pollution (Satsmadjis 1985): Calculation of p is based on several integrated equations. These equations are: $S' = s + t/(5 + 0.2s)$ $i_0 = (−0.017s^*2 + 2.63s' − 4)(2.20 − 0.0166h)$ $g' = i/(0.0124i + 1.63)$ $P = g'/[g(i/i_0)^{1/2}]$ where P = coefficient of pollution S' = sand equivalent S = percent sand t = percent silt i_0 = theoretical number of individuals i = number of individuals h = station depth g' = theoretical number of species g = number of species

B-IBI (Weisberg et al. 1997):
Eleven metrics are used to calculate the B-IBI—

- Shannon–Wiener species diversity index
- Total species abundance
- Total species biomass
- Percent abundance of pollution-indicative taxa
- Percent abundance of pollution-sensitive taxa
- Percent biomass of pollution-indicative taxa
- Percent biomass of pollution-sensitive taxa
- Percent abundance of carnivores and omnivores
- Percent abundance of deep-deposit feeders
- Tolerance score
- Tanypodinae to Chironomidae percent abundance ratio

The scoring of metrics to calculate the B-IBI is done by comparing the value of a metric from the sample of unknown sediment quality to thresholds established from reference data distributions.

Exergy index (Jørgensen and Mejer 1979, 1981; Marques et al. 1997):

$Ex = T * \sum \beta i * C_i$

where T is the absolute temperature, C_i is the concentration in the ecosystem of component i (e.g., biomass of a given taxonomic group or functional group), βi is a factor able to express roughly the quantity of information embedded in the genome of the organisms. Detritus was chosen as reference level, i.e., βi = 1 and Exergy in biomass of different types of organisms is expressed in detritus energy equivalents.

Specific Exergy (Jørgensen and Mejer 1979, 1981):
SpEx = Extot/Biomtot

Ascendency (Ulanowickz 1986):

$$A = \sum_i \sum_j T_{ij} \log\left[\frac{T_{ij}T_{..}}{T_{.j}T_{i.}}\right]$$

T_{ij} = trophic exchange from taxon i to taxon j

Thermodynamically Oriented and Network-Analysis-Based Indicators

Exergy and Specific Exergy: quantitative samples. Data on taxa (higher taxonomic groups) biomasses. Useful. Not sufficiently tested. Developmental phase.
Ascendancy: Quantitative benthic samples; taxa identification; species density (number of individuals and/or biomass). Knowledge of the food-web structure and system energy through flow.
Objective, powerful, most often impossible to apply because of lack of data.

TABLE 11.3

Application of Different Indicators as a Function of Data Requirements and
Data Availability

Data Availability	Indicators
Qualitative Data	
Metadata	
Rough Data	Shannon–Wiener
	Margalef
	Average taxonomic distinctness (Δ^+)
Quantitative Data	
Populations Numeric Density Data	AMBI
	BENTIX
	Bellan
	Bellan-Santini
	Shannon–Wiener
	Margalef
	Simpson
	Berger–Parker
	K-dominance curves
	Average taxonomic diversity index (Δ)
	Average taxonomic distinctness (Δ^+)
	Benthic index of environmental condition
	Coefficient of pollution
Numeric Density Data and Biomass Data	Individuals Identification up to Specific Level
	AMBI
	BENTIX
	Bellan
	Bellan-Santini
	Shannon–Wiener
	Margalef
	Simpson
	Berger–Parker
	K-dominance curves
	Average taxonomic diversity index (Δ)
	Average taxonomic distinctness (Δ^+)
	Benthic index of environmental condition
	Coefficient of pollution
	Method ABC
	Exergy
	Specific Exergy
	Ascendency
	Individuals Identification up to Family or Higher Taxonomic Levels
	Shannon–Wiener
	Margalef

TABLE 11.3

Application of Different Indicators as a Function of Data Requirements and
Data Availability (continued)

Data Availability	Indicators
	Simpson
	Berger–Parker
	K-dominance curves
	Benthic index of environmental condition
	B-IBI
	Method ABC
	Exergy index
	Specific Exergy
	Ascendency

Marques et al. 1997, 2003; Flindt et al. 1997; Lopes et al. 2000; Pardal et al. 2000; Martins et al. 2001; Cardoso et al. 2002).

Regarding the Mondego estuary case study, two different data sets were selected to estimate different ecological indicators: (a) The first one was provided by a study on the subtidal soft bottom communities, which characterized the whole system with regard to species composition and abundance, taking into account its spatial distribution in relation to the physicochemical factors of water and sediments. The infaunal benthic macrofauna was sampled twice during spring, in 1998 and 2000, at 14 stations covering the whole system (Figure 11.4). (b) The second one proceeded from a study on the intertidal benthic communities carried out from February 1993 to February 1994 in the south arm of the estuary (Figure 11.5). Samples of macrophytes, macroalgae, and associated macrofauna, as well as samples of water and sediments, were taken fortnightly at different sites, during low water, along a spatial gradient of eutrophication symptoms, from a noneutrophied zone, where a macrophyte community (*Zostera noltii*) is present, up to a heavily eutrophied zone, in the inner areas of the estuary, from where the macrophytes disappeared while *Enteromorpha* spp. (green macroalgae) blooms have been observed during the last decade. In this area, as a pattern, *Enteromorpha* spp. biomass normally increases from early winter (February/March) up to July, when an algal crash usually occurs. A second but much less important algal biomass peak may sometimes be observed in September, followed by a decrease up to the winter (Marques et al. 1997).

In both studies organisms were identified to the species level and their biomass was determined (g · m^{-2} [Ash Dry Free Weight] AFDW). Corresponding to each biological sample the following environmental factors were determined: salinity, temperature, pH, dissolved oxygen, silica, chlorophyll a, ammonia, nitrates, nitrites, and phosphates in water, and organic matter content in sediments. In addition, aiming specifically at estimating Ascendency, data on epiphytes, zooplankton, fish, and birds were collected from different

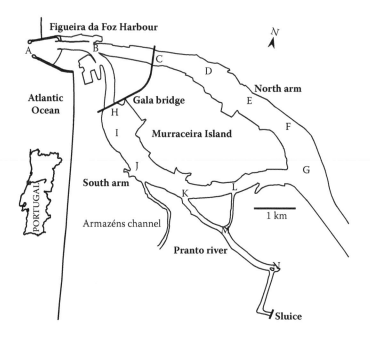

FIGURE 11.4
The Mondego estuary. Location of the subtidal stations in the estuary.

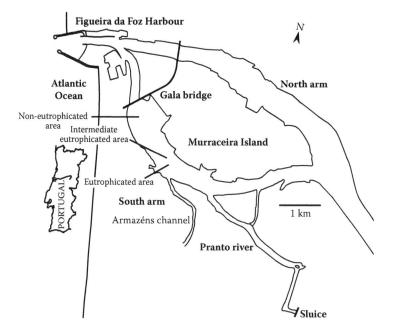

FIGURE 11.5
The Mondego estuary. Location of the intertidal stations in the south arm.

sources (e.g., Azeiteiro 1999; Jorge et al. 2002; Lopes et al. 2000; Martins et al. 2001) taken from April 1995 to January 1998.

Regarding the Mar Menor case study, a single data set was used. In this system, biological communities are adapted to more extreme temperatures and salinities than those found in the open sea. Furthermore, some areas in the lagoon present high levels of organic pollution proceeding from direct discharges, while other zones exhibit accumulations of organic materials originating from biological production of macrophytes meadows. Besides these areas, we can find other communities installed on rocky or sandy substrates that do not present any significant influence of organic matter enrichment.

To estimate different ecological indicators we used, in this case, data from Pérez-Ruzafa (1989), as they have the advantage of being a complete characterization of the benthic populations in the lagoon with the information needed for a study such as the present one. The subtidal benthic communities were sampled at six stations, located on soft substrates along the lagoon, representative of the different biocoenosis and main polluted areas (Figure 11.6). In station M3, samples were taken in July (A), February (B), and May (D).

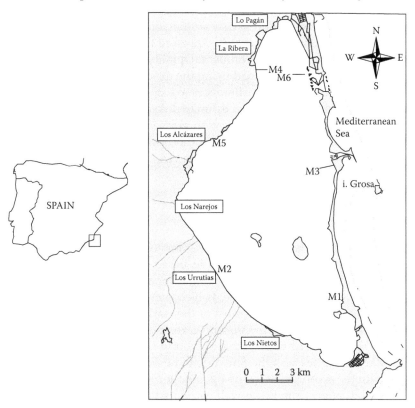

FIGURE 11.6
Location of the different stations in the Mar Menor.

Likewise with the Mondego estuary case study, organisms were identified to the species level and their biomass was determined ($g \cdot m^{-2}$ AFDW). The environmental factors taken into account were salinity, temperature, pH, and dissolved oxygen, as well as sediment particle size, organic matter, and heavy metal contents.

11.4.2 Selected Ecological Indicators

In each case we selected ecological indicators representative of each of the groups characterized above and capable of evaluating the system from different perspectives. The discussion with regard to their applicability in each system was based on the potential of each ecological indicator to react positively to different stress situations.

The following ecological indicators were used in both case studies: AMBI, polychaetes/amphipods ratio, Shannon–Wiener index, Margalef index, ABC method (by means of W-statistic), exergy, and specific exergy (Table 11.2). To estimate exergy and, subsequently, specific exergy from organisms' biomass, we used a set of weighing factors (β), as discussed in Chapter 2. For reasons of comparison between different case studies, all of them dated from the last 10 years, exergy estimations are still expressed taking into account the old β values. In fact, in terms of environmental quality evaluation, the relative differences between values obtained using the new β values or the old ones are minor, although the absolute differences are significant. Finally, only in the case of the Mondego estuary, we estimated Ascendency at three intertidal sampling areas along the eutrophication gradient in the south arm. Possible relations between values of the different indicators used and the ecological status of ecosystems are provided in Table 11.4.

11.4.3 Summary of Results

11.4.3.1 Mondego Estuary

We focused first on the analysis of the subtidal communities from both arms of the estuary (first data set). As a whole, based on the comparison between results from the 1998 and 2000 sampling campaigns, all the indicators estimated, with the exception of the Polychaetes/Amphipods index (which could not have been applied to most of the stations anyway), indicated in a few cases some changes in the system, corresponding to a different pattern of species spatial distribution (Table 11.5A and B).

The Margalef index was the only one to be significantly correlated to the others, with the exception of the AMBI and the Exergy indices. The Shannon–Wiener index, apart from being well correlated to Margalef's, showed a pattern of variation similar to the W-statistic index. The AMBI values appeared

TABLE 11.4

Possible Relations between Indicators Values and Environmental Quality Status of Ecosystems

Indices	Ecological Status
AMBI	Unpolluted: 0–1 Slightly polluted: 2 Meanly polluted: 3–4 Heavily polluted: 5–6 Extremely polluted: 7
Polychaetes/ amphipods ratio	≤1: nonpolluted >1: polluted
Shannon–Wiener Index	Values most often vary between 0 and 5 bits·individual^{-1}. Resulting from many observations, an example of a possible relation between values of this index and environmental quality status could be: 0–1: bad status 1–2: poor status 2–3: moderate status 3–4: good status >4: very good status This is, of course, subjective and must be considered with extreme caution.
Margalef Index	High values are usually associated with healthy systems. Resulting from many observations, an example of a possible relation between values of this index and environmental quality status could be: <2.5: bad to poor status 2.5–4: moderate status >4: good status This is, of course, subjective and must be considered with extreme caution.
W-statistics	The index can take values from +1, indicating a nondisturbed system (high status), to –1, which defines a polluted situation (bad status). Values close to 0 indicate moderate pollution (moderate status).
Exergy index and Specific Exergy	Higher values are usually associated with healthy systems, but there is no rating relationship between values and ecosystem status.
Ascendency	Higher values are usually associated with healthy systems, but there is no rating relationship between values and ecosystem status.

as negatively correlated with the Specific Exergy (Table 11.6). This suggests that most of the information expressed by specific exergy was related to the dominance of taxonomic groups usually absent in environmentally stressed situations. This uneven relationship between different indices can be recognized in the following cases:

1. Following the temporal variation of the communities at the different stations, although the diversity indices and the W-statistics show,

with regard to station A, that there is a worsening of the system between 1998 and 2000 (Table 11.5A and B), the AMBI, the Exergy index, and Specific Exergy suggest, on the contrary, an improvement. In fact, in 1998 the AMBI reveals co-dominance among species of group I (54.2%), group II (10.8%), and group III (35.0%), while in 2000 only group I (51.3%) and group II (48.7%) had been represented. The decrease in environmental quality described by the other indices is basically due to dominance of *Elminius* spp. in station A during 2000. Actually, although this species does not indicate any kind of pollution, its abundance caused a decrease in diversity values, as the Shannon–Wiener index depends on species richness and evenness. Also, the W-statistic was influenced by the dominance of *Elminius* spp. because, by coincidence, these species are very small in size. The increase in the values of the Exergy index and Specific Exergy was fundamentally due to the increase in the biomass of species from groups such as mollusks and equinoderms, which have higher β factors.

2. Additionally, according to the diversity indices and W-statistics, in stations B and C the environmental quality of the system should be improving (Table 11.5A and B), while AMBI shows a worsening. In the case of station B, the decline occurs drastically (from 1.90, in 1998, to 3.5 in 2000), changing from what could be considered an unbalanced community, in which species belonging to ecological group I prevailed (42.9%), to a transitional pollution state, revealed by the dominance of species of ecological groups III (43.8%) and IV (41.6%). Station C also changed to a transitional pollution or even meanly polluted situation (AMBI: 3.9) as a function of the dominance of ecological groups III (48.8%), IV (41.5%), and V (9.7%). With regard to the Exergy index and Specific Exergy, results point to an improvement in station B, this being coincident with the information provided by diversity measures and the W-statistic indices, while they revealed a worsening in station C, similarly to AMBI.

By applying a one-way analysis of variance (ANOVA) to the 1998 results (Table 11.7), we can verify that diversity indices and the W-statistic were efficient in distinguishing between stations from the north and south arms of the estuary, although values estimated for the south arm consistently indicated a higher disturbance, which is contradictory to our knowledge regarding the system reality. With regard to AMBI, Exergy index, and Specific Exergy, differences between both arms of the estuary were not statistically significant. On the other hand, regarding the 2000 results, none of ecological

TABLE 11.5A

Values of the Different Indices Estimated at the 14 Sampling Stations in the Mondego Estuary, Campaigns from 1998

Station	Polychaetes/ Amphipods Ratio	AMBI	Shannon– Wiener	Margalef	W-Statistics	Exergy Index	Specific Exergy
A	—	1.21	2.64	2.32	0.27	214.08	99.75
B	—	1.90	2.45	1.08	0.40	31.59	218.84
C	—	3.10	1.36	0.89	0.21	21.39	122.61
D	0.82	2.70	2.77	1.99	0.59	3416.39	230.27
E	—	1.70	2.14	1.26	0.30	59.59	59.13
F	—	1.60	2.61	1.55	−0.05	6.33	202.32
G	—	3.00	0.87	0.60	0.18	3.55	222.38
H	—	7.00	0.00	0.00	−1.00	5.76	450
I	—	2.00	1.43	0.94	−0.15	6.53	159.35
J	—	3.13	2.03	1.07	−0.06	33.29	165.58
K	—	2.02	1.91	1.25	0.22	15.31	10.98
L	—	3.00	1.66	0.81	−0.04	310.90	119.26
M	—	2.94	1.32	0.98	−0.20	72.35	179.68
N	—	3.00	0.63	0.72	−0.18	3.131	146.37

TABLE 11.5B

Values of the Different Indices Estimated at the 14 Sampling Stations in the Mondego Estuary, Campaigns from 2000

Station	Polychaetes/ Amphipods Ratio	AMBI	Shannon– Wiener	Margalef	W-statistics	Exergy Index	Specific Exergy
A	—	0.73	0.90	1.44	−0.19	3528.27	276.30
B	2.38	3.5	3.44	4.01	0.20	3424.53	217.43
C	—	3.9	2.40	1.52	0.23	4.52	50.90
D	—	2.3	1.84	0.89	0.39	1.95	220.86
E	—	2.4	0.65	0.27	−0.50	2.48	321.82
F	—	0.75	1.37	0.66	0.20	2.30	145.61
G	—	3	2.03	1.23	0.19	16.16	175.20
H	—	2.3	2.55	1.73	0.45	15.04	65.44
I	1.60	2.6	2.92	1.99	0.50	31.09	348.10
J	—	3	2.51	1.34	0.24	427.15	215.13
K	—	2.9	1.46	1.02	0.06	307.04	200.52
L	—	3	2.39	1.43	0.11	85.18	82.85
M	—	2.8	1.68	1.14	−0.09	7.22	69.52
N	—	3	1.38	0.79	0.24	1.67	1.82

TABLE 11.6

Pearson Correlations between the Values of the Different Indicators Estimated in 1998 and 2000 at the 14 Sampling Stations Located in the Two Arms of the Mondego Estuary

	AMBI	Shannon–Wiener	Margalef	W-Statistics	Exergy Index
Shannon–Wiener	+0.36				
Margalef	+0.20	+0.83**			
W-statistics	–0.18	+0.75**	+0.72*		
Exergy	+0.22	+0.46	+0.68**	+0.27	
Specific Exergy	–0.76**	–0.23	–0.46	–0.60**	+0.15

$* P \leq 0.05; ** P \leq 0.01.$

TABLE 11.7

Values Obtained after the Application of a One-Way ANOVA Test Considering the Sampling Stations Located in the Two Arms of the Mondego Estuary in 1998

	n	Mean	F	P
Shannon–Wiener				
North arm	6	2.32	10.47	0.007
South arm	8	1.23		
Margalef				
North arm	6	1.51	8.40	0.013
South arm	8	0.79		
W-statistics				
North arm	6	0.28	6.53	0.025
South arm	8	–0.15		
Exergy Index				
North arm	6	536.13	4.74	0.34
South arm	8	63.89		
Specific Exergy				
North arm	6	165.04	4.74	0.89
South arm	8	175.89		
AMBI				
North arm	6	2.03	2.65	0.13
South arm	8	3.38		

indicators were able to capture the differences between stations of both arms (Table 11.8).

With regard to the relationship between physicochemical factors and the variation of ecological indicators, we may observe that salinity and temperature were significantly correlated to the values of the Shannon–Wiener index

TABLE 11.8

Values Obtained after the Application of a One-Way ANOVA Test Considering the Sampling Stations Located in the Two Arms of the Mondego Estuary during 2000

	n	Mean	F	P
Shannon–Wiener				
North arm	6	1.76	0.65	0.43
South arm	8	2.15		
Margalef				
North arm	6	1.46	0.07	0.79
South arm	8	1.33		
W-statistics				
North arm	6	0.05	1.23	0.28
South arm	8	0.21		
Exergy Index				
North arm	6	997.17	1.84	0.20
South arm	8	124.91		
Specific Exergy				
North arm	6	201.16	1.16	0.30
South arm	8	140.48		
AMBI				
North arm	6	2.26	1.39	0.26
South arm	8	2.82		

($r = +0,81$; $P < 0.01$, with salinity), Margalef index ($r = +0.78$; $P < 0.05$, with salinity), and AMBI ($r = +0,9$; $P < 0.01$, with salinity, and $r = -0,93$; $P < 0.01$ with temperature).

Let us consider now the intertidal communities along the gradient of eutrophication symptoms in the south arm of the estuary (second data set) (Figure 11.7). In this case, despite different patterns of variation, with the exception of the AMBI and the polychaetes/amphipods ratio, the indicators used were able to differentiate between the three sampling areas along the south arm, as shown by a one-way ANOVA (Table 11.9). The Margalef index, as well as the Exergy index and Specific Exergy, behaved as expected, exhibiting higher values at the *Zostera noltii* beds and lower values in the inner areas of the south arm. However, contrary to expectations, the Shannon–Wiener and the W-statistics showed higher values in the most heavily eutrophied zone ($x = 1.69$, $x = 0.48$, and $x = 0.04$, respectively) than in the *Z. noltii* beds ($x = 0.78$, $x = 0.79$, and $x = -0.01$, respectively).

Regarding Ascendency, we could recognize a similar pattern of spatial variation along the gradient of eutrophication in the south arm of the estuary,

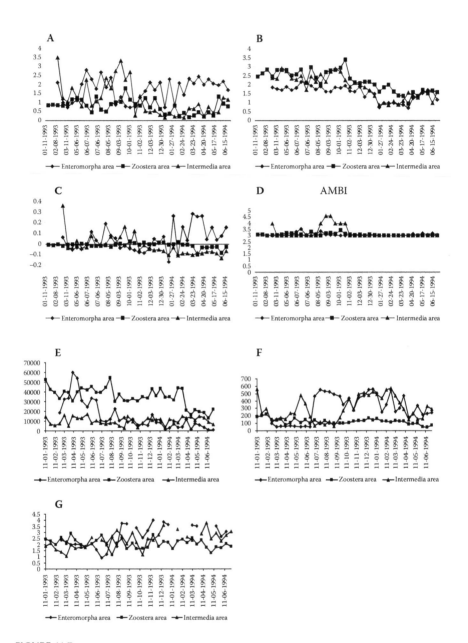

FIGURE 11.7
Temporal and spatial variation of the Shannon–Wiener index (A), Margalef index (B), W-statistic (C), AMBI (D), Exergy (E), Specific Exergy (F), and polychaetes/amphipods ratio (G) in the south arm of the Mondego estuary.

exhibiting a higher value in the noneutrophied area (16,549 g AFDW m^{-2} y^{-1}, bits; 42.3% of the total development capacity), followed by the heavily eutrophied zone (3976 g AFDW m^{-2} y^{-1}, bits; 36.7% of the total development

TABLE 11.9

Values Obtained after the Application of One-Way ANOVA Test Considering the Three Sampling Areas Located along the Spatial Gradient of Eutrophication Symptoms in the South Arm of the Mondego Estuary, in 1993–1994

	n	Mean	F	P
Shannon–Wiener				
Noneutrophicated area	35	0.78	17.12	0.00003
Eutrophicated area	35	1.69		
Intermedia area	35	1.14		
Margalef				
Noneutrophicated area	35	2.17	13.78	0.00004
Eutrophicated area	35	1.52		
Intermedia area	35	1.86		
Polychaetes/amphipods ratio				
Noneutrophicated area	35	2.08	6.46	0.0002
Eutrophicated area	35	2.67		
Intermedia area	35	2.39		
W-statistics				
Noneutrophicated area	35	−0.01	6.27	0.002
Eutrophicated area	35	0.04		
Intermedia area	35	−0.02		
Exergy Index				
Noneutrophicated area	35	14893.58	6.23	0.0006
Eutrophicated area	35	35048.9		
Intermedia area	35	10143.89		
Specific Exergy				
Noneutrophicated area	35	120.96	20.28	0.00008
Eutrophicated area	35	308.54		
Intermedia area	35	296.99		
AMBI				
Noneutrophicated area	35	3.07	3.36	0.06
Eutrophicated area	35	3.07		
Intermedia area	35	3.23		

capacity). The lowest values were found in the intermediate eutrophicated area (1731 g AFDW m^{-2} y^{-1}, bits; 30.4% of the total development capacity).

As mentioned above, the AMBI index was unable to distinguish those three areas since estimated values of the AMBI were close to 3, which indicates slightly polluted scenarios where species of the ecological group III are expected to dominate (Borja et al. 2000). Exceptionally, AMBI values between 4 and 5 were estimated from 22 July to 1 October at station B (Figure 11.7D),

TABLE 11.10

Pearson Correlations between the Values of the Different Indices Estimated in
1993–1994 at the Three Sampling Areas along the Spatial Gradient of Eutrophication
Symptoms in the South Arm of the Mondego Estuary

	AMBI	Shannon–Wiener	Margalef	W-Statistics	Exergy Index	Specific Exergy
Shannon–Wiener	+0.35*					
Margalef	−0.40**	+0.31*				
W-statistics	+0.22*	+0.82**	+0.21*			
Exergy	−0.16	−0.36**	+0.36**	−0.21*		
Specific Exergy	+0.21	+0.28**	−0.14	+0.14	−0.59**	
Polychaetes/ amphipods ratio	−0.02	+0.04	+0.17	+0.06	+0.16	−0.13

* $P \leq 0.05$; ** $P \leq 0.01$.

located in the intermediate eutrophied zone, which indicates a meanly polluted situation. Moreover, the AMBI showed an opposite pattern of variation
in relation to the other indicators used, as demonstrated by Pearson correlations estimated (Table 11.10). This was exactly the contrary of the observed
with regard to the subtidal communities, when AMBI showed a similar
response to the Shannon–Wiener index and Specific Exergy.

As for the polychaetes/amphipods ratio, it expressed the existence of a
eutrophication gradient, exhibiting lower values in the *Z. noltii* beds and
higher values in the intermediate and most eutrophied areas, but has not
been sensitive enough to distinguish between these ($P \leq 0.05$) (Table 11.9).
Finally, with regard to relationships with physicochemical factors, the
Shannon–Wiener index and W-statistic showed significant correlation with
ammonium, nitrite, and nitrate concentrations in the water column, while
the Margalef index, the Exergy Index, and Specific Exergy were significantly
correlated with phosphate concentrations levels (Table 11.11).

11.4.3.2 Mar Menor

The values of the different environmental parameters analyzed showed that
the areas mostly affected by organic enrichment correspond to stations M1
and M3, where organic matter content in sediments reaches values higher
than 5%, and they also have in common dominance of the polychaetes taxonomic group, *Heteromastus filiformis* being the most abundant species. We
should then expect there to be the occurrence of lower values of the Exergy
index and Specific Exergy, diversity measures, and W-statistic, as well as
higher ones for AMBI and the Polychaetes/Amphipods.

This fact was confirmed for all of them in station M3, but it was not in station M1, where only W-statistics and Margalef indices obtain the lower values (Table 11.12). Besides, the latter index is the only one capable of detecting

TABLE 11.11

Values Obtained after the Application of the Pearson Correlations between the Different Indices, and between the Indices of the Different Environmental Variables in South Arm of Mondego Estuary

	Temperature	NH_4^+	NO_2^-	NO_3^-	$PO_4^=$
Polychaetes/amphipods ratio	+0.13	−0.06	−0.15	−0.002	−0.06
AMBI	+0.03	−0.10	−0.20*	−0.15	−0.15
Shannon–Wiener	+0.11	−0.27**	−0.26**	−0.34**	+0.11
Margalef	−0.03	−0.19*	−0.33**	−0.04	−0.34**
W-statistics	−0.06	−0.31**	−0.15	−0.20*	−0.01
Exergy Index	−0.11	−0.16	−0.01	+0.10	−0.25*
Specific Exergy	0.05	+0.14	+0.10	−0.15	+0.22*

* $P \le 0.05$; ** $P \le 0.01$.

TABLE 11.12

Values of the Different Indices Estimated at the Sampling Stations in the Mar Menor

Station	Polychaetes/ Amphipods Ratio	AMBI	Shannon– Wiener (bits/indvs)	Margalef	W-Statistics	Exergy Index (g*m^{-2}det. energy equiv.)	Specific Exergy (ex/unit. biomass)
M1	—	2.10	2.75	5.34	−0.3	15762446	603402
M2	1.86	3.42	2.06	7.79	0.25	211020	1592
M3A	—	3.5	0	1.28	0.27	285182	14990
M3C	—	3.33	1.44	4.61	−0.15	94659	92702
M3D	2.20	3.32	1.9	8.49	−0.01	1555244	109065
M4	1.32	2.08	3.54	10.05	0.31	70381987	67160
M5	—	2	2.75	8.32	−0.11	2523455	14250
M6	—	3.45	3.75	12.16	0.24	28713101	78518

significant differences between polluted and nonpolluted areas (Table 11.13). AMBI values show similar values in all the stations, with slight disturbance in stations M1, M4, and M5 and moderate for the other stations. This similarity in results for all the stations has made it impossible to distinguish different situations of disturbance in that area. In spite of this, it shows a positive response to the organic matter content in the sediment, but in an insignificant manner ($r = +0.41$; $P > 0.05$). Meanwhile, the polychaetes/amphipods ratio could only be applied in stations M3 (during spring), M4, and M2. In the other stations, the absence of amphipods originated values tending to infinitum, therefore, without any meaning.

The W-statistic gives somewhat confusing results as station M1, which has organic matter content, presents lower values than M3, which is seen as the most polluted one ($W = −0.3$).

TABLE 11.13

Values Obtained after the Application of One-Way ANOVA Test
Considering the Nonpolluted and Polluted Stations in the Mar Menor

	n	Mean	F	P
Shannon–Wiener				
Nonpolluted area	4	1.52	4.69	0.07
Polluted area	4	3.02		
Margalef				
Nonpolluted area	4	4.93	6.98	0.04
Polluted area	4	9.58		
W-statistics				
Nonpolluted area	4	0.17	6.01	0.06
Polluted area	4	−0.15		
Exergy Index				
Nonpolluted area	4	4424382.75	1.57	0.25
Polluted area	4	25457390.8		
Specific Exergy				
Nonpolluted area	4	205039.75	1.47	0.27
Polluted area	4	40380		
AMBI				
Nonpolluted area	4	2.73	0.39	0.55
Polluted area	4	3.06		

In general terms, we could only recognize a similar pattern of variation
between diversity measures and the Exergy index, which showed positive
and significant correlations (Table 11.14). These indicators were also negative and significantly correlated ($P < 0.05$) with the organic matter content
in the sediment, as well as with other structuring factors of the system, such
as salinity or, in the case of the Margalef and Shannon–Wiener indices, also
with sediments particle size (Table 11.15A and B). Specific Exergy showed a
clear positive correlation with the presence of certain heavy metals such as
Pb ($r = +0.89$; $P \leq 0.05$) and Zn ($r = +0.71$; $P \leq 0.05$), which does not correspond
to what we should expect. For instance, stations M2D, which presented the
highest concentration of these two heavy metals, also exhibited the higher
value of Specific Exergy.

Regarding the Exergy index values, the influence of biomass losses or gains,
which are related to numerical changes in the dominant populations in environmentally stressed situations, is much more important than fluctuations
in the β factors, which are related to the quality of the biomass in the system.
In the case of Specific Exergy, the influence of such biomass fluctuations is
very much diluted, as the β factors related to the quality of the biomass take

TABLE 11.14

Pearson Correlations between the Values of the Different Indices Estimated in the Mar Menor

	AMBI	Shannon–Wiener	Margalef	W-Statistics	Exergy Index
Shannon–Wiener	−0.50				
Margalef	−0.15	+0.88**			
W-statistics	+0.33	+0.01	+0.25		
Exergy	−0.49	+0.66	+0.52	+0.40	
Specific Exergy	−0.44	+0.20	−0.16	−0.66	+0.05

* $P \leq 0.05$; ** $P \leq 0.01$.

TABLE 11.15A

Values Obtained after the Application of the Pearson Correlations between the Different Indexes and the Content of Organic Matter and Heavy Metals in Stations of the Mar Menor

	Organic Matter	Cd	Pb	Cu	Mn	Zn
Exergy Index	−0.49*	0.33	−0.36	−0.31	−0.17	−0.42
Specific Exergy	0.30	+0.35	+0.81**	+0.60	+0.50	+0.71**
Shannon–Wiener	−0.67*	0.14	−0.22	−0.09	0.06	−0.34
Margalef	−0.68*	0.11	−0.44	−0.17	−0.13	−0.48
AMBI	+0.35	+0.34	+0.07	+0.39	−0.29	+0.28
W-statistics	−0.48	+0.17	−0.52	−0.25	−0.27	−0.46

* $P \leq 0.05$; ** $P \leq 0.01$.

precedence. In this sense, the loss of the taxonomic groups affected by toxic substances, as a function of different degrees of tolerance, will deeply affect specific exergy values. Now, mollusks, namely bivalves, are known by their ability to bio-accumulate heavy metals, which is not the case in, for instance, polychaetes, crustaces, and equinodermes. Since the β factor for mollusks is higher than the other groups (see Chapter 2), it becomes immediately easy to understand why specific exergy was found to be higher in areas affected by heavy metals pollution.

11.5 Was the Use of the Selected Indicators Satisfactory in the Two Case Studies?

In order to compare the efficiencies of the ecological indicators used in each of the case studies, we considered it suitable to evaluate a basic property: their ability to reflect different stress situations.

TABLE 11.15B

Values Obtained after the Application of the Pearson Correlations between the Different Indices, and between the Indices of the Different Environmental Variables in Stations of the Mar Menor Sampling Stations

	Salinity	Temperature	Hydrodynamism	Dissolved Oxygen	Suspension Material	Gravel (%)	Sand (%)	Silt (%)	Clay (%)
Exergy Index	−0.60*	+0.02	+0.10	+0.06	−0.22	+0.02	+0.12	−0.39	−0.59
Specific Exergy	+0.27	+0.50	−0.47	+0.20	−0.38	+0.92*	+0.23	−0.32	−0.14
Shannon–Wiener	−0.61*	+0.40	+0.34	+0.09	+0.26	+0.27	+0.60	−0.70*	−0.76**
Margalef	−0.60*	+0.28	+0.51	+0.22	+0.50	+0.006	+0.57	−0.77*	−0.71**
AMBI	−0.47	+0.16	+0.21	+0.22	+0.29	−0.32	−0.36	+0.29	+0.43
W-statistics	−0.31	+0.16	+0.14	+0.09	+0.16	−0.49	+0.12	−0.38	−0.42

* $P \leq 0.05$; ** $P \leq 0.01$.

In light of our results, we may reasonably consider that all the indicators, with the exception of the polychaetes/amphipods ratio, worked satisfactorily when considered separately, although in several cases the information provided was contradictory.

11.5.1 Application of Indicators Based on the Presence versus Absence of Species: AMBI

As a whole, the AMBI worked reasonably well, although it was inefficient in discriminating among areas with clearly different eutrophication symptoms along the spatial gradient in the south arm of the Mondego estuary (e.g., dominance of *Z. noltii* versus *Enteromorpha* spp. as main primary producers), or between stations affected by organic enrichment (covered by *Cymodocea-Caulerpa* meadows) from those presenting sediments with low organic matter content in the Mar Menor. In the Mondego estuary case this fact may perhaps be explained if we consider that eutrophication effects at the level of primary producers, which are clearly visible, are still not strong enough at other trophic levels to be detected by AMBI. In fact, although a number of shifts in species composition are already recognizable, in qualitative terms, the benthic community structure in the three zones along the spatial gradient still exhibits, to a certain extent, a reasonably similar arrangement regarding the macrofaunal species (Marques et al. 2003). Nevertheless, regarding other impact sources, such as outfalls, oil platforms, etc., it has been demonstrated that AMBI clearly shows the stress gradient (Borja, Muxika, et al. 2003). The AMBI values estimated in the Mondego estuary were similar at the three sampling areas because of the common dominance of *Hydrobia ulvae*, which belongs to ecological group III. Besides, all the other indicators were strongly affected by large abundances of *H. ulvae* and *Cerastoderma edule*, the dominant species, although such dominance does not have anything to do with pollution, being related to higher resources availability instead (Pardal et al. 2000). In the case of Mar Menor, a sampling station such as M1, with a remarkable amount of organic matter in the sediment, was evaluated as slightly polluted because of co-dominance between species belonging to the genus *Bittium*, belonging to ecological group I, and polychaete species, belonging to group IV. The reason for this is that *Bittium* species have a herbivorous trophic strategy because of the available food resources provided by the presence of *Caulerpa-Cymodocea*, while polychaetes usually tend to be favored by sediment organic enrichment. Apart from that, in both study sites, AMBI did not vary with time, therefore being less influenced by seasonal changes in abundance than the other indicators.

11.5.2 Indices Based on Ecologic Strategies: Polychaetes/Amphipods Ratio

The polychaetes/amphipods ratio was able to reflect correctly the existence of a eutrophication gradient in the case of the intertidal communities in the south arm of the Mondego estuary, but in the other two studies it was impossible to apply it simply because of the absence of amphipods in the samples. The ratio in this case would reflect an extremely polluted scenario, which we knew for sure not to be the case. This indicator has been successfully used to detect the effects of organic and oil pollution on subtidal communities at the Bay of Morlaix (Mediterranean Sea) and at the Ría de Area and Betanzos (Atlantic Ocean), but in the present case studies the best results were provided when applying it to intertidal data. This indicator, as for instance the nematodes/copepods ratio used in relation to meiobenthic communities, is probably influenced by a large spectrum of ecological factors, in which perhaps some types of pollution are included, meaning that this simplistic ratio is inadequate and difficult to relate to environmental quality.

11.5.3 Biodiversity as Reflected in Diversity Measures: Margalef and Shannon–Wiener Indices

Regarding diversity measures, results showed that the old Margalef index, despite its simplicity, was the one with better performance, being more sensitive in distinguishing different eutrophication levels, in the case of the Mondego estuary, and in detecting organic enrichment situations, in the case of the Mar Menor marine lagoon. In fact, the more complex Shannon–Wiener index has been influenced too much by the dominance of certain species (e.g., *H. ulvae* in the Mondengo estuary, or *Bittium reticulatum* in the Mar Menor lagoon), whose presence has no relation with any type of disturbance or pollution phenomenon, being rather favored by abundant food resources.

11.5.4 Indicators Based on Species Biomass and Abundance: W-Statistic

Lastly, the W-statistic was capable, to a certain extent, of distinguishing between nondisturbed, slightly disturbed, and strongly disturbed situations. Moreover, it does not depend on reference values previously known. Nevertheless, the dominance of certain species with small size individuals which are characteristic of nonpolluted environments, is not unusual (as illustrated by the Mondego estuarine benthic community) will lead to erroneous evaluations regarding the environmental quality status. This problem was in fact perceived in several case studies (Ibanez and Dauvin 1988; Beukema 1988; Weston 1990; Craeymeersch 1991). The reason this indicator was not very successful in detecting organic pollution in the Mar Menor lagoon may rely on the fact that it was exclusively developed to evaluate the impact of organic pollution, since in Mar Menor, although sediment organic enrichment is a concern, there are other types of pollution (e.g., metallic pollution) and different types of environmental stress occurring.

11.5.5 Thermodynamically Oriented and Network-Analysis-Based Indicators: Exergy Index, Specific Exergy, and Ascendancy

11.5.5.1 *Exergy and Specific Exergy*

As a whole, our results suggest that the Exergy index is able to capture useful information about the state of the community. In fact, more than a simple description of the environmental state of a system, the spatial and temporal variations of the Exergy index may provide us a much better understanding of the system development in the scope of a broader theoretical framework. However, at the present stage, in simple snapshots, the Exergy index can hardly provide an explicit evaluation about disturbed (e.g., polluted) versus nondisturbed scenarios. For instance, in the case of the Mar Menor marine lagoon, despite responding to sediment organic enrichment, both the Exergy index and Specific Exergy were unable to distinguish between areas affected by organic pollution and areas that were not. Nevertheless, regarding the intertidal communities of the south arm of the Mondego estuary, both Exergy-based indicators were able to distinguish between areas with different eutrophication symptoms. Differences in efficiency in both case studies might have been due to the fact that in the Mar Menor lagoon the effects of organic pollution are to a certain extent diluted among other system-structuring factors, while in the south arm of the Mondego estuary eutrophication is indubitably the major driving force behind the ongoing changes.

Finally, it is interesting to observe that Specific Exergy appeared positively correlated to heavy metals contamination, such as Pb and Zn, while the Exergy index did not, which was because of their different response to biomass fluctuations in the community. In fact, the influence of such fluctuations on Specific Exergy values is much less important because the weighting factors (β) expressing the quality of biomass take precedence.

11.5.5.2 *Ascendancy*

Ascendancy was tested only with regard to the Mondego estuarine intertidal communities, but nevertheless the results suggest that it is able to capture useful information about the system, namely distinguishing between areas along the eutrophication gradient. It was possible to observe that the *Z. noltii*–dominated community clearly presented the highest value, which was in accordance with theoretical expectations. Nevertheless, no test of statistical significance can (until now) be applied to the differences between the values pertaining to different areas, because of the complexity of comparing information-theoretic combinations. Although this approach is a powerful theoretical tool, this inconvenience limits *a posteriori* demonstration that there are statistically significant differences to interpret.

11.5.6 Brief Conclusions

When selecting an ecological indicator, we must first account for its dependence on external factors that escape our control, such as the need for reference values that often do not exist, or particular characteristics regarding the habitat type. As a result, we may reasonably conclude that no indicator will be valid in all situations. Therefore, different ranges of values will have distinct significance in different scenarios, meaning that diverse classification schemes should be applied to different habitat types.

When evaluating the health status of an ecosystem our task can be greatly facilitated if we select indicators that do not depend on any reference conditions to establish the Ecological Quality Status. The polychaetes/amphipods ratio, as well as diversity measures, Exergy index or Specific Exergy, and Ascendancy, all tested by us, do not fulfill this requirement. But that is, for instance, the case with AMBI and W-statistic. This, and the fact that they are also independent from the type of habitat, make them at first sight very suitable indicators.

The inconvenience of AMBI is that the classification of species as indicators of different grades of pollution, which constitutes its base, often contains very subjective elements, and interpreting the meaning of the presence of a given species may be ambiguous. For instance, *Chaetozone setosa*, depending on the authors, is considered an indicator of moderate pollution (Bellan 1967; Solís-Weiss 1982) or of intense pollution (Glemarec and Hily 1981; Glemarec et al. 1982; Majeed 1987). Also, *Spiochaetopterus costarum* is considered by Bellan (1967) to be an indicator of slightly polluted environments, and by López-Jamar (1985) to be a characteristic of highly polluted areas. Similarly, *Nereis caudata* is considered to be an indicator of intense pollution by Bellan (1967), Zabala et al. (1983), and Lardicci et al. (1993), and simply tolerant by Glemarec and Hily (1981), Glemarec et al. (1982), and Majeed (1987). On the other hand, the W-statistic holds the inconvenience of being strongly affected by the dominance of certain small-sized species characteristic of nonpolluted environments, which leads to inevitable bias.

Our application of diversity measures, as expression of biodiversity, was only partially successfully, with the simpler Margalef index performing nevertheless better than the more complex Shannon–Wiener. We tested only two indicators, but anyway many other tries proposing new ways to estimate diversity couldn't provide any tangible conceptual progress (see Magurran 1988), and probably there is no conceivable "diversity index" capable of expressing the dynamics of mixed populations, exhibiting stabilized values through space and time. The difficulties may be summarized as follows (Marques et al. 2003):

1. The increase of diversity through time is inevitably gradual, more often than not associated with the emergence and transformation of an organized system, but the decrease of diversity is most frequently abrupt.

2. Looking to the spatial characteristics of ecosystems, we are forced to conclude that it is impossible to have stabilized variance, which may lead us to favor any kind of spectral expression taking into account the way diversity may shift as a function of the space considered. The problem in this case is that each spatial enlargement provides a different spectrum as a function of the characteristics of new sites added to the sample.

3. Since the biosphere is continuous, it is not adequate to set apart "local" diversity (called α diversity) from diversity estimated by pooling discontinuous patches (β diversity) or measured at larger spatial scales (γ diversity), although, to a certain extent, such descriptions may be useful in assessing the state of an ecosystem.

The Exergy index and Specific Exergy constitute theoretically more ambitious ecological indicators, aiming at integrating empirical biological data and ecological observations in terms of a comprehensive thermodynamic hypothesis (Jørgensen and Marques 2001), instead of interpreting results according to a number of nonuniversal generalizations. This point is important because, despite the little respect accorded to it by those in other fields of science, ecology deals with some of the most complex phenomena encountered in modern science. Ecosystem analyses must encompass several disciplines in a coordinated fashion to answer specific questions concerning how large, multidimensional systems work (Livingston et al. 2000; Jørgensen and Marques 2001). Besides, both indicators have been applied in structurally dynamic models of shallow lakes, appearing to represent a promising approach. Theoretically, Exergy storage is assumed to become optimized during ecosystem development, and ecosystems are supposed to self-organize toward a state of optimal configuration in this property (Marques et al. 1997). Since the Exergy index and Specific Exergy may respond differently to environmental stress, as well as to a system's seasonal dynamics, providing different spatial and temporal pictures, it is advisable to use both in complement. There is nevertheless an obvious need for the determination of more accurate (discrete) weighing factors to estimate the Exergy index from the biomass of organisms, which presently constitutes a very strong constraint to the application of these indicators.

We tested Ascendancy in the only case where the available data were enough to estimate it, the Mondego estuarine intertidal communities, which was obviously a very circumscribed application. Data difficulties notwithstanding, network analysis appeared to provide a systematic approach to apprehending what is happening at the whole-system level, which is obviously powerful from the theoretical point of view. Moreover, the current study on the Mondego estuarine ecosystem seems to have provided an example of how the measures coming out of network analysis can lead to an improved understanding of the eutrophication process itself (Patrício et al. 2004). Nevertheless, there is a major inconvenience regarding its use, that is, the extremely considerable amount of time and labor needed to collect all the data necessary to perform network analysis, which greatly limits its application.

Summarizing, we can say that a single approach does not seem appropriate because of the complexity inherent in assessing the environmental quality of a system. Rather, this should be evaluated by combining a suite of ecological indicators, which may provide complementary information. This very same message has intermittently been conveyed to the scientific community working on environmental quality assessment (e.g., Dauer et al. 1993), together with an increasing concern regarding the need for a deeper understanding of ecological processes and for the development of a theoretical network able to explain observations, rules, and correlations on the basis of an accepted pattern of ecosystem theories (Jørgensen and Marques 2001; Marques and Jørgensen 2002).

References

Aubry, A., and M. Elliot. 2006. The use of environmental integrative indicators to assess seabed disturbance in estuaries and coasts: Application to the Humer estuary. *Marine Pollution Bulletin* 53 (1–4): 175–85.

Azeiteiro, U. M. M. 1999. Ecologia pelágica do braço Sul do estuário do rio Mondego. PhD thesis, University of Coimbra.

Bellan, G. 1967. Pollution et peuplements benthiques sur substrat meuble dans la region de Marseille. 1.- Le secteur de Cortiou. *Rev Intern Oceanogr Med* 6:53–87.

———. 1980. Annélides polychétes des substrats solids de troits mileux pollués sur les côrtes de Provence (France): Cortiou, Golfe de Fos, Vieux Port de Marseille. *Téthys* 9 (3): 260–78.

Bellan-Santini, D. 1980. Relationship between populations of amphipods and pollution. *Marine Pollution Bulletin* 11:224–27.

Belsher, T. 1982. Measuring the standing crop of intertidal seaweeds by remote sensing. In *Land and its uses actual and potential: An environmental appraisal*, eds. F. T. Last, M. C. Hotz, and B. Bell, 453–56. Edinburgh: NATO Seminar on Land and Its Uses, Sept. 19–Oct. 1, 10.

Beukema, J. J. 1988. An evaluation of the ABC method abundance/biomass comparison as applied to macrozoobenthic communities living on tidal flats in the Dutch Wadden Sea. *Mar Biol* 99:425–33.

Bongers, T., R. Alkemade, and G. W. Yeates. 1991. Interpretation of disturbance-induced maturity decrease in marine nematode assemblages by means of the Maturity Index. *Marine Ecology Progress Series* 76:135–42.

Bonne, W., A. Rekecki, and M. Vincx. 2003. Impact assessment of sand extraction on subtidal sandbanks using macrobenthos. In *Benthic copepod communities in relation to natural and anthopogenic influences in the North Sea*, 207–26. PhD thesis of W. Bonne, Ghent University, Biology Department, Marine Biology Section, Ghent, Belgium.

Borja, Á., and D. Dauer. 2008. Assessing the environmental quality status in estuarine and coastal systems: Comparing methodologies and indices. *Ecological Indicators* 8 (4): 331–37.

Borja, Á., J. Franco, and I. Muxika. 2003. Classification tools for marine ecological quality assessment: The usefulness of macrobenthic communities in an area affected by a submarine outfall. *ICES CM 2003/Session J-02*, Tallinn, Estonia, September 24–28.

Borja, Á., J. Franco, and V. Pérez. 2000. A Marine Biotic Index to establish the ecological quality of soft-bottom benthos within European estuarine and coastal environments. *Marine Pollution Bulletin* 40 (12): 1100–14.

Borja, Á., I. Muxika, and J. Franco. 2003. The application of a Marine Biotic Index to different impact sources affecting soft-bottom benthic communities along European coasts. *Marine Pollution Bulletin* 46 (7): 835–45.

Brundtland, G.H. (Chair), 1987. Report on the World Commission on Environment and Development: *Our common future*. Transmitted to the General Assembly as an Annex to Document A/42/427 – Development and International Cooperation: Environment. UN Documents: Gathering a Body of Global Agreements; http://www.un-documents.net/wcedocf.htm.

Cardoso, P. G., A. I. Lillebø, M. A. Pardal, S. M. Ferreira, and J. C. Marques. 2002. The effect of different primary producers on *Hydrobia ulvae* population dynamics: A case study in a temperate intertidal estuary. *J Exp Mar Biol Ecol* 277:173–95.

Clarke, K. R., and R. M. Warwick. 1999. The taxonomic distinctness measure of biodiversity: Weighting of step lengths between hierarchical levels. *Mar Ecol Prog Ser* 184:21–29.

Cooper, J. A., T. D. Harrison, A. E. Ramm, and R. A. Singh. 1993. Refinement, enhancement and application of the Estuarine Health Index to Natal's estuaries. Tugela, Mtamvuna. Unpublished technical report. CSIR, Durban.

Craeymeersch, J. A. 1991. Applicability of the abundance/biomass comparison method to detect pollution effects on intertidal macrobenthic communities. *Hydrobiol Bull* 24 (2): 133–40.

Dauer, D. M., M. W. Luckenbach, and A. J. Rodi. 1993. Abundance-biomass comparison ABC method: Effects of an estuarine gradient, anoxic/hypoxic events and contaminated sediments. *Mar Biol* 116:507–18.

Dauvin, J. C., and T. Ruellet. 2007. Polychaete/amphipod ratio revisited. *Marine Pollution Bulletin* 55:215–24.

De Boer, W. F., P. Daniels, and K. Essink. 2001. Towards ecological quality objectives for North Sea benthic communities. National Institute for Coastal and Marine Management (RIKZ), Haren, the Netherlands. Contract RKZ 808, report no. 2001–11.

Deegan, L. A., J. T. Finn, S. G. Ayvazian, and C. Ryder. 1993. Feasibility and application of the Index of Biotic Integrity to Massachusetts Estuaries (EBI). Final report to Massachusetts Executive Office of Environmental Affairs. North Grafton, MA: Department of Environmental Protection.

Engle, V., J. K. Summers, and G. R. Gaston. 1994. A benthic index of environmental condition of Gulf of Mexico. *Estuaries* 172:372–84.

Fano, E. A., M. Mistri, and R. Rossi. 2003. The ecofunctional quality index (EQI): A new tool for assessing lagoonal ecosystem impairment. *Estuar Coast Shelf Sci* 56:709–16.

Feldmann, J. 1937. Recherches sur le vegetation marine de la Méditerranée. La côte des Albères. *Revue Algologique* 10:1–339.

Flindt, M. R., L. Kamp-Nielsen, J. C. Marques, M. A. Pardal, M. Bocci, G. Bendoricho, S. N. Nielsen, and S. E. Jørgensen. 1997. Description of the three shallow water estuaries: Mondego River (Portugal), Roskilde Fjord (Denmark) and the Lagoon of Venice (Italy). *Ecol Model* 102:17–31.

Fuliu, Xu. 1997. Exergy and structural exergy as ecological indicators for the state of the Lake Chalou ecosystem. *Ecol Model* 99:41–49.

Gee, J. M., R. M. Warwick, M. Schaanning, J. A. Berge, and W. G. Ambrose, Jr. 1985. Effects of organic enrichment on meiofaunal abundance and community structure in sublittoral soft sediments. *J Exp Mar Biol Ecol* 91 (3): 247–62.

Glemarec, M., and C. Hily. 1981. Perturbations apportées a la macrofaune benthique de la Baie de Concarneau par les effluents urbains et portuaires. *Acta Oecol Appl* 2 (2): 139–50.

Glemarec, M., E. Hussenot, and Y. Le Moal. 1982. Utilization of biological indicators in hypertrophic sedimentary areas to describe dynamic process after the Amoco Cádiz oil-spill. *International Symposium on Utilization of Coastal Ecosystems* 1–18.

Gómez Gesteira, J. L., and J. C. Dauvin. 2000. Amphipods are good bioindicators of the impact of oil spills on soft bottom macrobenthic communities. *Mar Poll Bul* 40 (11): 1017–27.

Gorostiaga, J. M., Á. Borja, I. Díez, G. Francés, S. Pagola-Carte, and J. I. Sáiz-Salinas. 2004. Recovery of benthic communities in polluted systems. In *Oceanography and marine environment of the Basque country*, Elsevier Oceanography Series 70, eds. Á. Borja and M. Collins, 549–78. Amsterdam: Elsevier.

Gray, J. S. 1979. Pollution-induced changes in populations. *Phil Trans R Soc London* 286:545–61.

Hale, S. S., and J. F. Heltshe. 2008. Signals from the benthos: Development and evaluation of a benthic index for the nearshore Gulf of Maine. *Ecol Indicat* 8:338–50.

Heip, C., M. Vincx, and G. Vraken. 1985. The ecology of marine nematodes. *Oceanography and Marine Biology: An Annual Review* 23:399–489.

Hughes, R. G. 1984. A model of the structure and dynamics of benthic marine invertebrate communities. *Mar Ecol Prog Ser* 15 (1–2): 1–11.

Ibanez, F., and J. C. Dauvin. 1988. Long-term changes 1977–1987 in a muddy fine sand Abra alba-Melina palmata community from the Western Channel: Multivariate time series analysis. *Mar Ecol Prog Ser* 19:65–81.

Jeffrey, D. W., J. G. Wilson, C. R. Harris, and D. L. Tomlinson. 1985. The application of two simple indices to Irish estuary pollution status. In *Estuarine management and quality assessment*, eds. J. G. Wilson and W. Halcrow. London: Plenum Press.

Jorge, I., C. C. Monteiro, and G. Lasserre. 2002. Fisg Community of the Mondego estuary: Space-temporam organisation. In *Aquatic ecology of the Mondego river basin. Global importance of local experience*, eds. M. A. Pardal, J. C. Marques, and M. A. S. Graça, 199–217. Coimbra: Imprensa da Universidade de Coimbra.

Jørgensen, S. E. 1994. Review and comparison of goal functions in system ecology. *Vie Mileu* 44 (1): 11–20.

Jørgensen, S. E., and J. C. Marques. 2001. Thermodynamics and ecosystem theory, case studies from hydrobiology. *Hydrobiologia* 445:1–10.

Jørgensen, S. E., J. C. Marques, and S. N. Nielsen. 2002. Structural changes in an estuary, described by models and using Exergy as orientor. *Ecol Model* 158:233–40.

Jørgensen, S. E., and H. Mejer. 1979. A holistic approach to ecological modelling. *Ecol Model* 7:169–89.

———. 1981. Exergy as a key function in ecological models. In *Energy and ecological modelling. Developments in environmental modelling*, eds. W. Mitsch, R. W. Bosserman, and J. M. Klopatek, 587–90. Amsterdam: Elsevier.

Lambshead, P. J. D., and H. M. Platt. 1985. Structural patterns of marine benthic assemblages and their relationship with empirical statistical models. In *Proceedings of the nineteenth European Marine Biology symposium*, ed., P. E. Gibbs, 16–21. Plymouth.

Lambshead, P. J. D., H. M. Platt, and K. M. Shaw. 1983. The detection of differences among assemblages of marine benthic species based on an assessment of dominance and diversity. *J Nat Hist* 17:859–47.

Lardicci, C., M. Abbiati, R. Crema, C. Morri, C. N. Bianchi, and A. Castelli. 1993. The distribution of polychaetes along environmental gradients: An example from the Ortobello Lagoon, Italy. *Mar Ecol* 14 (1): 35–52.

Livingston, R. J., F. G. Lewis, III, G. C. Woodsum, X. Niu, B. Galperin, W. Huang, J. D. Christensen, M. E. Monaco, T. A. Battista, C. J. Klein, R. L. Howell, IV, and G. L. Ray. 2000. Use of coupled physical and biological models: Response of oyster population dynamics to freshwater input. *Estuar Coast Shelf Sci* 50:655–72.

Lopes, R. J., M. A. Pardal, and J. C. Marques. 2000. Impact of macroalgal blooms and water predation on intertidal macroinvertebrates: Experimental evidence from the Mondego estuary Portugal. *J Exp Mar Biol Ecol* 249:165–79.

López-Jamar, E. 1985. Distribución espacial del poliqueto *Spiochaeterus costarum* en las Rías Bajas de Galicia y su posible utilización como indicador de contaminación orgánica en el sedimento. *Bol Inst Esp Oceanogr* 2 (1): 68–76.

Losovskaya, G. V. 1983. On significance of polychaetes as possible indicators of the Black Sea environment quality. *Ekologiya Moray Publ* 12:73–78.

Magurran, A. E. 1988. *Ecological diversity and its measurement*. London: Croom Helm Limited.

Majeed, S. A. 1987. Organic matter and biotic indices on the beaches of North Brittany. *Mar Pollut Bull* 18 (9): 490–95.

Marques, J. C. 2001. Diversity, biodiversity, conservation, and sustainability. *The Scientific World* 1:534–43.

Marques, J. C., A. Basset, T. Brey, M. Elliot, 2009. The ecological sustainability trigon— A proposed conceptual framework for creating and testing management scenarios. *Marine Pollution Bulletin*, 58: 1773–1779.

Marques, J. C., and S. E. Jørgensen. 2002.Three selected observations interpreted in terms of a thermodynamic hypothesis. Contribution to a general theoretical framework. *Ecol Model* 158:213–22.

Marques, J. C., P. Maranhão, and M. A. Pardal. 1993. Human impact assessment on the subtidal macrobenthic community structure in the Mondego estuary (Western Portugal). *Estuar Coast Shelf Sci* 37:403–19.

Marques, J. C., S. N. Nielsen, M. A. Pardal, and S. E. Jørgensen. 2003. Impact of eutrophication and river management within a framework of ecosystem theories. *Ecol Model* 166:147–68.

Marques J. C, M. A. Pardal, S. N. Nielsen, and S. E. Jørgensen. 1997. Analysis of the properties of exergy and biodiversity along an estuarine gradient of eutrophication. *Ecol Model* 102:155–67.

———. 1998. Thermodynamic orientors: Exergy as a holistic ecosystem indicator: A case study. In *Ecotargets, goal functions and orientors. Theoretical concepts and interdisciplinary fundamentals for an integrated, system based environmental management*, eds. F. Müller and M. Leupelt, Chapter 2.5, 87–101. Berlin: Springer-Verlag.

Marques, J. C., L. B. Rodrigues, and A. J. A. Nogueira. 1993. Intertidal macrobenthic communities structure in the Mondego estuary (Western Portugal): Reference situation. *Vie Mileu* 43 (2–3): 177–87.

Martins, I., M. A. Pardal, A. I. Lillebø, A. R. Flindt, and J. C. Marques. 2001. Hydrodynamics as a major factor controlling the occurrence of green macroalgal blooms in a eutrophic estuary: A case study on the influence of precipitation and river management. *Estuar Coast Shelf Sci* 52:165–77.

McGinty, M., and C. Linder. 1997. An estuarine index of biotic integrity for Chesapeake Bay tidal fish community. In *Biological habitat quality indicators for essential fish habitat*, ed. S. Hartwell. NOAA Technical Memorandum NMFS-F/SPO-32.

McLusky, D., and M. Elliot. 2004. *The estuarine ecosystem. Ecology, threats, and management*. Oxford: Oxford University Press.

Moreno, D., P. A. Aguilera, and H. Castro. 2001. Assessment of the conservation status of seagrass (*Posidonia oceanica*) meadows: Implications for monitoring strategy and the decision-making process. *Biol Cons* 102:325–32.

Muxika I., A. Borja, and J. Bald. 2007. Using historical data, expert judgement and multivariate analysis in assessing reference conditions and benthic ecological status, according to the European Water Framework Directive. *Marine Pollution Bulletin* 55 (1–6): 16–29.

Nelson, W. G. 1990. Prospects for development of an index of biotic integrity for evaluating habitat degradation in coastal systems. *Chem Ecol* 4:197–210.

Niell, F. X., and J. P. Pazó. 1978. Incidencia de vertidos industriales en la estructura de poblaciones intermareales. II. Distribución de la biomasa y de la diversidad específica de comunidades de macrófitos de facies rocosa. *In. Pesq* 42 (2): 231–39.

Nielsen, S. N. 1990. Application of exergy in structural-dynamical modelling. *Vehr Int Ver Limnol* 24:641–45.

———. 1992. Strategies for structural dynamics modelling. *Ecol Model* 63:91–101.

———. 1994. Modelling structural dynamic changes in a Danish shallow lake. *Ecol Model* 73:13–30.

———. 1995. Optimisation of exergy in a structural dynamic model. *Ecol Model* 77:111–112.

Nilsson, H. C., and R. Rosenberg. 1997. Benthic habitat quality assessment of an oxygen stressed fjord by surface and sediment profile images. *J Mar Syst* 11:249–64.

Odum, H. T. 1983. *Systems ecology: An introduction*. Toronto: John Wiley & Sons, Inc.

———. 1996. *Environmental accounting. Emergy and environmental decision making*. New York: John Wiley & Sons, Inc.

Orfanidis, S., P. Panayotidis, and N. Stamatis. 2001. Ecological evaluation of transitional and coastal waters: A marine benthic macrophytes-based model. *Mediterranean Mar Sci* 2:45–65.

———. 2003. An insight to the ecological evaluation index (EEI). *Ecol Indic* 3:27–33.

Pardal, M. A., J. C. Marques, I. Metelo, A. I. Lillebø, and A. R. Flindt. 2000. Impact of eutrophication on the life cycle population dynamics and production of *Amphitoe valida* (Amphipoda) along an estuarine spatial gradient (Mondego estuary, Portugal). *Mar Ecol Progr Ser* 196:207–19.

Patrício, J., R. E. Ulanowicz, M. S. A. Pardal, and J. C. Marques. 2004. Ascendency as ecological indicator: A case study on estuarine pulse eutrophication. *Estuar Coast Shelf Sci* 60 (1): 23–35.

Paul, J. F., K. J. Scott, D. E. Campbell, J. H. Gentile, C. S. Strobel, R. M. Valente, S. B. Weisberg, A. F. Holland, and J. A. Ranasinghe. 2001. Developing and applying a benthic index of estuarine condition for the Virginian Biogeographic Province. *Ecol Indicat* 1:83–99.

Pearson, T. H., and R. Rosenberg. 1978. Macrobenthic succession in relation to organic enrichment and pollution of the marine environment. *Oceanogr Mar Biol Ann Rev* 16:229–331.

Pérez-Ruzafa, A. 1989. Estudio ecológico y bionómico de los poblamientos bentónicos del Mar Menor (Murcia, SE de España). PhD thesis, University of Murcia, Spain.

Pérez-Ruzafa, I. 2003. Efecto de la contaminaciónsobre la vegetación submarina y su valor indicador. In *Perspectivas y herramientas en el estudio de la contaminación marina*, eds. A. Pérez-Ruzafa, C. Marcos, F. Salas, and S. Zamora, 137–47. Murcia, Spain: Servicio de Publicaciones, University of Murcia.

Pergent-Martini, C. 1998. *Posidonia oceanica*: A biological indicator of past and present mercury contamination in the Mediterranean Sea. *Mar Environ Res* 45:101–11.

Pergent-Martini, C., V. Leoni, V. Pasqualini, G. D. Ardizzone, E. Balestri, R. Bedini, A. Belluscio, T. Belsher, J. Borg, C. F. Boudouresque, S. Boumaza, J. M. Bouquegneau, M. C. Buia, S. Calvo, et al. 2005. Descriptors of *Posidonia oceanica* meadows: Use and application. *Ecol Indic* 5:213–30.

Perillo, G., M. C. Piccolo, and R. H. Freije. 2001. The Bahía Blanca estuary. *Ecol Stud* 144:205–17.

Petrov, A. N. 1990. Study on ecology of mollusks (the Black Sea bivalves) employing some relevant indices. PhD thesis, Sevastopol, Ukraine.

Pinto, R., J. Patrício, A. Baeta, B. D. Fath, J. M. Neto, and J. C. Marques. 2009. Review and evaluation of estuarine biotic índices to assess benthic condition. *Ecological Indicators* 9:1–25.

Rafaelli, D. G., and C. F. Mason. 1981. Pollution monitoring with meiofauna: Using the ratio nematodes/copepods. *Mar Pollut Bul* 12 (5): 158–63.

Rhoads, D. C., and J. C. Germano. 1986. Interpreting long-term changes in benthic community structure: A new protocol. *Hydrobiologia* 142:291–308.

Roberts, R. D., M. G. Gregory, and B. A. Foster. 1998. Developing an efficient macrofauna monitoring index from an impact study: A dredge spoil example. *Mar Pollut Bul* 36 (3): 231–35.

Rosenberg, R., M. Blomquist, H. C. Nilsson, H. Cederwall, and A. Dimming. 2004. Marine quality assessment by use of benthic species-abundance distribution: A proposed new protocol within the European Union water framework directive. *Marine Pollution Bulletin* 49:728–39.

Ruiz, J. L., M. R. Perez, and J. Romero. 2001. Effects of fish farm loadings on seagrass (*Posidonia oceanica*): Distribution, growth and photosynthesis. *Marine Pollution Bulletin* 42 (9): 749–60

Salas, F., J. M. Neto, Á Borja, and J. C. Marques. 2004. Evaluation of the applicability of a Marine Biotic Index to characterise the status of estuarine ecosystems: The case of Mondego estuary (Portugal). *Ecol Indic* 4:215–25.

Satsmadjis, J. 1982. Analysis of benthic fauna and measurement of pollution. *Rev Int Oceanogr Med* 66–67:103–107.

Satsmadjis, J. 1985. Comparison of indicators of pollution in the Mediterranean. *Marine Pollution Bulletin*, 16: 395–400.

Scanlan, C. M., J. Foden, E. Wells, and M. A. Best. 2007. The monitoring of opportunistic macroalgal blooms for the water framework directive. *Marine Pollution Bulletin* 55:162–71.

Shannon, C. E. and W. Weaver, 1963. *The Mathematical Theory of Communication*. Chicago: University of Illinois Press.

Shaw, K. M., P. J. D. Lambshead, and H. M. Platt. 1983. Detection of pollution-induced disturbance in marine benthic assemblages with special reference to nematodes. *Mar Ecol Progr Ser* 11:195–202.

Simboura, N., and A. Zenetos. 2002. Benthic indicators to use ecological quality classification of Mediterranean soft bottom marine ecosystems, including a new Biotic Index. *Med Mar Sci* 3 (2): 77–110.

Smith, R. W., M. Bergen, S. B. Weisberg, D. Cadien, A. Dalkey, D. Montagne, J. K. Stull, and R. G. Velarde. 2001. Benthic response index for assessing infaunal communities on the mainland shelf of southern California. *Ecol Appl* 11:1073–87.

Solís-Weiss, V. 1982. Aspectos ecológicos de la contaminación orgánica sobre el macrobentos de las cuencas de sedimentación en la bahía de Marsella Francia. *An Inst Cienc Mar Limnol* 9 (1): 19–44.

Sommerfield, P. J., and K. R. Clarke. 1997. A comparison of some methods commonly used for the collection of sublittoral sediments and their associated fauna. *Mar Environ Res* 43:145–56.

Straskraba, M. 1983. Cybernetic formulation of control in ecosystems. *Ecol Model* 18:85–98.

Ulanowickz, R. E. 1986. *Growth and development ecosystems phenpomenology*. New York: Springer-Verlag.

Ulanowickz, R. E., and J. S. Norden. 1990. Symmetrical overhead in flow and networks. *Intern J Syst Sci* 21 (2): 429–37.

Warwick, R. M. 1986. A new method for detecting pollution effects on marine macrobenthic communities. *Mar Biol* 92:557–62.

———. 1993. Environmental impact studies on marine communities: Pragmatical considerations. *Aust J Ecol* 18:63–80.

Warwick, R. M., and K. R. Clarke. 1994. Relearning the ABC: Taxonomic changes and abundance/biomass relationships in disturbed benthic communities. *Mar Biol* 118:739–44.

———. 1995. New "biodiversity" measures reveal a decrease in taxonomic distinctness with increasing stress. *Mar Ecol Prog Ser* 129:301–305.

———. 1998. Taxonomic distinctness and environmental assessment. *J Appl Ecol* 35:532–43.

Weisberg, S. B., J. A. Ranasinghe, D. M. Dauer, L. C. Schaffner, R. J. Díaz, and J. B. Frithsen. 1997. An estuarine benthic index of biotic integrity (B-IBI) for the Chesapeake Bay. *Estuaries* 20:149–58.

Weston, D. P. 1990. Quantitative examination of macrobenthic community changes along an organic enrichment gradient. *Mar Ecol Progr Ser* 61 (3): 233–44.

Word, J. Q. 1979. The infaunal trophic index. *Calif Coast Wat Res Proj Annu Rep,* 19–39.

Van Dolah, R. F., J. L. Hyland, A. F. Holland, J. S. Rosen, and T. R. Snoots. 1999. A benthic index of biological integrity for assessing habitat quality in estuaries of the southeastern USA. *Mar Environ Res* 48:269–83.

Vollenweider, R. A., F. Giovanardi, G. Montanari, and A. Rinaldi. 1998. Characterisation of the trophic conditions of marine coastal waters with special reference to the NW Adriatic Sea: Proposal for a trophic scale, turbidity and generalised water quality index. *Environmetrics* 9:329–57.

Zabala, K., A. Romero, and M. Ibanez. 1983. La contaminación marina en Guipuzcoa: I. Estudio de los indicadores biológicos de contaminación en los sedimentos de la Ría de Pasajes. *Lurralde* 3:177–89.

12

Application of Ecological and Thermodynamic Indicators for the Assessment of Ecosystem Health of Lakes

Fu-Liu Xu

CONTENTS

12.1 Introduction

12.1.1 Ecosystem Type and Problem

Lakes are extremely important storages of the earth's surface freshwater, with the important ecosystem service functions that can continue the sustainable development of society and the economy.[1,2] However, eutrophication, acidification, heavy metal, oil, and pesticide pollution caused by human activities have continuously deteriorated the healthy status of lake ecosystems. Water quality in over half of the lakes around the world has been seriously polluted. If this trend continues, it not only affects human health and social-economic development, but also may cause the breakup of lake ecosystems.[3,4] Therefore, the studies on lake ecosystem health have important and practical significance for the restoration of lake ecosystem health and the maintenance of their ecological service functions.

Since the mid-1980s, the studies on lake ecosystem health have begun to draw the attention of environmentalists and ecologists, with increasingly frequent use in academic and government publications, as well as the popular media.[5] More and more environmental managers consider the protection of ecosystem health a new goal of environmental management.[6–11] In the past few years, many national and international environmental programs have been established. One of the leading programs is "Assessing the State of Ecosystem Health in the Great Lakes," supported by the governments of Canada and the United States.[12] In the United States, important ongoing programs related to lake ecosystem health mainly include "Assessing Health State of Main Ecosystems"[11] and "Stresses on Ecosystem Health—Chemical Pollution."[13] In Canada, an ongoing key program related to lake ecosystem health is the "Aquatic Ecosystem Health Assessment Project."[14] Also, in China, special attention has been paid to lake ecosystem health. Two projects have been carried out, namely The Effects of Typical Chemical Pollution on Aquatic Ecosystem Health[15] and The Indicators and Methods for Lake Ecosystem Assessment.[16,17] Ongoing programs supported by the Natural Science Foundation of China (NSFC) include The Limiting Factors and Dynamic Mechanism for Lake Ecosystem Health, Regional Differentia and Its Mechanisms for the Ecosystem Health of Large Shallow Lakes, and Assessment and Management of Watershed Ecosystem Health.[18]

So far, a number of indicators have been proposed for lake ecosystem health assessment, e.g., gross ecosystem product (GEP),[19] ecosystem stress indicators,[20] the Index of Biotic Integrity (IBI),[21] thermodynamic indicators including exergy and structural exergy,[22,23] and a set of comprehensive ecological indicators covering structural, functional, and system-level aspects.[15,16] Some methods or procedures have also been proposed for assessing lake ecosystem health, e.g., a tentative procedure by Jørgensen (1995),[23] the direct measurement method (DMM), and the ecological model method (EMM) by Xu et al.[16,17] However, owing to the lack of criteria, there are two major problems with using present methods to assess lake ecosystem health. First, we can only assess the relative health status, and it is extremely difficult to assess the actual health status. Second, it is impossible to compare ecosystem health status for different lakes. In order to solve these problems, a new method, Ecosystem Health Index Method (EHIM), is developed in this chapter.

12.1.2 The Chapter's Focus

This chapter focuses on indicators and methods for assessing lake ecosystem health, which are followed by case studies. Also, a tentative theoretical frame or procedure for assessing lake ecosystem health is proposed. Finally, the discussions on indicators, methods, and the results of case studies are presented.

12.2 Methodologies

12.2.1 A Theoretical Frame

A tentative theoretical frame or procedure for assessing lake ecosystem health is shown in Figure 12.1. It can be seen from Figure 12.1 that there are five necessary steps, within which the development of indicators and the determination of assessment methods are two key steps. However, in order to develop sensitive indicators, first the anthropogenic stresses have to be identified, and the responses of lake ecosystems to the stresses have to be analyzed, since the stresses caused by human activities are mainly responsible for the degradation of lake ecosystem health.

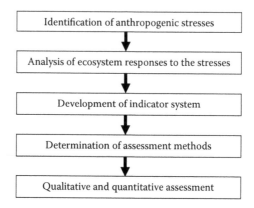

FIGURE 12.1
A tentative procedure for assessing lake ecosystem health.

12.2.2 Development of Indicators

12.2.2.1 The Procedure for Developing Indicators

The flow chart for developing indicators is shown in Figure 12.2. It can be seen that the identified anthropogenic stresses to the lake ecosystems include eutrophication and acidification, as well as heavy metals, pesticides, and oil pollution. The objected lake ecosystems should include actual and experimental ones. The response of lake ecosystems to the stresses should be composed of structural, functional, and system-level aspects.

12.2.2.2 Lake Data for Developing Indicators

The actual lake ecosystems including 29 Chinese lakes (Figure 12.3) and 30 Italian lakes (Table 12.1) are applied for eutrophication, while the 20 experimental lake ecosystems are applied for acidification as well as heavy metals, pesticides, and oil pollution (Table 12.2).

It can be seen from Figure 12.3 that 29 Chinese lakes are distributed in different regions in China. Their surface areas range from 3.7 km² (Lake Xuanwu-Hu) to 4200 km² (Lake Qinghai-Hu). Their trophic statuses are from oligotrophic (e.g., Lake Qinghai-Hu) to extremely hypertrophic (e.g., Lake Liuhua-Hu, Lake Dongshan-Hu, and Lake Dong-Hu).

Thirty Italian lakes are located in the island of Sicily. About 70% of the lakes are used for irrigation, while 30% of the lakes are used for drinking. Their mean depths are between 1.5 and 19 m. Their surface area ranges from 1 to 577 km² with average volume varying from 0.1 to 154 billion cubic meters.[24]

Experimental ecosystems, including microcosms, mesocosms, and experimental ponds, have been increasingly used in research on the toxicity and

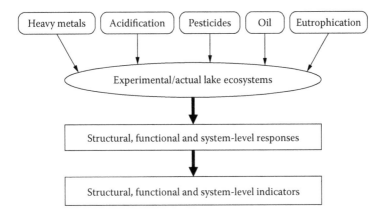

FIGURE 12.2
The flow chart for developing indicators for lake ecosystem health assessment.

FIGURE 12.3
Geographic locations of 29 Chinese lakes used for developing indicators. MX1: Lake Wulungu-Hu; MX2: Lake Beshiteng-Hu; MX3: Lake Wuliangshu-Hai; MX4: Lake Huashu-Hai; MX5: Lake Dai-Hai; MX6: Hulun-Hu; DB1: Lake Wudalianchi; DB2: Lake Jingbe-Hu; DB3: Lake Xiaoxingkai-Hu; DB4: Lake Daxingkai-Hu; QZ1: Lake Zhaling-Hu; QZ2: Lake Eling-Hu; QZ3: Lake Qinghai-Hu; YG1: Lake Erhai; YG2: Lake Fuxian-Hu; PY1: Lake Nanshi-Hu; PY2: Lake Hongzhe-Hu; PY3: Lake Chao-Hu; PY4: Lake Baoan-Hu; PY5: Lake Hong-Hu; PY6: Lake Tai-Hu; CS1: Lake Dian-Chi; CS2: Lake Liuhua-Hu; CS3: Lake Dongshan-Hu; CS4: Lake Lu-Hu; CS3: Lake Dong-Hu; CS6: Lake Xi-Hu; CS7: Lake Xuanwu-Hu; CS8: Lake Nan-Hu

TABLE 12.1

Basic Limnological Characteristics for 30 Italian Lakes

Lake Name	Cond. (mS/cm)	TP (μg/L)	N-NH₄ (μg/L)	N-NO₃ (μg/L)	SiO₂ (mg/L)
Ancipa	0.18	30.66	12	77	2.0
Arancio	0.72	166.65	667	676	4.8
Biviere di Cesaro	0.08	46.02	31	76	0.6
Biviere di Gela	2.72	45.15	22	78	2.3
Castello	0.96	109.88	775	263	2.9
Cimia	2.15	49.57	199	803	4.0
Comunelli	2.51	45.33	331	129	3.4
Dirillo	0.53	60.54	60	514	4.1
Disueri	1.21	1093.43	684	2226	3.6
Fanaco	0.53	54.34	199	1143	3.3
Gammauta	0.49	183.07	154	446	2.7
Garcia	0.77	51.36	22	1165	3.6
Gorgo	4.51	80.87	33	65	6.1
Guadalani	0.42	38.89	111	459	0.3
Nicoletti	1.42	35.18	46	66	1.5
Ogliastro	2.72	40.87	173	1710	2.9
Olivo	0.91	38.00	71	69	1.6
Piana degli Albanesi	0.37	46.77	349	412	0.4
Piana del Leone	0.41	46.85	160	546	2.4
Poma	0.74	51.11	73	994	1.4
Pozzillo	1.13	49.38	91	355	1.6
Prizzi	0.46	52.99	86	503	2.5
Rubino	1.05	28.94	18	711	1.0
San Giovanni	1.49	80.56	658	283	2.7
Santa Rosalia	0.42	55.81	125	279	3.4
Scanzano	0.50	61.65	300	1283	2.3
Soprano	1.85	2962.96	7671	57	12.7
Trinita	1.86	83.24	26	417	3.8
Vasca Ogliastro	0.32	106.69	28	177	3.4
Villarosa	2.27	64.06	524	276	1.0

impacts of chemicals on aquatic ecosystems during the last two decades. Experimental ecosystem perturbations allow us to separate the effects of various pollutants, to assess early effects of perturbations in systems with known background properties, and to quantitatively assess the result of known perturbations to whole ecosystems.[25,26] The experimental ecosystems for developing indicators include two microcosms, 14 mesocosms, and four experimental ponds, and the experimental perturbations include acidification, oil, copper, and organic chemical contamination (Table 12.2).

TABLE 12.2

Studies on the Responses of Lake Ecosystems to Experimental Perturbations

Stressors	Study Type[a]	Location	Duration (Days)	Reference[b]
Acidification	Meso.	West Virginia	75	54
Acidification	Meso.	California	35	55
Acidification	Meso.	Ohio	10	56
Acidification	Meso	Ohio	35	57
Copper	Meso.	Ohio	4	58
Copper	Meso.	Ohio	14	59
Oil	EP	Tennessee	420	60
Dursban	EP	California	90	61
2,4D-DMA	EP	Missouri	56	62
TCP	Meso.	Neuherberg	24	63
PCP	Meso.	Neuherberg	24	63
Trichloroethylene	Meso.	Southern Germany	44	64
TCB	Meso.	Southern Germany	22	65
Benzene	Meso.	Western Germany	26	66
Atrazine	Micro.	New Mexico	365	67
HCBP	Micro	New Mexico	365	67
Permethrin	Meso.	Tsukuba, Japan	30	68
Hexazinone	Meso.	Ontario	77	69, 70
Bifenthrin	EP	New Jersey	8	71
Carbaryl	Meso.	Ohio	4	58

[a] EP, experimental ponds; Meso., mesocosms; Micro., microcosms.
[b] For acidification also see Refs. 72–75; for oil pollution also see Refs. 76–79; for copper pollution also see Refs. 80–84; for pesticide pollution also see Refs. 85–90.
Source: Modified from Xu, F.-L. et al. 1999. Ecological indicators for assessing freshwater ecosystem health. *Ecol Model* 116:80, with permission.

12.2.2.3 Responses of Lake Ecosystems to Chemical Stresses

Xu et al. (1999)[15] examined the structural, functional, and ecosystem-level symptoms resulting from chemical stress; acidification; and copper, oil, and pesticide contamination in lake ecosystems, based on the above-mentioned data on experimental ecosystems. It could be concluded that the structural responses in freshwater ecosystems to chemical stresses were noticeable in terms of an increase in phytoplankton cell size, and in phytoplankton and microzooplankton biomass, and a decrease in zooplankton body size, zooplankton and macrozooplankton biomass, and species diversity, and in the zooplankton/phytoplankton and macrozooplankton/microzooplankton ratios. The functional responses included decreases in alga C assimilation, resource use efficiency, the P/B and B/E ratios, an increase in community production, and a departure from 1 for the P/R ratio. System-level responses included decreases in exergy, structural exergy, and ecological buffering capacities.[15,16]

Xu (1997)[27] investigated the structural responses of Lake Chao to eutrophication. He found that, with an increasing eutrophication gradient, algal cell number and biomass were increased, while algal biodiversity, zooplankton biomass, and the ratio of zooplankton biomass to algal biomass were decreased. Xu et al. (2001)[28] and Lu (2001)[29] studied the structural, functional, and system-level responses of 29 Chinese lakes and 30 Italian lakes to eutrophication. The results are summarized as Table 12.3. The results are very similar to the results from experimental lake ecosystems stressed by acidification and heavy metal, oil, and pesticide pollution, with the exception of zooplankton biomass and exergy for lakes with the trophic states from oligo-eutrophication to eutrophication.

12.2.2.4 Indicators for Lake Ecosystem Health Assessment

Ecological indicators for lake ecosystem health assessment resulting from chemical stress are important for both the early warning signs of ecosystem malfunction and confirmation of the presence of a significant ecosystem pathology.[9,20] Ecological indicators, as valid and reliable tools, should include

TABLE 12.3

The Structural, Functional, and System-Level Responses of Actual Lake Ecosystems to Eutrophication[a]

Responses Indicators		Oligo-eutrophication—Eutrophication	Eutrophication—Hyper-eutrophication
		Dynamics in lake trophic states	
Structural responses	Phytoplankton cell number[b,c]	Increase	Increase
	Phytoplankton biomass (BA)[b,c]	Increase	Increase
	Phytoplankton cell size[b,c]	Increase	Increase
	Phytoplankton diversity[b]	Decrease	Decrease
	Zooplankton biomass (BZ)[b,c]	Increase	Decrease
	Zooplankton body size[b,c]	Decrease	Decrease
	Zooplankton diversity[b]	Decrease	Decrease
	BZ/BA ratio[b,c]	Decrease	Decrease
	BZmacro./BZmicro. ratio[b,c]	Decrease	Decrease
Functional responses	Phytoplankton primary production[b]	Increase	Increase
	P/B ratio[b]	≈1	<0.5
	P/R ratio[b]	≈1	<1.0
System-level responses	Exergy[b,c]	Increase	Decrease
	Structural exergy[b,c]	Decrease	Decrease

[a] Please see Refs. 28 and 29 for the details.
[b] For the 29 Chinese lakes.
[c] For the 30 Italian lakes.
Note: P/R ratio = Gross production/community respiration
        P/B ratio = Gross production/standing crop biomass
        B/E ratio = Biomass supported/unit energy flow

structural, functional, and system-level aspects. According to the above-mentioned structural, functional, and system-level responses of actual and experimental lake ecosystems to chemical stress, a set of comprehensive ecological indicators, including structural, functional, and ecosystem-level aspects, for assessing lake ecosystem health can be derived (Table 12.4). Table 12.4 indicates that a healthy ecosystem can be characterized by small cell size in phytoplankton, large body size in zooplankton, high zooplankton and macrozooplankton biomass levels, low phytoplankton and micro-zooplankton biomass levels, a high zooplankton/phytoplankton ratio, a high macrozooplankton/microzooplankton ratio, high degrees of species diversity, high levels of algal C assimilation, high resource use efficiencies, low community production, high P/B and B/E ratios, a P/R ratio approaching 1, and high exergy, structural exergy, and buffer capacities.

TABLE 12.4

The Ecological Indicators for Lake Ecosystem Health Assessment

	Ecological Indicators	Relative Healthy State		Methods for Indicator Values
		Good	Bad	
Structural indicators	1. Phytoplankton cell size	Small	Large	Measure
	2. Zooplankton body size	Large	Small	Measure
	3. Phytoplankton biomass (BA)	Low	High	Measure
	4. Zooplankton biomass (BZ)	High	Low	Measure
	5. Macrozooplankton biomass (Bmacroz.)	High	Low	Measure
	6. Microzooplankton biomass (Bmicroz.)	Low	High	Measure
	7. BZ/BA ratio	High	Low	Calculate
	8. Bmacroz./Bmicroz. ratio	High	Low	Calculate
	9. Species diversity (DI)	High	Low	Measure and calculate
Functional indicators	10. Algal C assimilation ratio	High	Low	Measure
	11. Resource use efficiency (RUE)	High	Low	Measure and calculate
	12. Community production (P)	Low	High	Measure
	13. P/R ratio	≈ 1	> or <1	Measure and calculate
	14. P/B ratio	High	Low	Measure and calculate
	15. B/E ratio	High	Low	Measure and calculate
System-level indicators	16. Buffer capacities (β)	High	Low	Calculate
	17. Exergy (Ex)	High	Low	Calculate
	18. Structural exergy (Ex$_{st}$)	High	Low	Calculate

Source: Modified from Xu, F.-L. et al. 2001. Lake ecosystem health assessment: Indicators and methods. *Wat Res* 35 (1): 3159, with permission.

12.2.3 Calculations for Some Indicators

12.2.3.1 Calculations of Exergy and Structural Exergy

The definitions and calculations of exergy and structural exergy (or specific exergy) refer to Chapter 2, Section 2.5 and to references 22, 23, and 32–35.

12.2.3.2 Calculation of Buffer Capacity

The buffer capacity is defined as follows:[32,34,36]

$$\beta = \frac{1}{\delta(\text{state variable})/\delta(\text{forcing function})} \tag{12.1}$$

Forcing functions are the external variables that are driving the system, such as discharge of waste, precipitation, wind, solar radiation, and so on. State variables are the internal variables that determine the system, for instance, the concentration of soluble phosphorus or zooplankton in a lake, and so on. The concept should be considered multidimensionally, as all combinations of state variables and forcing functions may be considered. It implies that even for one type of change there are many buffer capacities corresponding to each of the state variables.

12.2.3.3 Calculation of Biodiversity

The definitions and calculations of diversity index (DI) for an ecosystem refer to Chapter 2, Section 2.5 and to references 37 and 38.

12.2.3.4 Calculations of Other Indicators

$$\text{RUE} = (\text{Zooplankton C assimilation rate})/ \\ (\text{Algal C assimilation rate}) *100\% \tag{12.2}$$

$$\text{P/R} = \text{Gross production (P)/Community respiration (R)} \tag{12.3}$$

$$\text{P/B} = \text{Gross production (P)/Standing crop biomass (B)} \tag{12.4}$$

$$\text{B/E} = \text{Standing crop biomass (B)/Unit energy flow (E)} \tag{12.5}$$

12.2.4 Methods for Lake Ecosystem Health Assessment

Three methods, direct measurement method (DMM), ecological model method (EMM) and ecosystem health index method (EHIM), have been applied to assess lake ecosystem health. The methods are reviewed in Chapter 2, Section 2.10 where the general methodology is mentioned. The indicators can be selected from Table 12.4.

12.3 Case Studies

12.3.1 Case 1: Ecosystem Health Assessment for Italian Lakes Using EHIM

12.3.1.1 Selecting Assessment Indicators

Assessment indicators are composed of basic and additional indicators. Basic indicators are crucial for lake ecosystem health assessment. Basic indicators have the consanguineous relationships to ecosystem health status, while additional indicators have less important relationships to ecosystem health status. A lake's ecosystem health status can be evaluated mainly on the basis of basic indicators; however, the assessment using additional indicators can be considered as the remedies of results by basic indicators.

In most lake ecosystems, the indicators that give the consanguineous relationships to ecosystem health status are phytoplankton biomass (BA) and Chl-a concentration. The higher the BA or Chl-a concentrations in a lake, the worse the lake ecosystem health status. Therefore, BA and Chl-a can serve as two basic indicators. According to data availability for Italian lakes, BA are selected as a basic indicator, while zooplankton biomass (BZ), BZ/BA, exergy (Ex), and structural exergy (Exst) are applied as additional indicators.

12.3.1.2 Calculating Sub-EHIs

The are two main steps to calculate sub-EHIs for all selected indicators. The first step is to calculate EHI(BA) for the basic indicator (BA). The second step is to calculate EHI(BZ), EHI(BZ/BA), EHI(Ex), and EHI(Exst) for the additional indicators, BZ, BZ/BA, Ex, and Exst, respectively. After the EHI(BA) for the basic indicator is obtained, the sub-EHIs including EHI(BZ), EHI(BZ/BA), EHI(Ex), and EHI(Exst) for the additional indicators can be deduced according to the relationships between the basic indicator (BA) and the additional indicators (BZ, BZ/BA, Ex, and Exst).

12.3.1.2.1 EHI(BA) Calculation

For the EHI(BA) calculation, it is assumed that EHI(BA) = 100 if BA is lowest, which means the best healthy state, and that EHI(BA) = 0 if BA is highest, which means the worst healthy state. Referring to Carlson's studies on trophic state index (TSI),[39] the relationship between ecosystem health status and phytoplankton biomass in a lake ecosystem can be described as logarithmic normal distribution. So, EHI(BA) can be calculated using the following equation:

$$EHI(BA) = 100 \times \frac{\ln C_x - \ln C_{min}}{\ln C_{max} - \ln C_{min}} \qquad (12.6)$$

where EHI(BA) is sub-EHI for basic indicator, BA; C_x is the measured BA value; C_{min} is the measured lowest BA value; C_{max} is the measured highest BA value.

Equation (12.6) can be predigested as the following format:

$$EHI(BA) = 10(a + b \ln C_x) \tag{12.7}$$

where a and b are constants, and they can be computed using the following equation:

$$\begin{cases} a = -10 \times \dfrac{\ln C_{min}}{\ln C_{max} - \ln C_{min}} \\ b = 10 \times \dfrac{1}{\ln C_{max} - \ln C_{min}} \end{cases} \tag{12.8}$$

According to the measured data for 30 Italian lakes,[24] $C_{min} = 0.004$ (mg/L), $C_{max} = 150$ (mg/L). Then, a = 5.2425, b = −0.94948. Thus, the expression for calculating EHI(BA) for Italian lakes can be obtained as follows:

$$EHI(BA) = 10 \times \left(5.2425 - 0.94948 \times Ln(BA)\right) \tag{12.9}$$

It can be seen that the equation for calculating EHI(BA) can be deduced from the BA measured data by logarithmic expression for differences between extremum values.

12.3.1.2.2 EHI(BZ), EHI(BZ/BA), EHI(Ex), and EHI(Exst) Calculations

The sub-EHIs for additional indicators, EHI(BZ), EHI(BZ/BA), EHI(Ex), and EHI(Exst), can be calculated according to the relationships between the basic indicator (BA) and the additional indicators (BZ, BZ/BA, Ex, and Exst). From Lu (2001),[29] there are very simple relationships between BA, and BZ/BA as well as Exst, while there are more complicated relationships between BA, and BZ as well as Ex. Thus, the different ways should be adopted to calculate EHI(BZ/BA), EHI(Exst), and EHI(BZ), EHI(Ex).

For the 30 Italian lakes, there are strongly negative relationships between BA, and BZ/BA as well as Exst. The following two expressions can be obtained by means of regression analysis:

$$\ln(BA) = 0.3878 - 0.7742 \times \ln(BZ / BA) \tag{12.10}$$

$$\ln(BA) = 5.1119 - 0.0688 \times (Exst) \tag{12.11}$$

Thus, the equations for calculating EHI(BZ/BA) and EHI (Exst) can be deduced from Equations (12.9), (12.10), and (12.11):

$$EHI(BZ / BA) = 10 \times \left(5.2425 - 0.94948 \times \left(0.3878 - 0.7742 \times \ln(BZ / BA)\right)\right) \tag{12.12}$$

$$EHI(Exst) = 10 \times \left(5.2425 - 0.94948 \times \left(5.1119 - 0.0688 \times (Exst)\right)\right) \tag{12.13}$$

According to Lu (2001),[29] there are three kinds of relationships between BA and BZ as well as Ex in the 30 Italian lakes, owing to the dynamics in phytoplankton community structure, the toxic effects of phytoplankton species, and the food sources of zooplankton. The first kind of relationship between BA and BZ as well as Ex is that BZ as well as Ex are apparently increased with the BA increase. The second kind of relationship between BA and BZ as well as Ex is that BZ as well as Ex are decreased with the BA increase. The third kind of relationship between BA and BZ as well as Ex is that BZ as well as Ex are slowly increased with the BA increase. The first and the third kinds of relationships between BA and BZ as well as Ex are more obvious than the second kind of relationship. However, the second kind of relationship between BA and BZ as well as Ex is less obvious, since there are many lakes, and BA is different in each lake when BZ as well as Ex start to be decreased. So, the second kind of relationship between BA and BZ as well as Ex can be considered as the transition from the first to the third kind of relationship. In order to better describe the relationships between BA and BZ as well as Ex, two linear expressions are used to simulate the first and the third relationships, respectively. By means of Fussy Mathematics' method, each data point in the second kind of relationship and in the first and the third kinds of relationships can be determined to belong to the first or to the third kind of relationship, through the comparison of its attributability to the first kind of relationship with its attributability to the third kind of relationship.

For the first and the third relationships between BA and BZ, two linear expressions can be obtained using regression analysis:

$$f_1: \ln(BA) = 0.1036 + 0.7997 \times \ln(BZ), \quad (N = 95, r = 0.702, p < 0.01) \quad (12.14)$$

$$f_2: \ln(BA) = 2.7359 + 0.6766 \times \ln(BZ), \quad (N = 19, r = 0.563, p < 0.01) \quad (12.15)$$

For the first and the third relationships between BA and Ex, two linear expressions are as follows:

$$f_3: \ln(BA) = -4.0256 + 0.8236 \times \ln(Ex), \quad (N = 95, r = 0.717, p < 0.01) \quad (12.16)$$

$$f_4: \ln(BA) = -2.5380 + 0.9899 \times \ln(Ex), \quad (N = 19, r = 0.829, p < 0.01) \quad (12.17)$$

Thus, four expressions for calculating EHI(BZ) and EHI(Ex) can be deduced from Equations (12.14), (12.15), (12.16), (12.17), and (12.9), respectively.

$$EHI(BZ)_1 = 10\left(5.2425 - 0.94948 \times \left(0.1036 + 0.7997 \times \ln(BZ)\right)\right) \quad (12.18)$$

$$EHI(BZ)_2 = 10\left(5.2425 - 0.94948 \times \left(2.7359 + 0.6766 \times \ln(BZ)\right)\right) \quad (12.19)$$

$$EHI(Ex)_1 = 10\left(5.2425 - 0.94948 \times \left(-4.0256 + 0.8236 \times \ln(Ex)\right)\right) \quad (12.20)$$

$$EHI(Ex)_2 = 10\left(5.2425 - 0.94948 \times \left(-2.538 + 0.9899 \times \ln(Ex)\right)\right) \quad (12.21)$$

Equations (12.18), (12.19), (12.20), and (12.21) can be synthesized as the following two comprehensive expressions:

$$EHI(BZ) = \begin{cases} EHI(BZ)_1 & (BA, BZ) \in \alpha \\ EHI(BZ)_2 & (BA, BZ) \in \beta \end{cases} \quad (12.22)$$

$$EHI(Ex) = \begin{cases} EHI(Ex)_1 & (BA, Ex) \in \gamma \\ EHI(Ex)_2 & (BA, Ex) \in \delta \end{cases} \quad (12.23)$$

In Equation (12.22), "α" and "β" are the attributability of measured data (BA,BZ) to the two linear expressions (12.14) and (12.15), which can be calculated using the following attributable functions:

$$\alpha(BA, BZ) = \begin{cases} 1 & 0 < BA \leq 2.5 \\ 1 - \dfrac{\left(\ln(BA) - f_1(BZ)\right)^2}{\left(\ln(BA) + f_1(BZ)\right)^2} & 2.5 < BA \leq 50 \\ 0 & 50 < BA \leq 150 \end{cases} \quad (12.24)$$

$$\beta(BA, BZ) = \begin{cases} 0 & 0 < BA \leq 2.5 \\ 1 - \dfrac{\left(\ln(BA) - f_2(BZ)\right)^2}{\left(\ln(BA) + f_2(BZ)\right)^2} & 2.5 < BA \leq 50 \\ 1 & 50 < BA \leq 150 \end{cases} \quad (12.25)$$

where BA is the measured values; $f_1(BZ)$ and $f_2(BZ)$ are the calculated BA values by Equations (12.14) and (12.15), respectively; 2.5 is the minimum BA value in the β set that expresses the third kind of relationship; 50 is the maximum BA value in the α set that expresses the first kind of relationship.

It can be see from Equations (12.24) and (12.25) that, for the measured data point (BA, BZ), if $\alpha(BA,BZ) \geq \beta(BA,BZ)$, then $(BA,BZ) \in \alpha$, its EHI(BZ) can be calculated using Equation (12.18); if $\alpha(BA,BZ) < \beta(BA,BZ)$, then $(BA,BZ) \in \beta$, its EHI(BZ) can be calculated using Equation (12.19).

In Equation (12.23), "γ" and "δ" are the attributability of actual data (BA,Ex) to the two linear expressions (12.16) and (12.17), which can be calculated using the following attributable functions:

$$\gamma(BA,Ex) = \begin{cases} 1 & 0 < BA \le 2.5 \\ 1 - \dfrac{\left(\ln(BA) - f_3(Ex)\right)^2}{\left(\ln(BA) + f_3(Ex)\right)^2} & 2.5 < BA \le 50 \\ 0 & 50 < BA \le 150 \end{cases} \quad (12.26)$$

$$\delta(BA,Ex) = \begin{cases} 0 & 0 < BA \le 2.5 \\ 1 - \dfrac{\left(\ln(BA) - f_4(Ex)\right)^2}{\left(\ln(BA) + f_4(Ex)\right)^2} & 2.5 < BA \le 50 \\ 1 & 50 < BA \le 150 \end{cases} \quad (12.27)$$

where BA is the measured values; $f_3(Ex)$ and $f_4(Ex)$ are the calculated BA values by Equations (12.16) and (12.17), respectively; 2.5 is the minimum BA value in the δ set that expresses the third kind of relationship; 50 is the maximum BA value in the γ set that expresses the first kind of relationship.

It can be seen from Equations (12.26) and (12.27) that, for the sample point (BA,Ex), if $\gamma(BA,Ex) \ge \delta(BA,Ex)$, then (BA,Ex) $\in \gamma$, its EHI(Ex) can be calculated using Equation (12.20); if $\gamma(BA,Ex) < \delta(BA,Ex)$, then (BA,Ex) $\in \delta$, its EHI(Ex) can be calculated using Equation (12.21).

12.3.1.3 Determining Weighting Factors (ω_i)

There are many factors that affect lake ecosystem health to different extents. It is therefore necessary to determine weighting factors for all indicators. Basic indicators have the consanguineous relationships to ecosystem health status, while additional indicators have the less important relationships to ecosystem health status. Thus, a lake's ecosystem health status can be evaluated mainly on the basis of basic indicators; however, the assessment by additional indicators can be considered as the remedies of results by basic indicators. So, the method of relation-weighting index can be used to determine the weighting factors for all indicators, i.e., the relation ratios between BA and other indicators can be used to calculate the weighting factors for all indicators. The equation is as follows:

$$\omega_i = \frac{r_{i1}^2}{\sum\limits_{i=1}^{m} r_{i1}^2} \quad (12.28)$$

where ω_i is the weighting factor for the ith indicator; r_{i1} is the relation ratio between the ith indicator and the basic indicator (BA); m is the total number of assessment indicators, here m = 5_o.

TABLE 12.5

Statistic Correlative Ratios between BA and Other Indicators

Relative Indicators	ln(BA)– ln(BA)	ln(BA)– ln(BZ)[a]	ln(BA)– ln(BZ)[b]	ln(BA–ln(BZ/BA)	ln(BA)– ln(Ex)[a]	ln(BA)– ln(Ex)[b]	ln(BA)– (Exst)
Sample number	114	95	19	114	95	19	114
r_{ij}	1	0.702	0.563	–0.731	0.717	0.829	–0.699
r_{ij}^2	1	0.4928	0.3170	0.5344	0.5141	0.6872	0.4886

[a] expresser the first kind of relationship between BA and BZ or Ex;
[b] expresser the third kind of relationship between BA and BZ or Ex.

The statistic correlative ratios between the basic indicator (BA) and other indicators are shown in Table 12.5. Considering two kinds of relationships between BA and additional indicators, BZ and Ex, there are two steps to calculate the weighting factors for BZ and Ex. First, which kind of relationship between BA and BZ or Ex has to be determined, and then the calculations of weighting factors can be done by using Equation (12.28) and the corresponding correlative ratios.

12.3.1.4 Assessing Ecosystem Health Status for the Italian Lakes

12.3.1.4.1 EHI and Standards for the Italian Lakes

According to the sub-EHI calculation equations for all selected indicators, the responding standards for all indicators to the numerical EHI in a scale of 0 to 100 can be obtained (Table 12.6).

12.3.1.4.2 Ecosystem Health Status

As case studies, the measured data in the summer of 1988 for the 30 Italian lakes, and the data in four seasons during 1987–1988 for Lake Soprano are used for assessing and comparing ecosystem health status. The results for the 30 Italian lakes and for Lake Soprano are presented in Tables 12.7 and 12.8, respectively.

It can be seen from Table 12.7 that the synthetic EHI in the summer of 1988 for the Italian lakes ranges from 60.5 to 12, indicating ecosystem health status from "good" to "worst." Ecosystem health state in Lake Ogliastro was "good" with maximum EHI 60.5, while that in Lake Disueri was "worst" with minimum EHI 12. Of 30 lakes, 20 lakes were with "middle" health status, six lakes were with "bad" health status, three lakes were with "worst" health status, and only one lake was with "good" health state.

TABLE 12.6

Ecosystem Health Index (EHI) and Its Associated Parameters as well as Their Standards for the Italian Lakes

EHI	Health Status	BA (mg/L)	BZ (mg/L)[a]	BZ (mg/L)[b]	BZ/BA	Ex (J/L)[a]	Ex (J/L)[b]	Exst (J/mg)
0		150		60.7	0.001319		3434.7	1.47
10	Worst	52.3		12.81	0.004576		1185.3	16.78
20		18.3	62.9	2.71	0.01588	8385.6	409.02	32.10
30	Bad	6.37	16.84	0.5713	0.0551	2334.3	141.15	47.42
40		2.22	4.512	0.1206	0.191	649.8	48.71	62.73
50	Middle	0.775	1.209		0.663	180.9		78.05
60		0.271	0.324		2.30	50.36		93.36
70	Good	0.094	0.0868		7.98	14.02		108.68
80		0.033	0.0233		27.7	3.9023		124.00
90	Best	0.011	0.00623		96.1	1.0863		139.31
100		0.004	0.00167		333	0.3024		154.63

[a] expresser the first kind of relationship between BA and BZ or Ex;
[b] expresser the third kind of relationship between BA and BZ or Ex.

Table 12.8 shows that, in Lake Soprano, the synthetic EHI ranges from 41.3–112.3, expressing ecosystem health status from "middle" to "worst." In the winter, the lake ecosystem had a "middle" health state, which followed with fall, and in the summer, the lake ecosystem had a "worst" health state.

12.3.2 Case 2: Ecosystem Health Assessment for Lake Chao Using DMM and EMM

Lake Chao is located in the central Anhui Province of southeastern China. It is characterized by a mean depth of 3.06 m, a mean surface area of 760 km^2, a mean volume of 1.9 billon m^3, a mean retention time of 136 days, and a total catchment area of 13,350 km^2. It provides the primary water resource for domestic, industrial, agricultural, and fishery use for a number of cities and counties, including Hefei, the capital of Anhui Province. As the fifth-largest freshwater lake in China, it was well known for its scenic beauty and the richness of its aquatic products before the 1960s. However, over the past decades, following population growth and economic development in the drainage area, nutrient-rich pollutants from discharge of wastewater and sewage, agricultural application of fertilizers, and soil erosion have been increasingly discharged into the lake, and the lake has been seriously polluted by nutrients. The extremely serious eutrophication has already caused severe negative effects on the lake ecosystem health, sustainable utilization, and management. Since 1980, some studies focusing on the investigation and assessment of pollution sources and water quality, eutrophication

TABLE 12.7

Assessment and Comparison of Ecosystem Health Status for the Italian Lakes in the Summer of 1988

Lake Name	EHI(BA)	EHI(BZ)	EHI(BZ/BA)	EHI(Ex)	EHI(Exst)	EHI	Health State	Order (Good–Bad)
Ogliastro	63.6	52.3	61.5	52.6	69.4	60.5	Good	1
Fanaco	60.4	49.6	61.7	49.8	69.9	58.6	Middle	2
Ancipa	60.6	56.1	55.0	56.5	52.8	56.9	Middle	3
Prizzi	55.3	43.2	64.1	43.2	74.7	56.0	Middle	4
Vasca	55.3	44.5	62.8	44.5	72.2	55.7	Middle	5
Comunelli	56.5	50.6	57.4	50.8	59.5	55.2	Middle	6
Nicoletti	54.2	50.0	56.0	50.2	55.6	53.4	Middle	7
Garcia	52.8	48.3	56.7	48.4	57.5	52.8	Middle	8
Cesaro	50.6	41.9	61.6	41.9	69.5	52.7	Middle	9
Poma	50.0	45.1	57.7	45.2	60.2	51.4	Middle	10
Pozzillo	51.4	52.4	51.2	52.5	42.0	50.2	Middle	11
Villarosa	45.6	42.5	56.7	42.5	57.4	48.4	Middle	12
Rosalia	48.6	52.9	48.2	53.0	33.8	47.6	Middle	13
Trinita	42.5	37.6	59.2	37.5	64.0	47.3	Middle	14
Dirillo	46.0	51.6	47.4	51.6	31.7	45.8	Middle	15
Gela	48.1	59.7	40.6	59.5	17.4	45.6	Middle	16
Olivo	47.3	57.0	42.7	56.9	21.3	45.5	Middle	17
Albanese	40.2	43.7	50.8	43.6	41.0	43.3	Middle	18
Castello	34.7	31.2	59.4	30.8	64.6	42.6	Middle	19
Rubino	41.4	51.6	43.5	51.3	22.8	42.1	Middle	20
Guadalani	34.9	36.3	54.2	36.1	50.6	41.3	Middle	21
Cimia	33.0	40.2	48.5	39.9	34.6	38.3	Bad	22
Scanzano	33.8	42.9	46.3	42.6	29.1	38.2	Bad	23
Giovanni	33.9	45.5	43.6	45.1	23.0	37.6	Bad	24
Leone	31.4	41.7	45.5	41.3	27.2	36.6	Bad	25
Gorgo	34.6	26.4	37.9	28.5	13.5	29.4	Bad	26
Gammauta	28.2	21.1	39.2	20.9	15.2	25.7	Bad	27
Arancio	11.8	29.6	14.6	21.9	2.0	15.3	Worst	28
Soprano	11.8	24.4	21.2	19.2	3.0	15.3	Worst	29
Disueri	6.2	23.2	18.0	15.2	2.4	12.0	Worst	30

mechanism, and ecosystem health, as well as on ecological restoration and environmental management have been carried out.[16,17,40–47]

12.3.2.1 Assessment Using DMM

The data measured monthly from April 1987 to March 1988 are used for the Lake Chao ecosystem health assessment. According to data availability, the

TABLE 12.8

Assessment and Comparison of Ecosystem Health Status for Lake Soprano in 1987–1988

Season	EHI(BA)	EHI(BZ)	EHI(BZ/BA)	EHI(Ex)	EHI(Exst)	EHI	Health State	Order (Good–Bad)
Winter	35.6	41.2	49.6	40.9	44.1	41.3	Middle	1
Fall	40.2	52.2	41.9	51.9	19.7	41.1	Middle	2
Spring	27.8	22.1	37.6	22.0	13.1	25.3	Bad	3
Summer	11.8	24.4	21.2	19.2	3.0	15.3	Worst	4

ecological indicators for the assessment were phytoplankton biomass (BA), zooplankton biomass (BZ), the BZ/BA ratio, algal primary productivity (P), algal species diversity (DI), the P/BA ratio, exergy (Ex), structural exergy (Exst), and phytoplankton buffering capacity ($\beta_{(TP)(Phyto.)}$). The values of these ecological indicators for different periods and the assessment results are presented in Table 12.9. A relative order of health states for the Lake Chao ecosystem proceeding from good to poor was obtained as follows: January to March 1988 > November to December 1987 > June to July 1987 > April to May 1987 > August to October 1987.

12.3.2.2 Assessment Using EMM

12.3.2.2.1 The Analysis of Lake Ecosystem Structure

In the early 1950s, the lake was covered with macrophytes appearing in order as floating plants → submerged plants → leaf floating plants → and emergent plants from the open waters to the shore. More than 190 species of zooplankton were identified. The lake was rich in large benthic animals and in fishery resources dominated by piscivorous fish. Phytoplankton populations were intensely suppressed to low densities by aquatic macrophytes, with diatoms as the dominant form. However, for the past few decades, the lake's ecosystem has been seriously damaged through eutrophication. From the early 1950s to the early 1990s the coverage of macrophytes decreased significantly from 30% to 2.5% of the lake's total area. Now, as a result of this reduction, more than 90% of the lake's primary productivity is from phytoplankton. At the same time, the fraction of large fish also dramatically decreased from 66.7% to 23.3%. Herbivorous fish also decreased from 38.4% to 3.5%, while carnivorous fish increased significantly from 32.6% to 83%.[45]

TABLE 12.9

The Ecological Indicators and Their Measured Values in Different Periods in Lake Chao (from April 1987 to March 1988)

Ecological Indicators[a]	Measured Indicator Values in Different Periods[b]					Relative Order of Health State in Different Periods (Good → Poor)
	A	B	C	D	E	
BA	4.5	1.31	21.82	0.60	0.58	E > D > B > A > C
BZ	0.33	0.34	1.76	4.15	13.54	E > D > C > B > A
BZ/BA	0.073	0.26	0.081	6.92	23.24	E > D > B > C > A
P	1.42	1.38	7.03	0.74	0.21	E > D > B > A > C
P/B	0.292	1.053	0.322	1.233	0.363	D > B > E > C > A
DI	1.59	1.62	0.28	1.83	1.97	E > D > B > A > C
Ex	112.0	98.5	606.3	1075.1	3350.9	E > D > C > A > B
Exst	25.33	52.8	48.0	213.6	238.6	E > D > B > C > A
β((TP)(Phyto.))	−0.014	6.45	0.04	0.92	−0.371	B > D > C > A > E
	Comprehensive results					E > D > B > A > C

[a] BA, phytoplankton biomass (g•m-3); BZ, zooplankton biomass (g•m-3); P, algal primary productivity (gC•m-2•d-1); DI, algal diversity index; Ex, exergy (MJ•m-3); Ex$_{st}$, structural exergy (MJ•mg-1); β((TP)(Phyto.)), phytoplankton buffer capacity to total phosphorus.

[b] A: Apr.–May 1987; B: Jun.–Jul. 1987; C: Aug.–Oct. 1987; D: Nov.–Dec. 1987; E: Jan.–Mar. 1988.

Note: The numbers are mean values of 31 sampling points' data measured monthly.

12.3.2.2.2 The Establishment of Lake Ecological Model

1. Conceptual diagram

   Given the ecosystem structure of Lake Chao, an ecological model describing nutrient cycling within the food web seemed reasonable. The model's conceptual framework is shown in Figure 12.4. The model contains six sub-models relative to nutrients, phytoplankton, zooplankton, fish, detritus, and sediments. The model's state variables include phytoplankton biomass (BA), zooplankton biomass (BZ), fish biomass (BF), the amount of phosphorus in phytoplankton (PA), the proportion of phosphorus in zooplankton (FPZ), the proportion of phosphorus in fishes (FPF), the amount of phosphorus in detritus (PD), the amount of phosphorus in the biologically active sediment layer (PB), the amount of exchangeable phosphorus in sediments (PE), the amount of phosphorus in interstitial water (PI), and the amount of soluble phosphorus in the lake's waters (PS). The model's forcing functions given as a time table (Table 12.10) include the inflow from tributaries (QTRI), the soluble inorganic P concentration in the inflow (PSTRI), the detritus P concentration in the inflow

(PDTRI), precipitation amounts to the lake (QPREC), outflow from the lake (Q), lake volume (V), lake depth (D), lake water temperature (T), and surface light radiation (I0).

2. Model equations

The equations for the state variables are presented in Table 12.11. Please see Xu et al. (1999, 2001)[17,44] for other equations for the process rates and limiting factors.

3. Model parameters

The parameters determined from literature, experiments, and calibrations are listed in Table 12.12.

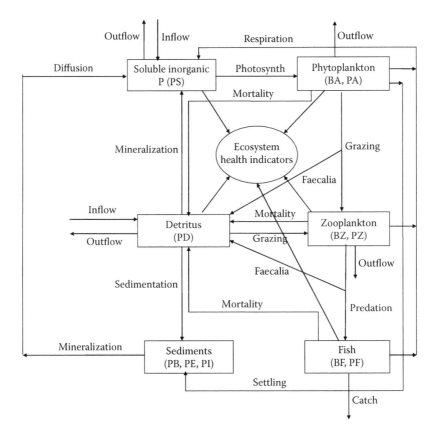

FIGURE 12.4
The conceptual diagram for the Lake Chao ecological model. *Source*: From Xu, F.-L. et al. 2001. Lake ecosystem health assessment: Indicators and methods. *Wat Res* 35 (1): 3160, with permission.

TABLE 12.10

The Model Forcing Functions during April 1987 to March 1998[a]

Month	T (°C)	I0 (Kcal/ m²)	D (m)	V (10⁸m³)	QTRI (10⁶m³/d)	PDTRI (mg/L)	PSTRI (mg/L)	Q (10⁶m³/d)	QPREC (10⁶m³/d)
1987									
April	23.80	4063.7	2.27	17.20	29.17	0.028	0.013	23.15	3.50
May	24.03	3794.2	2.28	19.20	13.15	0.022	0.022	24.48	1.82
June	27.40	4200.0	2.80	21.40	56.85	0.040	0.022	10.26	8.55
July	32.25	4500.0	4.30	33.70	39.36	0.067	0.022	17.73	4.43
August	28.90	3491.6	4.30	33.40	2.40	0.026	0.022	39.08	0.34
September	24.00	3506.7	3.37	25.90	13.29	0.024	0.022	31.94	2.99
October	18.08	2074.6	3.07	23.50	8.73	0.021	0.022	26.91	1.52
November	17.90	1788.9	2.28	17.20	0.87	0.018	0.022	24.32	0.00
December	6.21	2051.6	1.99	14.80	0.82	0.019	0.022	1.66	0.47
1988									
January	5.90	1480.5	1.99	14.80	7.59	0.024	0.024	0.90	2.32
February	5.20	1541.3	2.29	17.20	9.66	0.021	0.024	16.66	1.89
March	8.40	2244.6	2.11	15.70	3.96	0.021	0.024	8.13	0.82

[a] The model-forcing functions include inflow from tributaries (10⁶m³/d) (QTRI), soluble inorganic P concentration in inflow (mg/L) (PSTRI), detritus P concentration in inflow (mg/L) (PDTRI), precipitation to the lake (10⁶m³/d) (QPREC), outflow (10⁶m³/d) (Q), lake volume (10⁸m³) (V), lake depth (m) (D), temperature of lake water (°C) (T), light radiation on the surface of lake water (kcal/m².d) (I0).

TABLE 12.11

Differential Equations for State Variables of the Lake Chao Model

(1) $\dfrac{d}{dt}\left(BA = GA - MA - RA - SA - \dfrac{GZ}{Y0} - \dfrac{Q}{V}\right)* BA$

(2) $\dfrac{d}{dt} PA = AUP * BA -\left(MA + RA + SA + \dfrac{GZ}{Y0} + \dfrac{Q}{V}\right)* PA$

(3) $\dfrac{d}{dt} BZ =\left(MYZ - RZ - MZ - \dfrac{Q}{V}\right)* BZ -\left(\dfrac{PRED1}{Y1}\right)* BF$

(4) $\dfrac{d}{dt} FPZ = MYZ *\left(FPA - FPZ\right) = MYZ *\left(\left(\dfrac{PA}{BA}\right) - FPZ\right)$

(5) $\dfrac{d}{dt} BF =\left(GF - RF - MF - CATCH\right)$

(6) $\dfrac{d}{dt} FPF =\left(\dfrac{PREDY1}{Y1}\right)*\left(FPZ - FPF\right)$

TABLE 12.11 (continued)

Differential Equations for State Variables of the Lake Chao Model

$$(7) \quad \frac{d}{dt}PD = \left(\frac{1}{Y0}-1\right)*GZ*PA-\left(\frac{1}{Y0}-1\right)*PRED1*PZ+MA*PA+$$
$$MZ*PZ+MF*PF+QPDIN-\left(KDP+SD+\frac{Q}{V}\right)*PD$$

$$(8) \quad \frac{d}{dt}PB = \left(QSED*D\Big/(DB*DMU)\right)-QBIO-QDSORP$$

$$(9) \quad \frac{d}{dt}PE = \left((D*KEX*(SA*PS-QSED+SD*PD))\Big/(LUL*DMU)\right)-KE*PE$$

$$(10) \quad \frac{d}{dt}PI = \left(\frac{AE}{AI}\right)*KE*PE-\left(\frac{QDIFF}{AI}\right), \quad AI = LUL*(1-DMU)\Big/D$$

$$(11) \quad \frac{d}{dt}PS = RA*PA+RZ*PZ+RF*PF+QPSIN+KDP*PD+QDIFF+$$
$$\left(\left(\frac{DB}{D}\right)*DMU\right)*(QBIO+QDSORP)-AUP*BA-\left(\frac{Q}{V}\right)*PS$$

Notes:

(1) BA—Phytoplankton biomass (g/m³), GA—phytoplankton growth rate (1/d), MA—phytoplankton motality rate (1/d), RA—phytoplankton respiration rate (1/d), SA—phytoplankton motality rate (1/d), GZ—zooplankton grazing rate (1/d), Y0—assimilation efficiency for zooplankton grazing, Q—outflow(m³/d), V—lake volume(m³);

(2) PA—PA in phytoplankton (g/m³), AUP—phosphorus uptake rate (1/d);

(3) BZ—zooplankton biomass (g/m³), MYZ—zooplankton growth rate (1/d), RZ—zooplankton respiration rate (1/d), MZ—zooplankton motality rate (1/d), PRED1—fish predation rate (1/d), Y1—assimilation efficiency for fish predation;

(4) FPZ—P proportion in zooplankton (kg P/kg BZ);

(5) BF—fish biomass (g/m³), GF—fish growth rate (1/d), RF—fish respiration rate (1/d), MF—fish motality rate (1/d), CATCH—catch rate of fish (1/d);

(6) FPF—P proportion in fish (kg P/kg BF);

(7) PD—phosphorus in detritus (g/m³), QPDIN—PD from inflow (mg/L), KDP—PD decomposition rate (1/d), SD—PD settling rate (1/d);

(8) PB—P in biologically active layer (g/m³), QSED—sediment material from water, D—lake depth (m), DB—depth of biologically active layer in sediment (m), DMU—Dry matter weight of upper layer in sediment (kg/kg), QBIO—demineralization rate of PB (1/d), QDSOPD—sorption and desorption of PB (1/d);

(9) PE—exchangeable P (g/m³), KEX—ratio of exchangeable P to total P in sediments, LUL—depth of unstable layer in sediments (m), KE—PE mineralization rate (1/d);

(10) PI—P in interstitial water (g/m³), QDIFF—diffusion coefficient of PE;

(11) PS—Soluble inorganic P (g/m³), QPSIN—PS from inflow (mg/L).

TABLE 12.12
Parameters for the Lake Chao Ecological Model

Symbol	Description	Unit	Literature Range	Value Used	Sources
Phytoplankton submodel					
GAmax	Maximum growth rate of phytoplankton	1/d	1–5	4.042	Measurement
MAmax	Maximum mortality rate of phytoplankton	1/d	—	0.96	Measurement
RAmax	Maximum respiration rate of phytoplankton	1/d	0.005–0.8	0.6	Measurement
AUPmax	Maximum P uptake rate of phytoplankton	1/d	0.0014–0.01	0.003	Calculation
TAopt	Optimal temperature for phytoplankton growth	°C	—	28	Measurement
TAmin	Minimum temperature for phytoplankton growth	°C	—	5	Measurement
FPAmax	Maximum kg P per kg phytoplankton biomass	—	0.013–0.03	0.013	Ref. 91
FPAmin	Minimum kg P per kg phytoplankton biomass	—	0.001–0.005	0.001	Ref. 91
KI	Michaelis constant for light	kcal/m².d	173–518	400	Ref. 91
KPA	Michaelis constant of P uptake for phytoplankton	mg/L	0.0005–0.08	0.06	Measurement
SVS	Settling velocity of phytoplankton	m/d	0.1–0.8	0.19	Ref. 91
α	Extinction coefficient of water	1/m	—	0.27	Ref. 92
β	Extinction coefficient of phytoplankton	1/m	—	0.18	Ref. 92
θ	Temperature coefficient for phytoplankton settling	—	—	1.03	Ref. 92
Zooplankton submodel					
MYZmax	Maximum growth rate of zooplankton	1/d	0.1–0.8	0.35	Ref. 91
MZmax	Maximum basal mortality rate of zooplankton	1/d	0.001–0.125	0.125	Ref. 91
TOXZ	Toxic mortality rate	1/d	—	0.075	Calibration
Ktoxz	Toxic mortality adjustment coefficient	—	—	0.5	Calibration
RZmax	Maximum respiration rate of zooplankton	1/d	0.001–0.36	0.02	Ref. 91
PRED1max	Maximum feeding rate of fish on zooplankton	1/d	0.012–0.06	0.04	Calibration
TZopt	Optimal temperature for zooplankton growth	°C	—	28	Measurement
TZmin	Minimum temperature for zooplankton growth	°C	—	5	Measurement
KZ	Michaelis constant for fish predation	mg/L	—	0.75	Ref. 93

Symbol	Description	Unit	Range	Value	Method	Reference
KSZ	Threshold zooplankton biomass for fish predation	mg/L	—	0.75		Ref. 94
KA	Michaelis constant for zooplankton grazing	mg/L	0.01–2	0.5		Ref. 91
KSA	Threshold phytoplankton biomass for zooplankton	mg/L	0.01–0.2	0.2		Ref. 95
KZCC	Zooplankton carrying capacity	mg/L	—	30		Calculation
Y0	Assimilation efficiency for zooplankton grazing	—	0.5–0.8	0.63		Ref. 91
Fish submodel						
GFmax	Maximum growth rate of fish	1/d	—	0.015		Measurement
MFmax	Maximum basal mortality rate of fish	1/d	—	0.003		Ref. 93
Ktoxf	Toxic motality rate	1/d	—	0.05		Calibration
TOXF	Toxic motality adjustment coefficient	—	—	0.015		Calibration
RFmax	Maximum respiration rate of fish	1/d	0.00055–0.0055	0.002		Calculation
TFopt	Optimal temperature for fish growth	°C	—	22		Measurement
TFmin	Minimum temperature for fish growth	°C	—	5		Measurement
CATCH	Catch rate of fish	1/d	—	0.001		Calibration
KFCC	Fish carrying capacity	mg/L	—	40		Calculation
Y1	Assimilation efficiency for fish predation	—	—	0.5		Calibration
Detritus, sediments, and soluble inorganic phosphorus submodel						
DB	Depth of biologically active layer in sediment	m	—	0.005		Measurement
LUL	Depth of unstable layer in sediments	m	—	0.16		Measurement
DMU	Dry matter weight of upper layer in sediment	kg/kg	—	0.3		Measurement
KE20	Mineralization rate of PE at 20 C	1/d	—	0.0673		Measurement
KDIFF	Diffusion coefficient of P in interstitial water	—	—	1.21		Ref. 91
KEX	Ratio of exchangeable P to total P in sediments	—	—	0.18		Measurement
SVD	Settling velocity of detritus	m/d	—	0.002		Ref. 91
KDP10	Decomposition rate of detritus P at 10 C	1/d	—	0.1		Calculation
Φ	Temperature coefficient for detritus degradation	—	—	1.072		Ref. 91
θ	Temperature coefficient for PE decomposition soluble inorganic P	—	—	1.03		Ref. 92

Source: Modified from Xu, F.-L. et al. 1999. Modeling the effects of ecological engineering on ecosystem health of a shallow eutrophic Chinese lake (Lake Chao). *Ecol Model* 117:348, with permission.

12.3.2.2.3 The Calibration of Ecological Model

The comparisons of the simulated and the observed values of important state variables and process rates are presented in Figure 12.5, including phytoplankton rates for growth, respiration, mortality and settling, internal

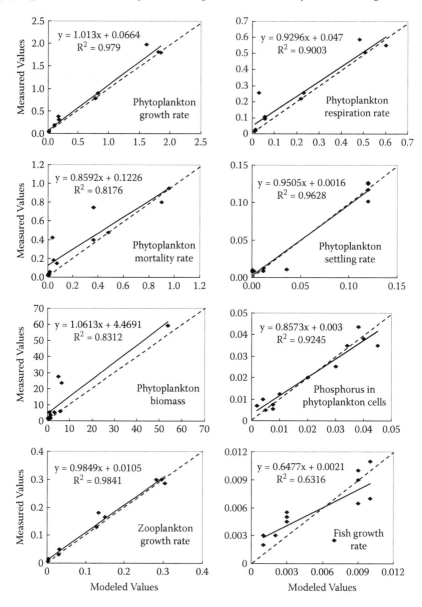

FIGURE 12.5
The comparisons of the modeled and measured state variables and process rates. (The solid lines are trend lines; the dashed lines are "1:1" lines.)

phosphorus concentration in phytoplankton cells, phytoplankton biomass, zooplankton, and fish growth rates. It can be seen from Figure 12.7 that there were very good agreements between observations and simulations of the growth rates, respiration rates, mortality rates, settling rates, internal phosphorus and biomasses of phytoplankton, as well as zooplankton growth rate, with R^2 being over 0.8. There were also good agreements between the simulated and the observed values for fish growth rates, with R^2 being 0.6316.

The results of model calibration suggested that the model could reproduce most of the important state-variable concentrations and process rates using model equations and coefficients, and would represent pelagic ecosystem structure and function in Lake Chao. Therefore, it can be applied for the calculation of ecological health indicators.

12.3.2.2.4 *The Calculation of Ecosystem Health Indicators*

The ecosystem health indicators used in the model include phytoplankton biomass (BA), zooplankton biomass (BZ), zooplankton/phytoplankton ratio (R_{BZBA}), exergy (Ex), and structural exergy (Exst). The calculation results of ecosystem health indicators are presented in Table 12.13.

12.3.2.2.5 *The Assessment of Lake Ecosystem Health*

Relative to contaminated ecosystems, a healthy ecosystem will have a higher zooplankton biomass, lower phytoplankton biomass, higher zooplankton/ phytoplankton ratio, and higher exergy and structural exergy (see Table 12.4 and Ref. 15). According to these principles, and the calculated values for the ecological health indicators in Table 12.13, the results of the lake ecosystem health assessment of Lake Chao are presented in Table 12.13. The relative health states in terms of time span have been arranged from good to poor as follows: January to March 1988 > November to December 1987 > June to

TABLE 12.13

The Modeled Values of Ecological Indicators and Relative Health State in Different Periods in the Lake Chao Ecosystem (from April 1987 to March 1988)

Ecological Indicators	Time Periods[a]					Relative Health State (Good → Poor)
	(A)	**(B)**	**(C)**	**(D)**	**(E)**	
BA (mg•m^{-3})	4.21	4.00	35.31	1.34	0.01	E > D > B > A > C
BZ (mg•m^{-3})	0.90	0.91	3.52	12.52	12.77	E > D > C > B > A
BZ/BA	0.21	0.23	0.10	9.34	127.70	E > D > B > A > C
Ex (MJ•m^{-3})	1312.88	1413.52	1951.99	3256.01	3358.11	E > D > C > B > A
Exst (MJ•mg^{-1})	152.52	158.20	57.97	184.01	196.02	E > D > B > A > C
	Comprehensive results					E > D > B > A > C

[a] A: Apr. to May 1987; B: Jun. to Jul. 1987; C: Aug. to Oct. 1987; D: Nov. to Dec. 1987; E: Jan. to Mar. 1988.

Source: From Xu, F.-L. et al. 2001. Lake ecosystem health assessment: Indicators and methods. *Wat Res* 35 (1): 3165, with permission.

July 1987 > April to May 1987 > August to October 1987. These results are the same as the results using DMM.

12.4 Discussions

12.4.1 About Assessment Results

12.4.1.1 Assessment Results for Lake Chao

The results obtained using two assessment methods, the direct measurement method (DMM) and the ecological modeling method (EMM), are very similar. Namely, the relative health states from good to poor in Lake Chao during April 1987 to March 1988 are as follows: January to March 1988 > November to December 1987 > June to July 1987 > April to May 1987 > August to October 1987. This means that the worst health state occurred between August and October of 1987, followed by April to May of 1987, while the best health state happened between January and March 1988. These results were good correspondence with the observed eutrophic states and the results of eutrophication assessment at Lake Chao.

In terms of the observations made from April 1987 to March 1988, the most serious algal bloom occurred between August and October of 1987 (summer–autumn bloom). Another algal bloom occurred between April and May 1987 (spring bloom). These two algal blooms are typical symptoms of the eutrophication condition at Lake Chao. Both events severely impaired the lake's ecosystem health.

The results for the lake's trophic state index (TSI) calculations are illustrated in Figure 12.6. The calculations were carried out using the same time

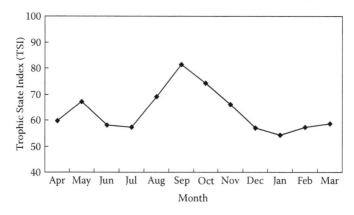

FIGURE 12.6
Dynamics of trophic state in Lake Chao during April 1987 to March 1988.

period data and six indicators: total phosphorus, total nitrogen, chemical oxygen demand, Secchi disk depth, chlorophyll-a concentration, and phytoplankton biomass (see Ref. 27 for details). The average TSI levels were 56.7 for January to March 1988; 61.4 for November to December 1987; 57.7 for June to July 1987; 63.4 for April to May 1987; and 74.4 for August to October 1987. This indicates that the most serious eutrophication event occurred between August and October of 1987, followed by the April to May 1987 period. The lowest eutrophication levels happened between January and March of 1988. The assessment results of the lake's ecosystem health obtained using both the DMM and EMM procedures correspond closely with the lake's existing trophic states.

12.4.1.2 Assessment Results for the Italian Lakes

The relationships between Ecosystem Health Index (EHI) and TSI for the 30 Italian lakes and for Lake Soprano are demonstrated in Figures 12.7 and 12.8, respectively. The TSI calculations were made using the same time period data and the different indicators (total phosphorus, Secchi disk depth, and chlorophyll-a concentration) with the EHI calculations.

It can be seen from both Figures 12.7 and 12.8 that, for the 30 Italian lakes and for Lake Soprano, there are strongly negative relationships between EHI and TSI with R^2 being over 0.72. This means that the lakes with higher TSI have the lower EHI, i.e., the lake ecosystem healthy states become worse with increasing trophic states. For instance, in the 30 Italian lakes, the TSI values for the Disueri, Soprano, and Arancio lakes in the summer of 1988 are 92.3, 78.5, and 78.5, respectively, which are the higher values, and correspondingly, the EHI values for these three lakes in same time period are 12.0, 15.3, and 15.3 respectively, which are the lower values. This means that the more serious the eutrophication, the worse the health state. The same situation can also be found in Lake Soprano in the summer of 1988

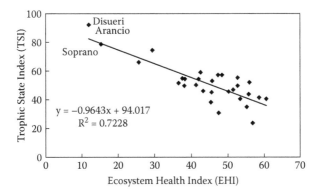

FIGURE 12.7
Relationships between EHI and TSI in 30 Italian lakes in the summer of 1988.

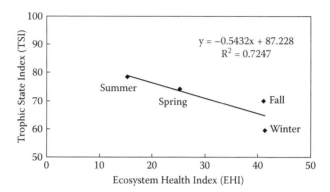

FIGURE 12.8
Relationships between EHI and TSI in Lake Soprano in the fall 1987 to summer 1988.

(see Figure 12.8). Therefore, it can also be concluded that the assessment results of ecosystem health states for the 30 Italian lakes and for Lake Soprano obtained by EHIM procedures accord closely with the lake's existing trophic states.

12.4.2 About Assessment Indicators

The ecological and thermodynamic indicators presented in this chapter cover structural, functional, and system-level aspects of lake ecosystem health. In order to provide a fully informative assessment of the health condition of a lake ecosystem, it is also necessary to apply the indicators simultaneously. Structural changes represent the first response of a lake ecosystem to external anthropogenic stresses. These are then followed by the functional and system-level changes. Structural and functional changes can be described using structural and functional ecological indicators, while system-level changes can be described using thermodynamic indicators, exergy, and structural exergy.

It has been proven that exergy as the ecosystem health indicator consists of (1) homeostasis; (2) absence of disease, partly; (3) diversity or complexity; (4) stability or resilience; (5) vigor or scope for growth of Costanza's definition of ecosystem health; and (6) balance between system components. Structural exergy as an ecosystem health indicator consists of biodiversity (point 3) and the balance between system components (point 6). Ecological buffering capacity as an ecosystem health indicator coincides with description point (4) in Costanza's concept definition of ecosystem health. The combination, therefore, of exergy, structural exergy, and ecological buffering capacity provides an appropriate system-level description of lake ecological health.[22,23,27,30,31,33,35,48] It is suggested that more lake biological components such as bacteria, phytoplankton, zooplankton, benthos, macrophyte, and fish should be used to calculate exergy and structural exergy if the data are available, so that the more reasonable and practical results can be obtained.

However, in the case studies presented in this chapter, only phytoplankton and zooplankton were used to calculate exergy and structural exergy, since the data for biological components other than phytoplankton and zooplankton are very limited.

The indicators are also applicable to the other sites, since the indicators are retrieved from the worldwide experimental and actual research on the responses of lake ecosystems to chemical stresses, including eutrophication, acidification, as well as heavy metal, oil, and pesticide contamination, not just from a single research.

12.4.3 About Assessment Methods

Three methods, DMM, EMM, and EHIM, have been suggested in this chapter for the assessment of ecosystem health of lakes. Combinations of the three methods can be applied for the relative and absolute diagnosis as well as prediction of lake ecosystem health states. The results from the case studies show that the DMM can be used only for the relative assessment of a single lake, while the EHIM can be used for the absolute assessment of both a single lake and different lakes.

The DMM can be used for the rough assessment of a lake's ecosystem health in the case of insufficient data. However, owing to the lack of criteria, there are two major problems with using DMM to assess lake ecosystem health. First, it can only assess the relative healthy status, and it is extremely difficult to assess the actual health status. Second, it is impossible to make comparisons of ecosystem health status for different lakes.

The results from the case studies show that the EHIM can solve the above-mentioned two problems. This method offers a numerical scale from 0 to 100, which can easily make the quantitative assessment and comparison of ecosystem health states for single and different lakes. The criteria for different healthy states can also be obtained using EHIM (see Table 12.6). The EHIM is a valuable method with the advantages of an uncomplicated principle, handy calculation, and reliable and intuitive results. It is expected that the EHIM can be widely used for the quantitative assessment and comparison of ecosystem health states.

It is quite new to apply the EMM for ecosystem health assessment, although ecological modeling has gone through a long history since Lotka–Volterra and Streeter–Phelps models were developed in the 1920s.[49] This may partly be because ecosystem health is a relatively new research field with only about 10 years' history. Compared with the experiment and monitoring method, the EMM is less time-consuming and laborious. However, only a tentative procedure for the practical assessment of ecosystem health using EMM was suggested by Jørgensen (1995).[22] It is necessary to give more documentation and more case studies for ecosystem health assessment using EMM to promote the development of this new field. The key problems in the EMM approach to ecosystem health assessment are how to develop a reliable lake ecological model and how to integrate it with ecosystem health indicators. Previous efforts to develop and apply lake ecological models have indicated that the most important steps are a pre-examination of the lake ecosystem, a determination of the proposed model's complexity, an estimation of the model's parameters, and a calibration of the model's results.[34,50]

In order to predict the changes of lake ecosystem health following changes in environmental conditions, lake models have to be either validated or designed with a dynamic structure. In the first instance, different data sets will have to be used for the model validations in order for the most suitable parameters to be found.[51-53] However, these validated models can only be applied to the ecological health prediction of lakes whose biological structures have remained unchanged, or only slightly changed, since such models have both a given structure and a set of fixed parameters. In the second case, the models can be called Ecological Structural Dynamic Models (ESDM), where goal functions are used to determine how to change the current parameters to express changes in the lake's biological structure following changes in environmental conditions.[51-53] The most appropriate use of such ESDMs would seem to be in predicting changes to lake ecosystem health as a result of changes in environmental conditions. If accomplished, this would represent an important step in ecosystem health assessment through EMMs.

12.5 Conclusions

A tentative theoretical frame, a set of ecological and thermodynamic indicators, and three methods have been proposed for the assessment of ecosystem health of lakes in this chapter. The tentative theoretical frame is composed of five necessary steps: (1) the identification of anthropogenic stresses; (2) the analysis of ecosystem responses to the stresses; (3) the development of indicators; (4) the determination of assessment methods; and (5) the qualitative and quantitative assessment of lake ecosystem health.

In order to develop indicators for lake ecosystem health assessment, five kinds of anthropogenic stresses including eutrophication and acidification, as well as heavy metals, pesticides, and oil pollution were identified, and 59 actual and 20 experimental lakes served as the sample lake ecosystems. A set of ecological and thermodynamic indicators covering lake structural, functional, and system-level aspects were developed, according to the structural, functional, and system-level responses of lake ecosystems to the five kinds of anthropogenic stresses. The structural indicators included phytoplankton cell size and biomass, zooplankton body size and biomass, species diversity, macro- and microzooplankton biomass, the zooplankton-to-phytoplankton ratio, and the macrozooplankton-to-microzooplankton ratio. The functional indicators encompassed the algal C assimilation ratio, resource use efficiency, community production, gross production-to-respiration (i.e., P/R) ratio, gross production-to-standing crop biomass (i.e., P/B) ratio, and standing crop biomass-to-unit energy flow (i.e., B/E) ratio. The ecosystem-level indicators consisted of ecological buffer capacities, exergy, and structural exergy.

Three methods, the Direct Measurement Method (DMM), the Ecological Modeling Method (EMM), and the Ecosystem Health Index Method (EHIM), are proposed for lake ecosystem health assessment. The DMM procedures were designed to (1) identify key indicators; (2) measure directly or calculate indirectly the selected indicators; and (3) assess ecosystem health on the basis of the indicator values. The EMM procedures were designed to (1) determine the structure and complexity of the ecological model according to the lake's ecosystem structure; (2) establish an ecological model by designing a conceptual diagram, establishing model equations, and estimating model parameters; (3) compare the simulated values of important state variables and process rates with actual observations; (4) calculate ecosystem health indicators using the ecological model; and (5) assess lake ecosystem health according to the values of the ecological indicators. The EHIM, which is based on the four-season measured data from 30 Italian lakes, possessed three major steps. First, a numerical EHI in a scale of 0 to 100 was developed. Second, in order to calculate the specific and synthetic EHI, phytoplankton biomass (BA) was selected to serve as a basic indicator, while zooplankton biomass (BZ), the ratio of BZ to BA (BZ/BZ), exergy, and structural exergy were used as additional indicators. Third, the specific and synthetic EHI were calculated based on these indicators, and then the quantitative assessment results for lake ecosystem health could be obtained according to the synthetic EHI values.

The results from Lake Chao demonstrated that both DMM and EMM provided very similar results. A relative order of health states from poor to good was found: August to October 1987 > April to May 1987 > June to July 1987 > November to December 1987 > January to March 1988. This result reflected well the actual situation of Lake Chao. Also, the EHI method was successfully applied to the assessment and comparison of ecosystem health for the Italian lakes.

Acknowledgments

This chapter is supported by the National Science Fund for Distinguished Young Scholars (No. 40725004), the National Natural Science Foundation of China (No. 40671165), and the National Basic Research Program (973 Project) (No. 2007CB407304, No. 2006CB403304).

References

1. Costanza, R., et al. 1997. The values of the world's ecosystem services and natural capital. *Nature* 387:253.
2. Westman, W. E. 1997. What are nature's services worth? *Sciences* 197:960.
3. Cairns, J., Jr. 1997. Sustainability, ecosystem services, and health. *Int J Sustain Dev World Exol* 4:153.
4. Daily, G. C. 1997. Nature's services: Societal dependence on natural ecosystems. Washington, DC: Island Press, 18.
5. Xu, F.-L., and S. Tao. 2000. On the study of ecosystem health: The state of the art. *J Envir Sci (China)* 12 (1): 33.
6. Schaeffer, D. J., E. E. Herricks, and H. W. Kerster. 1988. Ecosystem health: 1. Measuring ecosystem health. *Environmental Management* 12:445.
7. Costanza, R., B. G. Norton, and B. D. Haskell, eds. 1992. *Ecosystem health: New goals for environmental management.* Washington, DC: Island Press.
8. Haskell, B. D., B. G. Norton, and R. Costanza. 1992. What is ecosystem health and why should we worry about it? In *Ecosystem health: New goals for environmental management*, eds. R. Costanza and B. G. Norton, Haskell. Washington, DC: Island Press, 3.
9. Rapport, D. J. 1995. Ecosystem health: Exploring the territory. *Ecosystem Health* 1 (1): 5.
10. Rapport, D. J., L. Richard, and R. E. Paul, eds. 1998. *Ecosystem health.* Malden, MA: Blackwell Science Inc.
11. Rapport, D. J., R. Costanza, and A. J. McMichael. 1998. Assessing ecosystem health. *Trends in Ecology and Evolution* 13 (10): 397.
12. Shear, H. 1996. The development and use of indicators to assess ecosystem health state in the Great Lake. *Ecosyst Health* 2:241.
13. WRI (Water Research Institute). 2002. Stresses on ecosystem health—Chemical pollution. http://www.wri.org/health/ecohealt.html
14. NWRI (National Water Research Institute). 2002. Aquatic Ecosystem Health Assessment Project. http://www.cciw.ca/nwri/aecb/aquatic-eco-health.html
15. Xu, F.-L., S. E. Jørgensen, and S. Tao. 1999. Ecological indicators for assessing freshwater ecosystem health. *Ecol Model* 116:77.
16. Xu, F.-L., et al. 2001. Lake ecosystem health assessment: Indicators and methods. *Water Research* 35 (13): 3157.
17. Xu, F.-L., et al. 2001. A method for lake ecosystem health assessment: An Ecological Modeling Method and its application. *Hydrobiol* 443 (1–3):159–75.
18. Xu, F.-L., et al. 2004. Marine coastal ecosystem health assessment: A case study of Tolo Harbour, Hong Kong, China. *Ecol Model* 173:355–70.
19. Hannon, B. 1985. Ecosystem flow analysis. *Can Bul Fish and Aqu Sci* 213:97.
20. Rapport, D. J., H. A. Regier, and T. C. Hutchinson. 1985. Ecosystem behavior under stress. *American Naturalist* 125:617.
21. Karr, J. R., et al. 1986. Assessing biological integrity in running waters: A method and its rationale. Champaign, Illinois, Natural History Survey, Special Publication 5.
22. Jørgensen, S. E. 1995. Exergy and ecological buffer capacities as measures of ecosystem health. *Ecosystem Health* 1 (3): 150.

23. Jørgensen, S. E. 1995. The application of ecological indicators to assess the eco-logical condition of a lake. *Lakes and Reservoirs: Research and Management* 1:177.
24. Calvo, S., et.al. 1993. Limnological studies on lakes and reservoirs of Sicily. ISSN 0394-0063.
25. Lundgren, A. 1985. Model ecosystem as a tool in freshwater and marine research. *Arch Hydrobiol* Suppl. 70:157.
26. Schindler, D. W. 1987. Detecting ecosystem responses to anthropogenic stress. *Can J Fish Aquat Sci* 44:6.
27. Xu, F.-L. 1997. Exergy and structural exergy as ecological indicators for the development state of the Lake Chao ecosystem. *Ecol Model* 99:41.
28. Xu, F.-L., et al. 2001. Lake ecosystem health assessment: Indicators and methods. *Wat Res* 35 (1): 3157–67
29. Lu, X.-Y. 2001. The dynamics of lake ecological structure and its effects on lake ecosystem health. Master's thesis, Peking University.
30. Jørgensen, S. E., and H. Mejer. 1977. Ecological buffer capacity. *Ecol Model* 3:39.
31. Jørgensen, S. E., and H. F. Mejer. 1979. A holistic approach to ecological model-ing. *Ecol Model* 7:169.
32. Jørgensen, S. E. 1992. *Integration of ecosystem theories: A pattern*. 1st ed. Ch. 8. London: Kluwer Academic Publishers.
33. Jørgensen, S. E. 1992. Parameters, ecological constraints and exergy. *Ecol Model* 62:163.
34. Jørgensen, S. E. Review and comparison of goal functions in system ecology. *Vie Milieu* 44 (1): 11.
35. Jørgensen, S. E., S. N. Nielson, and H. F. Mejer. 1995. Emergy, environ, exergy and ecological modelling. *Ecol Model* 77:99.
36. Jørgensen, S. E. 1982. Exergy and buffering capacity in ecological system. In *Energetics and systems*, eds. W. Mitsch, 61. Ann Arbor, MI: Ann Arbor Science Publishers.
37. Washington, H. G. 1984. Diversity, biotic and similarity indices. *Water Res* 18 (6): 653.
38. Margalef, R. 1961. Communication of structure in plankton population. *Limn Oceanogr* 6:124.
39. Carlson, R. E. 1977. A Trophic State Index for lakes. *Limnol Oceanogr* 22 (2): 361.
40. Tu, Q. Y., et al. 1990. *The researches on the Lake Chao eutrophication*. Heifei: The Publisher of University of Science and Technology of China.
41. Wang, S. Y., et al. 1995. Environmental research for Lake Chao in Anhui Province. In *Lakes in China (Vol. 1)*, ed. X. C. Jin, 189. Beijing: China Ocean Press.
42. Xu, F.-L. 1994. Scientific decision-making system for environmental manage-ment of the Lake Chao watershed. *Environ Protection* 21 (5): 36.
43. Xu, F.-L. 1996. Ecosystem health assessment of Lake Chao, a shallow eutrophic Chinese lake. *Lakes and Reservoirs: Research and Management* 2:101.
44. Xu, F.-L, Jørgensen, S. E., and S. Tao. 1999. Modelling the effects of ecological engineering on ecosystem health of a shallow eutrophic Chinese lake (Lake Chao). *Ecol Model* 117:239.
45. Xu, F.-L., S. Tao, and Z. R. Xu. 1999. The restoration of riparian wetlands and macrophytes in the Lake Chao, an eutrophic Chinese lake: Possibility and effects. *Hydrobiol* 405:169.
46. Xu, F.-L., et al. 2001. A GIS-based method of lake eutrophication assessment. *Ecol Model* 144 (2–3): 231–44.

47. Xu, F.-L., et al. 2003. The distributions and effects of nutrients in the sediments of a shallow eutrophic Chinese lake. *Hydrobiol* 492 (1–3): 85.
48. Xu, F.-L., S. Tao, and R. W. Dawson. 2002. System-level responses of lake ecosystems to chemical stresses using exergy and structural exergy as ecological indicators. *Chemosphere* 46 (2): 173.
49. Xu, F.-L., S. Tao, and R. W. Dawson. 2002. Lake ecological model: History and development. *J Environ Sci (China)* 14 (2): 255.
50. Jørgensen, S. E. 1983. Ecological modelling of lakes. In *Mathematical modelling of water quality: Streams, lakes, and reservoirs,* ed. G. T. Orlob, 116. New York: J. Wiley.
51. Jørgensen, S. E. 1986. Structural dynamic model. *Ecol Model* 31: 1.
52. Jørgensen, S. E. 1988. Use of models as experimental tools to show that structural changes are accompanied by increased exergy. *Ecol Model* 41:117.
53. Jørgensen, S. E. 1992. Development of models able to account for changes in species composition. *Ecol Model* 62:195.
54. Havens, K. E., and J. DeCosta. 1987. Freshwater plankton community succession during experimental acidification. *Arch Hydrobiol* 111:37.
55. Barmuta, L. A., et al. 1990. Responses of zooplankton and zoobenthos to experimental acidification in a high-evaluation lake (Sierra Nevada, California, USA). *Freshwater Biology* 23:571.
56. Havens, K. E. 1992. Acidification effects on the algal-zooplankton interface. *Can J Fish Aquat Sci* 49:2507.
57. Havens, K. E., and R. T. Heath. 1989. Acid and aluminium effects on freshwater zooplankton: An in situ mesocosm study. *Envir Pollu* 62:195.
58. Havens, K. E. 1994. An experimental comparison of the effects of two chemical stressors on a freshwater zooplankton assemblage. *Envir Pollu* 84:245.
59. Havens, K. E. 1994. Structural and functional responses of a freshwater community to acute copper stress. *Envir Pollu* 86:259.
60. Giddings, J. M., et al. 1984. Effects of chronic exposure to coal-derived oil on freshwater ecosystems: II. Experimental ponds. *Environ Toxicol and Chem* 3:465–88.
61. Hurlbert, S. H., M. S. Mulla, and H. R. Wilson. 1972. Effects of organophosphorous insecticide on the phytoplankton, zooplankton, and insect populations of freshwater ponds. *Eco Mono* 42:269.
62. Boyle, T. R. 1980. Effects of the aquatic herbicide 2,4-DMA on the ecology of experimental ponds. *Envir Pollu* 21:35.
63. Schauerte, W., et al. Influence of 2,4,6-trichlorophenol and pentachlorophenol on the biota of aquatic system. *Chemosphere* 11:71.
64. Lay, J. P., W. Schauerte, and W. Klein. 1984. Effects of trichloroethylene on the population dynamics of phyto- and zooplankton in compartments of a natural pond. *Environ Pollu* 33:75.
65. Lay, J. P., et al. 1985. Influence of benzene on the phytoplankton and on *Daphinia pulex* in compartments of an experimental pond. *Ecotoxicol and Environ Safety* 10:218.
66. Lay, J. P., et al. 1985. Long-term effects of 1,2,4-trichlorobenzene on freshwater plankton in an outdoor-model-ecosystem. *Bull Environ Contam Toxicol* 34:761.
67. Lynch, T. R., H. E. Johnson, and W. J. Adams. 1985. Impact of atrazine and hexachlorobiphenil on the structure and function of model stream ecosystem. *Environ Toxi and Chemi* 4:399.

68. Yasuno, M., et al. 1988. Effects of permethrin on phytoplankton and zooplankton in an enclosure ecosystem in a pond. *Hydrobiol* 159:247.

69. Thompson, D. G., et al. 1993. Impact of hexazinone and metilsulfuron methyl on the phytoplankton community of a boreal forest lake. *Envir Toxi and Chem* 12:1695.

70. Thompson, D. G., et al. 1993. Impact of hexazinone and metilsulfuron methyl on the zooplankton community of a boreal forest lake. *Envir Toxi and Chem* 12:1709.

71. Drenner, R. W., et al. 1993. Effects of sediment-bound bifenthrin on gizzard shad and plankton in experimental tank mesocosms. *Envi Toxi and Chem* 12:1297.

72. Havens, K. E., and T. Hanazato. 1993. Zooplankton community responses to chemical stress: A comparison of results from acidification and pesticide contamination research. *Environ Pollu* 82:277.

73. Schindler, D. W., et al. 1985. Long-term ecosystem stress: The effects of years of experimental acidification of a small lake. *Science* 228:1395.

74. Schindler, D. W. 1990. Experimental perturbations of whole lakes as tests of hypotheses concerning ecosystem structure and function. *Oikos* 57:25.

75. Webster, K. E., et al. 1992. Complex biological responses to the experimental acidification of Little Rock Lake, Wisconsin, USA. *Environ Pollut* 78:73.

76. Giddings, J. M. 1982. Effects of the water-soluble fraction of a coal-derived oil on pond microcosms. *Arch Environ Contam Toxicol* 11:735.

77. Cowser, K. E. 1982. Life sciences synthetic fuels semi-annual progress report for the period ending December 31, 1981. ORNL/TM-8229. Oak Ridge, TN: Oak Ridge National Laboratory.

78. Cowser, K. E. 1982. Life sciences synthetic fuels semi-annual progress report for the period ending June 30, 1982. ORNL/TM-8441. Oak Ridge, TN: Oak Ridge National Laboratory.

79. Franco, P. J., et al. Effects of chronic exposure to coal-derived oil on freshwater ecosystems: I. Microcosms. *Environ Toxicol and Chem* 3:447.

80. McNight, D. M. 1981. Chemical and biological processes controlling the response of a freshwater ecosystem to copper stress: A field study of the $CuSO_4$ treatment of Mill Pond Reservoir, Burlington, Massachusetts. *Limnol Oceanogr* 26:518.

81. McNight, D. M., S. W. Chisholm, and D. R. F. Harleman. 1983. $CuSO_4$ treatment of nuisance algal blooms in drinking water reservoirs. *Environ Managem* 7:311.

82. Moore, M. V., and R. W. Winner. 1989. Relative sensitivities of *Ceriodaphnia dubia* laboratory tests and pond communities of zooplankton and benthos to chronic copper stress. *Aquatic Toxicol* 15:311.

83. Taub, F. B., et al. Effects of seasonal succession and grazing on copper toxicity in aquatic microcosms. *Verh Int Ver Limnol* 24:2205.

84. Winner, R. W., and H. A. Owen. 1991. Seasonal variability in the sensitivity of freshwater phytoplankton communities to a chronic copper stress. *Aqu Toxicol* 19:73.

85. Past, M. H., and M. G. Boyer. 1980. Effects of two organophosphorous insecticides on chlorophyll a and pheopigment concentrations of standing ponds. *Hydrobiol* 69:245.

86. Hughes, D. N., et al. 1980. Persistence of three organophosphorus insecticides in artificial ponds and some biological implication. *Arch Environ Contam Toxicol* 9:269.

87. Kaushik, N. K., et al. 1985. Impact of permethrin on zooplankton communities in limnocorrals. *Can J Fish Aquat Sci* 42:77.
88. Hanazato, T., and M. Yasuno. 1987. Effects of a carbonate insecticide, Carbaryl, on summer phyto- and zooplankton communities in ponds. *Environ Pollut* 48:145.
89. Hanazato, T., and M. Yasuno. 1990. Influence of time of application of an insecticide on recovery patterns of a zooplankton community in experimental ponds. *Arch Environ Contam Toxicol* 19:77.
90. Hanazato, T. 1991. Effects of repeated application of Carbaryl on zooplankton communities in experimental ponds with or without the predator Chaoborus. *Environ Pollut* 73:309.
91. Jørgensen, S. E. 1976. A eutrophication model for a lake. *Ecol Model* 2:147.
92. Chen, C. W., and G. T. Orlob. 1975. Ecological simulation of aquatic environments. In *Systems analysis and simulation in ecology, vol. 3.* ed. B. C. Patten, 476. New York: Academic Press.
93. Jørgensen, S. E., S. N. Nielsen, and L. A. Jørgensen. 1991. *Handbook of ecological parameters and ecotoxicology.* Amsterdam: Elsevier Press, 380.
94. Steele, J. H. 1974. *The structure of the marine ecosystems.* Oxford, UK: Blackwell Scientific Publication, 128.
95. Biermen, V. J., et al. 1974. Multinutrient dynamic models of algal growth and species competition in eutrophic lakes. In *Modeling the eutrophication process*, eds. E. Middlebrooks, D. H. Falkenberg, and T. E. Maloney, 89. Ann Arbor, MI: Ann Arbor Science.

13

Application of Ecological Indicators in Forest Management in Africa

Kouami Kokou, Adzo Dzifa Kokutse, and Kossi Adjonou

CONTENTS

13.1 Introduction

The forests in Africa cover about 650 million hectares and represent more than 17% of the world's forests. They are important for cultural, socioeconomic development, and environmental services. According to the United Nations Environmental Programme (UNEP; 2000) forests contribute up to 6% of gross domestic product (GDP) in rich forest countries. Despite this immense status, Africa's natural forest share of the global trade is a mere 2% and its people are among the world's poorest.

In order to ensure sustainable health and productivity of these forest ecosystems, it is necessary to understand the components, and to improve their management. In many cases, the biological systems are threatened by a certain number of factors (Stuart et al. 1990). MacKinnon and MacKinnon (1986) indicate that about 65% of the original habitat of the flora and fauna in South Saharan Africa (SSA) has been lost because of human activities. The deteriorations come from population growth and economic development needs, but also from the transfer of European systems of production that are not adapted to the local conditions, and from the long-term ecological consequences in the African context that are not taken into account.

But in Africa, the socioeconomic and cultural contexts are also more important than the demographic factor to understand the loss or the conservation of forest. However, the roles of the local populations are also part of the models of the biodiversity conservation in SSA. In many traditional societies, local populations are responsible for the maintenance of portions of ecosystem (e.g., sacred forests or riparian forests) (Kokou et al. 2008).

The perspective of climate changes represents another potential threat to the productivity of the African ecosystems. Global warming could cause an important reduction of crop performance in the tropical regions. In some African countries, agricultural performance could decrease from 30% to 50% before 2060 in spite of the introduction of new technologies to mitigate climate changes. Then, Africa becomes more and more vulnerable to the changes that are beyond its control (Monastersky 1992).

The precise threats to African forest ecosystems—and the necessary actions to remedy them—vary from one country to another and from one place to another. Basically, the tendencies of inadequate methods of production and uncontrolled transformation of the landscape are extensively spread in SSA. The structural factors have an important influence on the persistence of these tendencies. The conservation of biodiversity and the management of the forest resources in Africa can be measured, at a considerable scale, by applying ecological indicators (EI) (Appendix 13.1).

These EI serve as tools to assess the current condition and the progress or trend in the intended directions. Some key forest indicators are historical conditions, present conditions, and the projected trend in conditions. It also concerns the development of policies, exploitation with reduced impact, forest planning and inventories, and involvement of local communities in the management, restoration of the forests, and conservation of biodiversity. These measurable factors are integrated into the Principles, Criteria, and Indicators (PCI) at international, national, and especially Forest Management Units (FMU) scale to ensure Sustainable Forest Management (SFM) (McDonald and Lane 2004; International Tropical Timber Organization [ITTO] 1998, 2002).

The objective of this chapter is to present the progress achieved in application of EI to SFM in SSA countries. More experiences are being implemented for EI application in the area. Virtually all countries have enacted supportive policies, legislations, institutional instruments, and reforms. This study highlights the application of EI for enhancing social justice and economic, environmental, institutional, and human capital in forest management in African countries.

13.2 International and National Policies Impacting EI Application to SFM in SSA

The setting up of equitable and extensively widespread international standards influences EI application in SSA. The recent tendencies underline the need to ensure a more equitable distribution, but also to more economically encourage the local populations to conserve natural resources (McNeely 1988). The Convention on the Biologic Diversity (CBD) also explicitly recognizes the rights of genetic and intellectual properties, and underlines the questions of equity and national sovereignty with regard to property and compensation concerning biological resources.

The system of international trade constitutes an important source of pressure on African biodiversity. A set of subsidies (the Convention of Lomé, the agricultural subsidies of the industrialized countries), of commercial agreements (the statute of most favored nation, the General Agreement on the customs Tariffs and Trade [GATT], the Preferential Trade Agreement [PTA], the United Nations Conference on the Trade and Development [UNCTAD], the unions of customs and currencies, the African Financial

Community [AFC]), and the trade barriers exercise important influences on the structure of the imports and exports, at the regional and world levels. The Convention on the International Trade of the Extinction Species (CITES) governs the world trade of the threatened species by forbidding or regulating the trade of some species or populations, based mainly on the level of threat that they undergo. Throughout the world, 113 countries had signed this treaty (Groombridge 1992), including 34 SSA countries (Stuart et al. 1990). The Programs of Structural Adjustment (PSA) initiated by the International Monetary Fund and the World Bank had some negative effects on forest management in Africa. Forest has been damaged more in some countries because of economic uneasiness after the PSA. The response to the market indications caused more fragmentation of the natural ecosystems by intensive agriculture. Agricultural production increased the levels of use of pesticides and pollution (Clark and Juma 1991). In sustainable management of protected areas, the international support influences the national politics. For example, during the last years, the development of the buffer zones around the protected areas was growing according to the belief that total exclusion of humans from the system of protected areas is neither feasible, nor desirable. This approach is currently implemented in several countries through projects.

At the national level, government policies in natural resources management play an important role in EI application. Regarding forest countries in West Africa (Ghana, Côte d'Ivoire, Nigeria), the exploitation of wood had provoked a high level of deforestation during the 1970s and 1980s (Gillis 1988). The governments of these countries tended to considerably underestimate their natural forest resources by making little control on the export. These governments also adopted wood logging systems that create waste on a large scale, and that also harmed the noncommercial species. Currently similar tendencies are being developed in Central Africa for the same reasons; the negative consequences of an extended deforestation could be experienced there more seriously than in West Africa (World Resources Institute [WRI] 1993).

Land tenure and regional development are also important domains where a national environment policy can contribute to create destructive models of landscape. Local suitable practices of land lease have been ignored or replaced by laws and policies coming from outside (Talbott and Furst 1991). In many SSA countries, land management is at rudimentary levels, and the local expertise to carry it out as well as the necessary databases are nonexistent. Evidence of the non-application of the EI is the agricultural overlapping, which often follows the new roads created by the loggers and hunters in forest (Wells and Brandon 1992). Although such opportunist conversion of forest area into agricultural land is an important factor in African deforestation, it is rarely supervised in the management plan (WRI 1993).

The absence of suitable policy on demographic questions such as growth rate, urbanization, and domestic migration are EI in forest management. In some countries, the movements and the settlement of refugees can influence biodiversity conservation. For example, in the Democratic Republic of the Congo (DRC), the migration of the populations of the Kivu region produced negative effects on the forest of Ituri (WRI 1993). On the other hand, in Nigeria and Kenya, the increasing density of the population caused an increase of the invasion of the land, the deforestation being necessary to the production of combustible wood, fodder, fruits, and honey (Cline-Cole et al. 1990).

Another important aspect of EI application in forest management in the SSA is governance. The inability of the governments to pay appropriate salaries to the foresters or the custom services leads them to obtain supplementary incomes.

13.3 EI Application in African Forest Management

13.3.1 Assessing Forest Health, Ecological Role, and Biodiversity Conservation with EI

All African countries possess precious ecosystems. Countries such as the DRC and Gabon, which possess large extents of moist tropical forests, present large numbers of plants and animal species, and the island of Madagascar shelters numerous endemic species. The mountainous regions in Cameroon, Tanzania, Ethiopia, and Rwanda also contain important concentrations of endemism (Stuart et al. 1990). The dry tropical forests are more spread out in SSA than the moist forests, and they are more ecologically complex, although they contain fewer species (Janzen 1988). SSA also contains deserts as well as mountain forests, mangroves, interior lakes, coral reefs, and marshes. These different ecosystems contain plant and animal communities that vary extensively and are exploited in different manners by human beings. SSA counts about 167 national parks and more than 500 forest reserves (MacKinnon and MacKinnon 1986). In Botswana, the protected surface areas are over 17%, and in Malawi, Namibia, Rwanda, Senegal, Tanzania, and Togo this number varies from 11% to 14%. On the other hand, there are no protected areas in Equatorial Guinea, Guinea-Bissau, and Mozambique (Groombridge 1992). Most of these protected areas are currently submitted to pressures, due to the fast growth of the population, and also due to the lack of required resources to manage them appropriately (Hannah 1992). Buffer zones for multiple purposes have been created around these protected areas in order to balance the local needs in resources and for conservation priorities (Wells and Brand 1992). This improvement of land management at the national, provincial, or local level is an important action of the EI application (Kiss 1990).

Local communities use EI in different ways to protect many components of the environment such as forests and biodiversity. Many Masaï groups ban the extraction of plants and wildlife hunting according to the belief that they are the gifts of God and that they must be respected. These people consider that the well-being of an individual lineage depends on how its members maintain a respectful relationship toward a particular animal species. Therefore they abstain from killing or eating such animals; they avoid the place where such animals died or have been killed. These practices are deteriorated nowadays, but the statement of ecological truth, according to which human well-being depends on the well-being of the plants and the animals, is underlined as EI application in forest management.

Small patches of sacred groves were kept close to the villages. Such forests exist in many countries in Africa (Ghana, Togo, Benin, etc.). They are less than one hectare, often containing a tree, a stone, a rock, etc., considered like being a god (Dorm-Adzobu et al. 1991). Such forests may not be important in terms of conservation of the habitat, flora, and fauna, but a unique tree in a sacred site can represent a precious source of genetic material for the reproduction of some plants. They contain species of wild animals considered as sacred, totem, or taboo. Traditionally, such species were strictly protected, and in some cases it was prohibited to eat them or to kill them. A certain number of sacred forests take their origin from a historic event related to the culture of a community (Lieberman 1979; Dorm-Adzobu et al. 1991). These patches of forest are part of the protected zones, filling the vital function to preserve concentrations of biodiversity, and to provide many supplementary advantages (Kokou et al. 2008).

Many rivers and streams that provide the main source of drinking water to a community were considered sacred, and the neighboring forestlands were protected because people believed that the spirit of the river resided in the forest. The taboos associated with such places led to the prohibition of cultivating, cutting the trees, and exploiting the forestlands along the banks of the river. Thus, although the protection of the forests around the rivers was based on religious and cultural beliefs, it also represented a case of management of the riverbed and surrounding vegetation.

13.3.2 EI Integration in Principles, Criteria, and Indicators (PCI) for Forest Management

The concept of SFM (Rametsteiner 1999) is implemented today through important efforts to use criteria and indicators (C&I) as well as tests of their applicability to the conditions in SSA. The follow-up and monitoring of national standards involve the stakeholders in order to be sure that all verifiers have the possibility to be measured at national, local, and FMU levels. Technicians of the public and private sectors as well as those of the nongovernmental organization (NGO) involved directly or indirectly in the management, exploitation, and conservation of the forests are more familiar with the C&I and the relevant issues on the SFM in several countries (Congo Basin, Côte d'Ivoire, and Ghana). The implementation of PCI permitted the restoration and improvement of the social, economic, ecological, biologic, and environmental functions of the forests. It also made possible appreciation of the sense of changes to SFM in SSA.

Manuals of C&I for forest planning have been elaborated in some countries like Gabon and Cameroon, and the C&I are integrated in the management plans of forest concessions. The Basin of Congo is currently being endowed with regional standards. In addition, the national manuals of the C&I serve the National Working Groups (NWG) to promote SFM and forest certification.

In the ITTO member countries (Figure 13.1), the role of the African Timber Organization (ATO) in the development and implementation of PCI is significant. The ATO has been efficient in promoting SFM in African forests by the application and the implementation of ATO/ITTO PCI in 10 countries such as Gabon, Cameroon, Côte d'Ivoire, Central Africa Republic, DRC, Congo Republic, Liberia, Togo, Ghana, and Nigeria, by instituting (1) the mechanism of national consultation; (2) the development of national PCI; (3) the spread of information on the progress accomplished to SFM; (4) the development of regional harmonized PCI ATO/ITTO; (5) the establishment of a consultative forum at the regional level; (6) training in forest audits; (7) capacity building, etc.

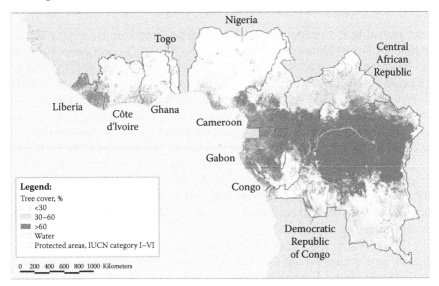

FIGURE 13.1
African countries that are members of ITTO.

13.3.3 Monitoring Forestland Area Security and Change Over Time with EI

Most African forest administrations use remote sensing and the Geographical Information System in forest development, protected areas, and forest concessions management (in Cameroon, Gabon, Congo). Spatial analysis permits one to determine and localize the geographical boundaries of the FMU, the yearly felling plan, the protected areas, the public and forest roads, the dwellings, and the hydrographic network. It is also possible to discriminate between geomorphology (high slopes, swamps, etc.) and soil occupation (dwellings, savannas, plantations, degraded forest, etc.) and follow management instructions, various vocations such as conservation, protection, timber production and community development. Thanks to these techniques, current maps are available for forest management. Forest stratification map development permits one to optimize the political and technical decisions in resources management, the adjustment of investments, the mastery of the relative costs of management, the facilitation of boundaries management from their implementation, control, and conflict management. The spatial analysis also allows forest administrations in some countries in the Congo Basin to identify the behavior of the concessioners and the administrators of protected areas with respect to forest legislation.

13.3.4 EI Application in Forest Zoning, Conservation Easements or Restrictions, Natural Heritage Sites, and Forests Enrolled in High Conservation Value

Land planning is in progress in West Africa where countries have limited Permanent Forest Domains (PFD). In Ghana, the PFD covers about 1.6 million hectares, among which 1 million is production forests and the rest are protected areas. The forests of production have been divided into 52 FMU. All FMU are managed on the basis of management plans elaborated by the forest administration that concedes rights of exploitation to private companies. The first generation of management plans are subject to review. The maximal surface for exploitation assigned to a concessioner is 50,000 ha and the average size is 20,000 ha. Nevertheless, the most important part of timber production in Ghana comes from the Non-permanent Forest Domain (NPFD) represented especially by trees in farmlands. Plantations play a more important role and currently thousands of hectares of Teak plantations are already being exploited.

In Côte d'Ivoire, the PFD covers about 6 million hectares (forests of production and protected areas). The surface of the PFD effectively under management for timber production is estimated to be 2.5 million hectares. SODEFOR, which is responsible for the management of the production forests (classified forests), manages 4 million hectares of natural forests and 150,000 ha of plantations that should spread in fine over 700,000 ha. From 219 classified forests, 85 have management plans and other plans are in preparation. However, only nine classified forests are under timber exploitation. As in Ghana, the most important part of produced timber comes from the NPFD. The exploitation is achieved by private operators selected on call for bids. As Côte d'Ivoire is a country of heavy agricultural activity, with a high population density, its forests face a problem of invasion by the population. More than 80,000 farmers are operating inside the classified forests and 30% of the classified forests were invaded.

Contrary to West Africa, the Congo Basin is, in general, a domain of big natural forest concessions. In Gabon, the PFD covers 12 million hectares, of which 8 million hectares are production forests. Concessions managed by private companies cover 600,000 ha. These companies have elaborate management plans for their concessions three years after signing a temporary convention with the forest administration. Some small and average forest exploitations are managed by local loggers. These local loggers face numerous financial and technical difficulties. They could not implement SFM and therefore could not correctly apply the EI in the field.

In the Republic of Congo, the production forests cover a surface estimated between 20 and 21 million hectares. These forests have been divided into 34 FMU, out of which some have surfaces of more than a million hectares. The forest concessions are managed by private companies that sign agreements with the forest administration, obliging them to achieve forest inventories and management plans. In the DRC, some actions are undertaken for a partial definition of PFD on 60 million hectares. In spite of the abundance of resources, the logging is low because of the instability of the sociopolitical environment. The assignment of the logging is done by mutual agreement between the forest administration and the operators on previously identified forest blocks inventoried by the administration.

In the Republic of Central Africa (RCA), the production forests cover about 5 million hectares and are distributed between the massif of the southwest that covers 3.5 million hectares and the massif of the southeast that covers 1.5 million hectares. The production of lumber varies between 600,000 and 700,000 m³/an. Half of the production in volume is exported and the other half is transformed on the spot. The surfaces to be exploited are conceded by the ministry in charge of the forests after survey of the run book submitted by the concerned company. The forests identified to undergo SFM were subdivided in 12 Permits of Exploitation and Planning (PEP) of surfaces larger than 150,000 ha. The assignments are made for the life span of the company. But besides the PEP, special and artisanal licenses are also assigned for smaller forest areas. The special and artisanal permits don't quite follow sustainable management logics. The RCA benefits from important support of international cooperation for the implementation of SFM in the framework of the project Forest Ecosystems of Central Africa (ECOFAC) and the project *Appui à la Réalisation des Plans d'Aménagement Forestier* (PARPAF).

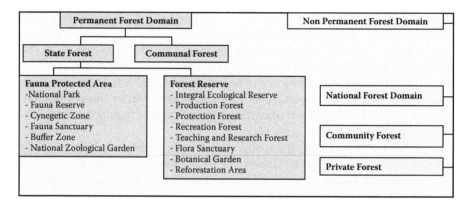

FIGURE 13.2
Example of forest zoning in Cameroon.

In Cameroon, the situation of forest management is intermediate between West Africa and the Congo Basin. Cameroon has important forest resources. The total surface of the production forests is estimated to be 17.5 million hectares. Efforts have been made to delimit the PFD by zoning plan since 1995 (Figure 13.2). After the adoption of the forest code in 1994, the Ministry of Forests began to allow new types of titles for logging activities and progressively eliminated the system of licenses used before. The FMU were assigned through a call of bids and were governed by a procedure of selection. The plan of zoning proposed a field boundary of 90 UFA whose surfaces vary from 30,000 to 200,000 ha. Besides the forests of production of the PFD, Cameroon implemented communal forests (CF) managed by local communities according to a simple management plan approved by the administration. The CF are organized in a sustainable management logic and can therefore be interested in forest certification. But their small size (less than 5,000 ha) and the level of information of the communities are important constraints for such evolution.

13.3.5 EI Application by Stakeholders Involved in Forest Management

The involvement of the stakeholders in forest management is fundamental. The policy of "Top to Down," widespread during the last two decades, is abandoned for the more participative approach of "Down to Top" (Arnold 1995). Encouraged by progress reported from traditional natural resources management initiatives and results of participatory approaches, NGOs, donors, and government agencies initiated pilot trials of participatory forestry on degraded forests in the early 1990s (Adams and Hulme 1999). Much effort has been centered on how to harness sustainable forest management in development. New community-based forest management systems (CBFM) models continue to emerge, and more than one model may be practiced in one country such as joint ventures, leases, consultation, contracts, consigned management, loose confederation, co-management, community-based forest management, etc. By 2002, CBFM was under way in over 35 countries in Africa (Wily 2002; Forestry Outlook Study for Africa [FOSA] 2003) through more than 100 projects, involving about 5,000 communities (FOSA 2003; Sarrazin 2002; Wily 2002). By the end of 2006, virtually all countries had promulgated pro-CBFM policies and legislations, and established units/sections in the forest service responsible for CBFM (World Rainforest Movement [WRM] 2007). Some countries, such as Benin, Burundi, Cameroon, the Democratic Republic of the Congo, and Ghana, had more than 20% of their total forest areas under some form of CBFM (Wily 2002). Rights in CBFM, ranging from temporary agreements or contracts in combination with a management plan for 5 to 15 years, had been implemented in Lesotho, Mozambique, Cameroon, Benin, Gambia, and Ghana (Sarrazin 2002). In Cameroon, the collaboration of the local populations was especially useful to develop the communal forests in order to restore and maintain forest ecosystem in some places. In the framework of ITTO projects on forest concessions management and planning (Congo, Gabon), restoration of degraded forests (Ghana), reforestation in the classified forests (Togo), and struggle against bush fires (Côte d'Ivoire), local communities have been involved. In return, local populations gain socio-collective infrastructures (schools, well, clinic). These projects also ensured the training and the organization of the populations in foresters, beekeepers, and pisciculturists groups or NGO. In Madagascar, the local community accepted the creation of the reserve of Beza Mahafaly, with the intention to improve their means of subsistence (Wells and Brandon 1992). In the program titled Administrative Design for Game Management Areas (ADMADE) in Zambia, local communities participate directly in the protection and management of natural resources and part of the income is returned to them (Kiss 1990). Wily (2002) also reported that CBFM has built a recognizable base during the last two decades, and is gaining the confidence of communities as a promising route for securing SFM. Currently, the process is taking root in all countries, capturing the attention of communities, NGOs/the civil society, governments, and the private sector. The majority of CBFM initiatives begin under the patronage of donor support, often with NGO backstopping. Many of the pilot initiatives that started about two decades ago have paved the way for policies and laws that have in turn introduced the practice in the national forest development agenda, in virtually all countries in the continent (Food and Agriculture Organization [FAO] 2003).

Giving this responsibility to the populations is a better ownership of the development programs. Therefore, CBFM in Cameroon may only be established in nonclassified forests, apart from a few pilot exceptions; restricted to a maximum size of 5,000 ha on 10-year agreements (Egbe 1997). By contrast, Uganda, South Africa, Ethiopia, and Guinea Bissau allow CBFM in forest reserves, including those with high conservation value (Wily 2002). But even in such occasions, wide gaps occur between policies and practice (Barrow et al. 2002). In Kenya, a latecomer into the process, the new Forest Act of 2005 and the draft policy show inclinations to a co-management system, under a buffer zone approach. A common practice is for the state to retain most or all control over licensing, live felling, and enforcement. Despite this limitation, a growing number of community forest associations involved in co-management have emerged and are operating ahead of the policy and forest management rules. In Gambia and Tanzania, pioneer leaders in the CBFM experience in Africa (Sonko and Camara 2000), progress had been made, including wording of by-laws, and other countries are considering the possibilities of doing so (Wily 2002). Few countries, except Gambia, have moved into national planning. Overall, countries continue to limit the CBFM to community forests and joint forest management (JFM) to state forests (Iddi 2002), although official guidelines for nationwide application increasingly exist, e.g., Cameroon, Tanzania, and Senegal (Sarrazin 2002; Wily 2002). Communities in southern Africa, Malawi, and Tanzania are involved in industrial plantation programs under outgrower contract programs (Wily 2002), addressing forest degradation and selling of forest products in Botswana (Mogaka et al. 2001), Mozambique (Mansur and Cuco 2002), and Niger and Mali (Fries and Heemans 1992). Malawi has articulated supportive forest policies and a forest act that specifies community rights and mechanisms for achieving CBFM (Jones and Mosimane 2000). Uganda, Lesotho, and Namibia are also developing along the same line (Wily 2002).

An important case study of collaboration experience with the populations in Ghana is the emphasis put on the role of the women. Indeed, the 31st December Women's Movement (DWM, Ghana) in collaboration with the Forest Office supported communal reforestation in degraded forest and improved the life of women in rural areas. This experience plans how to share the profit in the communal forests through an agreement for 40% to the Forest Office, 40% to the communities, 15% to the traditional landowners, and 5% to the riparian communities.

More complex collaboration has been experimented between the government, local NGO, and populations to promote the forestry and the sustainable management of the natural resources. The acceptance of and adherence to the objectives of the projects by these actors minimize the costs of investment and maximize the results. In Congo, this partnership is associated with the Ministry of Forest Economy, the International NGO Wildlife Conservation Society (WCS), and the enterprise *Congolaise Industrielle des Bois* (CIB) to ensure the long-term integrity to the national park Nouabalé Ndoki, while allowing the CIB to pursue its logging activities.

13.3.6 EI Application in Natural Forests Restoration

The first task achieved by the colonial administrations in African forests was botanical and forest inventories, which permitted the creation of experimental stations to put in place silvicultural methods. Today, these exploratory works are very advanced in most countries. The services entered in an era of convenient forest researches. In Ghana, e.g., the technicians of the Forest Research Institute of Ghana (FORIG) tried to solve the problem of the bud miner of African mahoganies (Meliaceae), which are very precious tropical timber in exports in the international market. The continuous provision and conservation of the mahogany are threatened by over-logging in natural forests and the inability to plant this local species because of the bud miner *Hypsipyla*, which destroys the main stem of the young trees. FORIG has developed technologies to multiply clones of resistant mahoganies. These results provide information that can serve in an integrated management of *Hypsipyla* and contribute to the restoration and conservation of African mahogany in West Africa. Also, Côte d'Ivoire put in place cubage tariffs for 24 forest species of natural forests and plantations. These tools appreciate the capacity of the forests to provide timber for which the tariffs have been established. These cubage tariffs ensure reliability of the data on the potential availability of concerned species in order to plan their sustainable use.

In moist forest areas, the problems are to stop or compensate their impoverishment and increase their economic value. Two main opinions exist. The oldest takes into account the possibilities to exploit some species, which are well known and marketed already. It recommends an intensive silviculture entirely in charge of the forest administration that assigns the execution to a competent technical staff, which concentrates its efforts on very small surfaces called forest reserves of enrichment. The second follows the existent works of old foresters of the Far East and Africa that recommended an extensive silviculture, based on the use of a very large number of species. In this case, it is therefore necessary that a large volume of a very important range of species be marketed, or at least used in place.

The techniques of reforestation in the savanna, nonexistent at the beginning of the twentieth century, has made great progress since 1930, thanks to the experimental stations that determined robust and fast growth species adapted to the climate of these areas. Exotic species such as senna, teak, Albizia, etc., have been grown successfully. But the main enemy of the plantations is the grass that is prone to bush fires in the dry season and can choke the young plantations in the rainy season during the first years.

In the arid regions, the populations are generally nomadic. They possess livestock and harvest products such as gum. The herds graze the savannas and destroy the natural regeneration. Pruning trees to feed the livestock is practiced in an excessive way. With regard to the gum trees, in some regions like Sudan the bleed is practiced correctly, but in a lot of other countries like Mauritania and Senegal it is excessive and too frequent. But in each region possessing forest services, methodical reserves of the most beautiful gum tree populations have been made. These reserves have been purged of all use rights or have been regulated. In some regions, to produce wood for domestic or industrial needs, it was necessary to conduct real reforestations. The results were honestly mediocre; almost always, it was necessary to water the seedlings and the young saplings during the first dry season and so the expenses were out of proportion with the achieved results.

13.3.7 EI Education and Training in Forest Management

The colonial governments introduced models of teaching from Europe. The independence wave that swept through the continent in the early 1960s coincided with the need to find nationals to replace expatriate staff leaving the public service, including forestry. Available local forestry professionals were very few and all were trained outside the continent. The Food and Agricultural Organization (FAO) of the United Nations spearheaded expert and consultative meetings on the need to produce professional foresters (FAO 1962). The inception of forestry education in SSA was largely patterned and shaped after models that were already in place in Europe and North America. These put much emphasis on biophysical aspects of timber production as the main end product of forest management and underemphasized economic, social/cultural, and ecological/environmental issues that also impacted on forestry. The conceptual framework was a vibrant public forestry sector raising and managing forests to feed into public and private wood and fiber industries, and also conserving forestry for multiple benefits (Wyatt-Smith 1970), but the latter was taken as a spillover benefit rather than a mainstream purpose of managing forests. However, soil and water conservation were taken seriously. In this objective, forestry education was structured to produce vocational workers, technicians, and professionals.

Investment in education has varied considerably from country to country and over different time periods. Reidar (2003) sketches the different phases of forestry assistance in the past four decades, focusing on industrial forestry (predominant in the late 1960s and 1970s), social forestry (1980s), environmental forestry (1980s–1990s), and the more recent focus on natural resource management. In most cases, funding has been from public and donor resources.

There was considerable bilateral donor financing especially during the program's (institutional setup) inception. Such funding usually involved establishing physical infrastructure (classrooms, laboratories, computer labs, field stations, vehicles, etc.) and paying salaries to expatriate faculty for a defined time frame during which national staff capacity was developed. Other resources are collaborative research projects with other universities, individual projects to international organizations. This kind of funding is usually very limited and used for research and for extremely limited equipment support to educational institutions in the form of laboratory equipment, computers, copiers, etc. The volume of such funds depends heavily on leadership and the creativity of individual lecturers in the institution, as well as institutional policy and practice in managing grants. Such funds have played an important role in advancing forest science, in maintaining professional interest and contacts among educators, and in supporting postgraduate programs. Thus they contribute to faculty retention and stability.

The shrinking capacity for postgraduate training is pushing interested students to foreign universities. However, studying overseas is expensive and very few find the necessary resources. Besides, few of them return to serve their countries. This has negatively impacted overall scientific and especially research capacity. A regional approach to postgraduate education is recommended. Griffin (1982) questioned the rationale of graduate training of personnel from developing countries in developed countries where young scientists are exposed to sophisticated equipment and experimental conditions that are way beyond what is available to them when they return home. Thus, although there is a need for rigorous exposure to research tools and methodologies, care must be taken to ensure the relevance of such training.

The expert consultative meeting held in Rabat (FAO 2001) identified regional networking and inter-institutional exchange of knowledge and experience as one concrete way of supporting and strengthening forestry education. One such initiative is the RIFFEAC (*Réseau des institutions de Formation Forestière et Environnementale d'Afrique Centrale or Forestry Schools in Central Africa*) network. The network was created by eight forestry schools and research institutions in October 2001, with a view to improving the quality of forestry training to meet the needs of sustainable management of forest ecosystems in the Congo Basin. Among other objectives, the network seeks to promote exchanges between the members, particularly in teaching and research. The facilitation role of the International Union for Conservation of Nature (IUCN) helps RIFFEAC to build strong collaborative ties among its members and develop synergies with other regional initiatives.

The FAO (2003) advocated that curricula at all levels must be updated to include such topics as role of trees outside forests, collaborative management, gender equity, access and benefit sharing, the potential impact of certification schemes on forest practices, and participatory learning. Although it would be nice to have a curriculum that addresses all these and other aspects of forestry (including "traditional core forestry" courses), in reality such a program will be impossible to implement as it is likely to be amorphous and lead to no definable competency. Inadequacies of forestry education and emerging issues have been addressed through short courses addressing specific topics. For example, the International Center for Research in Agroforestry (ICRAF) runs training courses in agroforestry. The International Training Centre (ITC) conducts courses in social forestry, participatory forest management, and natural resource management in which forest managers, extension workers, and those teaching in forestry schools have benefited. The Oxford Forestry Institutes have also given these types of courses in the past. Several universities in Africa offer short courses in agroforestry, social forestry, community forestry, and some aspects of mainstream forestry subjects. In 1995–1996, a Global Environment Facility (GEF)–funded biodiversity project made it possible for university academic staff from Kenya, Uganda, and Tanzania to attend intensive field courses on biodiversity resources assessment techniques including the use of participatory methods. Short courses on ethno-botany have been supported by the World Wildlife Fund for Nature (WWF); United Nations Educational, Scientific, and Cultural Organization (UNESCO); Kew Royal Botanical Gardens; and Center for International Forestry Research (CIFOR). These institutions also support the regular publication of the *People and Plants* handbook. Training workshops related to formulation and project management have also become a common phenomenon, especially among NGOs and for many donor-funded projects. Egerton University (Kenya) is well known for short courses on Participatory Rural Appraisal (PRA), a participatory approach used in all sectors of rural development including in the field of natural resources. All these are critical aspects of continuing education and it can be correctly argued that much of the professional awareness created in emerging issues of tree and forest resource management has been achieved through issue-specific and targeted short courses offered at a variety of institutions. However, in most cases, the efforts are anecdotal and highly dependent on external support. There is a need to establish fitting regional and/or sub-regional mechanisms to recognize the needs and to design and manage such programs. The emerging African Forest Forum (AFF) could provide an excellent platform for this.

13.3.8 EI Application and Forest Products Management

In West Africa the most important product is fuel wood, which represents more than 85% of the total consumption of energy. During the last 20 years, the part of firewood in the total production of wood was already very high (Table 13.1).

The production of industrial wood in West Africa increased over the past years. The European markets are important to Côte d'Ivoire and Ghana, with the first being the main supplier in terms of both value and volume. In general, the trend has been in a decline with trade value falling from 2001 to 2007. Guinea Conakry was an exception, approximately doubling the value of exports in 2005 and 2006 (Table 13.2). Trade is mainly in sawn-timber and molded products, but veneer is also an important trade component for Côte d'Ivoire and Ghana. Prior to the UN ban on exports, Liberia was a major supplier of logs. Nigeria, with limited forest remaining and a large internal market, no longer supplies substantial volumes of timber products except for charcoal exports, which have almost quadrupled since 2001.

TABLE 13.1

Evolution of Fuel Wood Consumption in West Africa

Country	1980 (×1000 m³)	1990 (×1000 m³)	2000 (×1000 m³)
Benin	5,261	5,977	6,453
Burkina Faso	8,655	10,393	12,660
Cap-Vert	106	130	194
Côte d'Ivoire	7,636	8,132	9,284
Gambia	407	571	777
Ghana	12,228	18,424	26,725
Guinea-Bissau	1,637	1,996	2,395
Guinea Conakry	8,744	10,443	12,248
Liberia	2,451	3,750	5,173
Mali	3,086	3,942	4,731
Niger	4,466	6,698	9,356
Nigeria	45,863	56,749	67,789
Senegal	4,095	4,687	5,114
Sierra Leone	5,257	5,115	6,018
Togo	4,055	5,049	6,168
Total	**113,948**	**142,057**	**175,086**

TABLE 13.2

Exports to Europe (Tons)

Country	2001	2002	2003	2004	2005	2006	2007	Average
Côte d'Ivoire	385,770	314,389	260,240	291,182	277,196	250,662	255,558	290,821
Liberia	245,813	234,595	153,540	67	0	0	35	159,208
Ghana	166,031	150,928	142,778	130,880	124,103	101,453	99,278	130,933
Nigeria	58,246	37,723	47,613	61,699	71,717	78,619	96,841	64,908
Guinea Conakry	9,810	4,231	2,743	7,565	9,813	11,486	8,740	7,980
Benin	3,480	2,740	2,561	2,729	3,920	3,147	2,022	3,040
Togo	2,342	1,340	2,333	1,832	1,172	775	985	1,665
Sierra Leone	275	722	227	395	272	400	1,158	786
Total	871,766	746,668	612,034	496,348	488,192	446,542	464,615	659,340

Source: Eurostats.

Exports to Europe and the United States have been declining in both volume and value, but Europe has consistently been the major market (Table 13.3). Meanwhile markets in Africa are increasing with the most significant destination being Nigeria. Exports to Europe in 2007 accounted for only 30% by volume but 42% by value, and Europe remains the prime destination for higher-value products.

Other important destinations for timber exports from West Africa are the United States, China, and India with buyers from the latter two nations

TABLE 13.3

Export Volume (m³ × 1000)

Region	2005 Volume	%	2006 Volume	%	2007 Volume	%
Europe	209	45	168	37	160	30
Africa	71	15	116	26	142	27
Asia/Far East	86	19	82	18	139	26
America	69	15	53	12	55	10
Middle East	27	6	29	6	29	5
Oceania	3	1	3	1	3	1
Total	466	100	452	100	529	100

Source: Forestry Commission, Timber Industry Development Division, 2007.

becoming increasingly active in the region. Indian buyers are particularly focused on teak lumber and poles, while Chinese interest is in logs and lumber of a wide variety of species including camwood and the false-teak or vene.

In Central Africa, forest logging contributes to 0.7% (DRC) to 10% to 13% Republic of Central Africa (RCA) of the GDP. It provides about 20% of the employment opportunities and occupies the second position after mining and/or oil exploitation. The surfaces assigned to forest exploitation during the last decades in the region reached 49.4 million hectares, representing 36% of the total area of the production forests and 27% of the total area of the moist forests. In Equatorial Guinea, Gabon, RCA, and Republic of Congo, 77% to 93% of forests have been allocated (Table 13.4). In DRC, the assignments covered 18% of the production forests because a lot of licenses have been canceled. In the same way, the production has increased and reached 8.5 million cubic meters for the whole region. Gabon was the first country, followed by Cameroon and Republic of Congo. In DRC, the production remains proportionally very weak. On average, 35% of production is exported as logs or lumber. In Equatorial Guinea this proportion reaches 85%, but in Cameroon it is only 6% because of the heavy legal restrictions to export lumbers. In absolute volume, Gabon is the main exporter of lumber. On average, 19% of production is exported after a first transformation. This percentage is the lowest in Equatorial Guinea (5%) and the highest in Cameroon (32%). In Cameroon, since 1986, the export of woods and derivative products (e.g., plywoods) plays an increasing role in the economy. The forest sector, which contributes about 6% to the GDP, currently generates about 45,000 jobs, out of which half are in the informal sector.

13.3.9 EI Application by Private Sector to Forest Management

The role of the private sector in forest management is increasing in African countries, and many multilateral and bilateral programs have been instituted to encourage companies, often accompanied by a reduction of the public agencies

TABLE 13.4

Timber Exportation in Central Africa

Country	Total Forest Area (ha)	Forest Production Areas (ha)	Area Allocated in 2004	Production (m³)	Export Lumber (m³)	Exported Processed Wood (m³)
Cameroon	19,639,000	12,000,000 (61%)	5,400,000 (45%)	2,375,000	141,000 (6%)	758,000 (32%)
Equatorial Guinea	1,900,000	1,500,000 (79%)	1,400,000 (93%)	513,000	438,293 (85%)	27,000 (5%)
Gabon	22,069,999	17,000,000 (77%)	13,800,000 (80%)	3,700,000	1,517,000 (41%)	515,000 (14%)
RCA	6,250,000	3,500,000 (56%)	3,000,000 (86%)	570,000	194,000 (34%)	57,000 (10%)
Republic of Congo	22,263,000	13,000,000 (58%)	10,000,000 (77%)	1,300,000	659,000 (50%)	284,000 (22%)
DRC	108,339,000	90,000,000 (83%)	16,000,000 (18%)	90,000	58,000 (64%)	15,000 (17%)
Central Africa	**180,460,999**	**137,000,000 (76%)**	**49,400,000 (36%)**	**8,548,000**	**3,007,293 (35%)**	**1,656,000 (19%)**

activities. However, in some countries, the potential of initiative toward the application of the EI by the private sector is regulated.

In West Africa, the private sector is very active in the domain of forest exploitation, the trade of the products, the wood processing, etc. It was less attracted by the investments in plantation, for various reasons, e.g., the problems of land property, the long periods of gestation of the investments, the uncertainties of the markets, the prices linked to other risks, and the instability of the policies and legislations (Contreras-Hermosilla 2001). Nevertheless, there is an increasing intervention of the private sector in arboriculture, often on a small scale. Some countries such as Ghana make efforts to encourage the private sector to invest in plantations by creating a specific fund. In Benin, the private forests that applied for the support of the government can be organized in FMU endowed with a management plan.

In Central Africa, several legislations authorize private forests, except DRC where concessions of natural forests are recognized as property of the concessioners. Four countries (Cameroon, Equatorial Guinea, Gabon, and DRC) adopted communal forests. But only Cameroon foresaw that local populations can have a private forest.

In southern Africa, several countries carry out reforms or elaborated laws aimed at transferring a part of the forest resources to the civil society. The holders of private forests can be individuals, villager councils (Tanzania), non-governmental organizations (NGOs), local administrations (Uganda), funds, or associations of communal heritage management (South Africa) or private societies (Zambia). A great part of the new laws in these countries encourages the privatization of production forest belonging to the state by direct sales, rents, or concession.

13.3.10 EI as Basis of Forest Certification and Forest Law Enforcement, Governance, and Trade Process

Many forest companies are involved in the certification process in Africa, especially in the Congo Basin. In Gabon, a forest concession of 580,000 ha managed by CEB, a company of the Thanry group, has been certified. In Congo, CIB, a company of the same group, was certified by Keurhout for a concession of 680,000 ha. Three other enterprises in the sub-region already entered into the Keurhout process. They are Rougier Gabon (288,000 ha), Leroy Gabon (300,000 ha), and IFB (Forest Industry of Batalimo), which manages 425,000 ha of forest in Central Africa Republic. The companies HFC and SIBAF in Cameroon got involved in the forest stewardship council (FSC) system. TRC, Pallisco, Wijma, etc., in Cameroon were already certified. Currently, FSC shows a particular interest in the continent. It has established its regional office in Ghana since April 2005, which could be moved in the coming months to the Congo Basin. The last General Assembly was organized in Cape Town in South Africa in November 2008. FSC adopted an Africa strategy and encouraged the development of Congo Basin regional standards. At the end of 2008, the forest areas certified in the Congo Basin are estimated at 3,011,293 ha (4%).

European markets are of continuing importance to Africa, and this puts the European Commission (EC) through the Forest Law Enforcement, Governance, and Trade (FLEGT) initiative in a unique position to influence reform by voluntary partnership agreement (VPA) negotiation and legality assurance systems (LAS) development. By resisting any push for the exclusive participation of governments only in the negotiating process, the EC is able to ensure wider debate and greater participation involving the experience and aspirations of NGOs and the timber industry. Such participation is essential if reforms that are unpalatable to government, such as industry downsizing, are to be given any proper consideration. The EC's position of influence also enables it to ensure that the highest levels of government are informed and aware of the issues.

The main importance of the EC is that it is in a position to raise international awareness and bring wider attention to the need for reforms. This is particularly relevant with respect to China and India as the timber industries and traders of both nations are extending the search for timber raw material, often with scant regard for the environmental consequences and disregard for legality. Activities are being supported by Indian government subsidies of €100/m^3 for imported raw material. To avoid undermining the FLEGT initiative it is crucial that the EC use its position of influence to encourage the governments of China and India to play an active role in ensuring responsible practice by the nationals of both nations, and explore ways that this may be done through greater cooperation. Actions to be encouraged should include the abolition of environmentally damaging subsidies and support for local industry development through policies prescribing only the import of at least semiprocessed timber products.

While the European Union and the United States are supporting initiatives to eradicate illegal logging, limited concern is being shown by either Indian or Chinese buyers who are allegedly involved in some of the illegal trade. Failure to curtail such activity will undermine efforts being made under the FLEGT initiative. The EC's intention of establishing an economic partnership agreement (EPA) for the entire region envisages open borders and free movement of goods in order to encourage trade, investment, and development. This would remove any control of timber movement currently imposed by Customs authorities. In theory, this means one less barrier to trade in illegal timber, but a widespread opinion is that the role played by Customs authorities is ineffective anyway. The abolition of border controls, along with the need to ensure appropriate monitoring of timber harvesting and movement, make the need for an effective system to track the source of all timber an absolute imperative.

13.4 Progress Assessment in EI Application to SFM

In Africa, assessment of the progress accomplished toward the EI application appears under the following aspects:

Planning and use of the forest resources: Conscious of forest deterioration by human activities and taking into account the international requirements concerning SFM, notably the principles contained in the declaration of Rio de Janeiro in 1992 and the new socioeconomic context, several countries defined a new forest policy in the framework of their National Forest Action Plan (NFAP). In the Congo Basin, forest concessions were allocated on the basis of a convention that requires the development of management plans. The criteria of the development of these management plans integrate the ecological, social, and economic aspects based on the C&I and strengthened by the national laws of forest management. Thanks to the use of the ecological indicators, the spatial organization of the FMU is elaborated on the basis of socioeconomic, ecological, and environmental studies and on forest stratification by remote sensing. The mapping supports detail studies to better value homogeneous forest units. Thanks to this technical integration, the FMU are managed in a series of production, conservation, protection, communal development, and research. In the same way, the rights and duties of the stakeholders were clearly stated.

The data of forest production (number of stems inventoried, logs, volumes by species, processed volume, exported volume, etc.) are recorded following the fluxes of production, from the logging places to the consumer. The spaces opened annually are mastered, the sources of the wood are known (comfortable identification of the illegal wood), and the volumes produced at different stages of the exploitation are well determined. This implies a better application of the various taxes linked to wood exploitation. In the forest companies of Cameroon, units of management are put in place to work with the local populations to study the possibilities of peaceful cohabitation.

The enterprises that wish to certify FSC are obliged to define a policy of preservation of the traditional uses of the forest. Some special measures are taken to insure that the operator's activities do not have a negative impact. The logging of competitive resources (that interest the local populations and the operators), such as some plant species that provide wood and nontimber forest products like Sapelli and Moabi, are regulated. The diameter of logging of some threatened species is reviewed in rise by the managers of the enterprises. Other improvements are the ban of logging at 30 m near the rivers, the creation of a nursery of forest trees for the reforestation of some areas in the forest after logging, the installation of permanent plots of research in the forest concessions, and the respect of environmental norms at the time of construction of the roads and bridges.

The use of the C&I clarifies the domains of intervention and the needs for human and material resources in sustainable management in the FMU. One notices a progressive involvement of the local populations, NGO, and private sector in forest management. The will to develop forest research is expressed in the new policy of several countries. However, many domains remain in which all the African countries need to achieve some goals, notably training, information, forest management funding, knowledge of different forest types and their vegetation, lack of qualified human resources, weak involvement of the rural populations in forest management, and promotion of refresher courses for officials in the forest sector.

The development of the national initiative (NI) in Africa has been one of the objectives of FSC since the establishment of the regional office in Ghana in April 2005. There are currently 14 national initiatives in Africa; most of them are FSC contact persons working in Burkina Faso, Cameroon, Côte D'Ivoire, Democratic Republic of Congo, Ethiopia, Gabon, Morocco, Mozambique, Republic of Congo, Senegal, South Africa, Tanzania, and Zambia.

Economic and fiscal policies aspects: Since the end of the 1990s, forest countries such as Gabon have gotten involved in the promotion and development of wood industries. Efforts have been made to

develop new legislation, taking into account socioeconomic development. In Cameroon, the legislation creates the possibility for the
populations to manage forest resources by creating communal forests and for remittance of a part of the income from the sale of the
forest products for the benefit of the riparian communities. In Côte
d'Ivoire, discussions are in progress on the incentives or financial
advantages to give to pioneer enterprises involved in SFM. Also, discussions have been undertaken between the operators and the government in order to propose solutions to the difficulties faced by the
enterprises. With certification in some countries, data on forest production are mastered, notably forest inventories; the production of
lumber, sawn-timber, veneers, etc.; and the delivery to destinations.
The certification requires a chain of custody of the forest products,
which yields a lot of data that are more reliable for the calculation of
different taxes and for mastering the flow of product.

Institutional aspects: All the countries adopted modern forest legislations favorable to the SFM. The most detailed and complete legislations are those of Cameroon and Ghana. In these two countries,
the laws on forest management go from broad constructs to detailed
norms for implementing management in the field. Most of the countries have a ministerial department in charge of the management
of the forest resources that replaces the approach adopted just after
the independence that conceded forest resource management to the
department of agriculture. Besides the ministerial departments that
are funded mainly with government subsidies, countries such as
Cameroon, Republic of Central Africa, and Congo use forest funds
supplied by a proportion of the revenue from the forest exploitation
and intended to directly fund forest planning operations. In West
Africa, the general picture is that the state is more liable to the forest management through public companies, e.g., SODEFOR in Côte
d'Ivoire and the Forest Commission in Ghana. These public bodies write the management plans for the forests of production in the
permanent domain and implement them themselves. That means
that if the forest certification is considered in these countries, it is
the management of the state that will be evaluated and therefore
national public bodies cannot easily play the role of certifiers and
accredited bodies.

Most countries have regular training facilities for forest technicians. On the other hand, forest research is weak in most countries,
except in Ghana, which creates some problems in refinement of forest planning and forest certification.

Decentralization and devolution strategies have created essential
frameworks and instruments that allow communities access to forest
resources and tenure rights. An increasing number of countries are

gradually devolving authority and ownership to community institutions within whose geographic areas the forests fall and who have the interest and capacity to protect and enforce SFM. Shackleton and Campbell (2001) observed that devolving authority directly to the community level simplifies management and minimizes ambiguities on rights and responsibilities. Wily (2002), concurred, advising against decentralizing powers to line departments or to district councils that have neither the interest nor the capacity for forest management, instead of local communities.

Wily (2002) and Kajembe et al. (2003) have observed that local participation is more meaningful and effective where the local population is involved not as cooperating users but as forest managers. Kajembe et al. (2003) have referred to such CBFM systems as "forest management by consent." Wells and Brandon (1992) stressed that a combined effect of enforcement and participation in resource management is essential for keeping communities from destroying forests. Under the CBFM, this is willingly implemented by communities. Despite imperfection, available positive trends are revealing that participation in forest management is emerging when communities are assigned managerial roles.

Social aspects: The enterprises that look for the FSC certification pay the annual forest royalties, which is a tax by logging area half of which is used for local development projects through the villages. In Cameroon, the companies organize the populations by groups of common initiatives and associations and train them to develop micro projects. There are also campaigns to inform and sensitize the populations on the forest laws.

Facing the threat of European countries closing their borders to timber of illegal or doubtful origin, and because of the implementation of the FLEGT process, forest companies in Cameroon no longer had alternatives and conformed to the norms. That is why they became involved in internal changes with the objective of acquiring a stamp of certification, especially the one of the FSC. Companies such as WIJMA, SEFAC, TRC, PALLISCO, and SFIL/GV in Cameroon made efforts in this respect. With the requirements of the FSC, the forest companies' employees are systematically recorded at the retirement insurance process. The verbal contracts of the temporary workers are adapted into contracts of specific time length. The subcontractors are obliged to formalize the contracts with their employees and treat them similarly to workers in the mother companies. The employees receive necessary safety equipment in their workplace (work clothes, sturdy shoes, helmets, safety glasses, comforters, earplugs). Particular attention is paid to the employees exposed to toxic chemicals; they observe safety measures and get systematic biannual medical checkups.

With FSC certification, communication increased between forest companies and residents. Before logging begins, there is a meeting of the administrative authorities, forest administration, local NGOs, and the economic operator to inform the villages concerned. During this meeting, the stakeholders debate the damages of exploitation, the recruitment of the local communities, the number of available job openings in the companies, the various tasks likely to be required of individuals: illuminating the tracks for the saws, containment, and setting the limits of the logging area or of the FMU.

13.5 Main Problems to EI Application in Africa

13.5.1 Policy, Law, and Governance

Most African countries are regarded as deficient in the implementation of international conventions of which they are members. The poor implementation of strategies on biodiversity makes it impossible to hold back illegal logging. This lack of policy implementation could be explained in some countries by the fact that they are undergoing a Structural Adjustment Program (SAP). For example, in Cameroon, out of the 109 FMU allocated and in use, 66 have an approved and implemented plan, and 41 are classified and managed on definite boundaries and areas. Six management plans have been operational for more than five years without being subject to an evaluation or review of the management plan. The permanent domains are not yet permanent because of illegal logging and farming in FMU. The management of buffer zones in forest-agriculture, forest-livestock, and forest-mining areas is a main problem. The access to land, which is an important condition for the implementation of forestry policy, remains a major concern for many countries. The sustainability of community forest management is held back by the land tenure on which these forests are located. The size of the FMU is sometimes too small, varying between 15,000 and 149,000 ha. Small areas are unsuitable for sustainable management considering natural tropical forests where the production per hectare is very low.

13.5.2 Participation of Local Communities and Stakeholders in Forest Management

Local communities participate efficiently in resources management when they possess a certain degree of control over them. According to Renard (1992), one of the biggest obstacles to the creation of community-based systems of management is the lack of enthusiasm of the government ministries to transfer responsibilities to the communities.

In Africa, land management plans do not take the socioeconomic aspect much into account. The different types of management do not consider indicators such as poverty. Sustainable forest management is incompatible with poverty so far as if the income does not benefit the local population, they will never become interested in the SFM. So poverty accounts for the unfavorable behavior of local populations toward sustainable management, resulting in illegal exploitation of resources.

There is low involvement of the stakeholders such as national NGOs and associations in the forestry sector. The forestry administration does not support local communities for the implementation of simple management plans. There is a lack of sensitization and information of local communities on the issues of sustainable management, and the failure of these communities to organize themselves into viable entities on which the capacity-building programs can be implemented. Concepts of SFM, biodiversity, bio-energy, carbon, legality, certification, etc., are new concepts that come from outside.

13.5.3 Control of Illegal Activities in Forest Management

The lack of an efficient industrial policy in the forestry sector led to the emergence of the informal sector. The wood processing capacity is greater than the forest supply. Few are the factories owners who do not resort to the informal sector and who generally operate illegally.

The inventory of trees before the logging is not systematic. Governance and transparency principles are deficient in the practices. Forestry control is only exercised on forest logging, and the implementation of the management plans is neglected. The proposed penalties do not match with the damages caused to forest potential and to the countries' economy; they do not deter offenders. The consequence of this situation is widespread illegal logging activity in all areas and especially in smaller owned areas. Cases of abuse of influence and corruption are present in the system. Many politicians are loggers and wood sellers. PCI, which serve as management guides within the FMU, are not implemented. Corruption and poor performance practices are regular in public officials.

13.5.4 Funding Forest Management, Administration, Research, and Human Resource Development

The lack of financial support from the state does not allow the forest service to perform their scheduled missions. This situation is the consequence of the Structural Adjustment Program (SAP), which does not make it possible to recruit competent foresters from technical schools or universities, therefore having a negative effect on the technical quality of the work.

This situation leads to the freezing of staff recruitment, aging of the staff, and retiring employees who are not replaced. The workers are not well paid, so there is no motivation, resulting in corruption. The capacity of management, monitoring and controlling the forest resources is limited. Community forests face financial difficulties in carrying out the inventory, management plans, and monitoring of administrative records. The timber economy in Cameroon does not rely on domestic capital; 99% of businesses are foreign-owned.

13.5.5 Structure and Staffing Responsible for SFM

Institutional instability makes the departments in charge of forests unable to implement SFM. The instability of the department implies the instability of the staff. The ministries in charge of forest resources management change often in SSA countries. The new staff receive no information about the previous regime's programs and works in progress. New recruits know almost nothing about what has been done before. Meanwhile, the SFM must be expressed in terms of staff sustainability in the field. The competition between the departments in charge of the rural development policy and forestry sector creates an unhealthy situation. Sometimes the basic infrastructures are nonexistent. In general, human resources are qualitatively and quantitatively insufficient. Most supervisors spend all their time attending meetings and do not have time to devote to the basics. The selection of these officials is sometimes carried out not on the basis of competence, but rather on political or ethnic considerations. Experts in a given field are assigned tasks for which they have no qualifications. Forestry administration is characterized by cumbersome bureaucracy. The capacity building of staff is not a priority. Very few managers have attended refresher courses in their field of expertise. This is a major handicap, a destruction of expert EI application in the field.

13.5.6 Training and Research Institutions

The training institutions that are supposed to provide support to the ministries in charge of forest resources are short of modern teaching materials in the different fields of forestry study (such as compass reading in the forest, biometrics, photo interpretation, remote sensing, GIS, and data processing). This has resulted in less competent technicians in the various fields of the forestry sector. There is a lack of training curricula adapted to the needs in terms of sustainable forest resources management (continuous training and refresher courses). The trainers don't have links with forest administration. Therefore, the collaboration between institutions is informal. The research results are not validated or not available to forest administration. Projects carried out and successfully completed have not had their results validated.

The unemployment rate for graduates of forestry training institutions is high in Africa.

13.6 Conclusion

A new approach to SFM in SSA is urgent and should incorporate the values, systems of knowledge, and points of view of Africans. It should involve the local populations in the management and exploitation of forest resources. It should also consider the conservation of biodiversity and economic development as integral parts of the same sustainable development process. This study synthesizes the experiences acquired through EI application in SFM in SSA. It presents the important progress achieved during the two decades of EI application in African countries. But the application of EI in African countries is very unequal. Progress is more important with regard to the development of legal and institutional aspects of forest resources management. Concerning the implementation of management in the field, the evolution of the effort is positive. In the countries of West Africa (Côte d'Ivoire and Ghana), resources are more limited and the governments directly manage the production forests. For instance, forests are under more pressure because of agricultural activities, and plantations play a more important role in timber provision to the industries. The forest certification should therefore take into account the role of the plantations in West Africa. In the Congo Basin (DRC, Congo, Gabon, Cameroon, RCA), important resources are available and the management of the natural forests for timber production is the most important option. Forest companies that possess big concessions are more involved in SFM and are more likely to be certified. The weak technical capacities of forest administrations in private enterprises is a major constraint to implementing SFM through EI application.

References

Adams, W. M., and D. Hulme. 1999. Conservation and community: Changing narratives, policies and practices in African conservation. Paper for the Conference on African Environments: Past and Present, Oxford, UK.

Arnold, J. E. M. 1995. Community forestry: Ten years review, CF-Note 7, FAO.

Barrow, E., C. Jeanette, I. Grundy, J. Kamungisha–Ruhombe, and Y. Tessema. 2002. Analysis of stakeholder power and responsibilities in community involvement in forest management in Eastern and Southern Africa. *Forest and Social Perspectives in Conservation* 9. NRIT IUCN.

Clark, N., and C. Juma. 1991. *Biotechnology for sustainable development: Policy options for developing countries.* Nairobi, Kenya: African Centre for Technology Studies (ACTS) Press.

Cline-Cole, R. A., H. A. C. Main, and J. E. Nichol. 1990. On fuelwood consumption, population dynamics, and deforestation in Africa. *World Development* 18 (4): 513–28.

Contreras-Hermosilla, A. 2001. Forest institutional issues. Forestry Outlook Study for Africa report.

Dorm-Adzobu, C., O. Ampadu-Agyei, and P. G. Veit. 1991. *Religious beliefs and environmental protection: The Malshegu sacred grove in Northern Ghana.* Washington, DC: World Resources Institute, and Nairobi: African Centre for Technology Studies (ACTS) Press.

Egbe, S. 1997. Forest tenure and access to forest resources in Cameroon: An overview. No. 6, Forest Participation Series. London: IIED Forestry and Land Use Programme.

Food and Agriculture Organization (FAO). 1962. Education and training of foresters. *Unasylva* 16 (1), 64.

———. 2001. Report of the expert consultation on forestry education. Rabat, Morocco, October 17–19. 2001.

———. 2003. *Proceedings of Second International Workshop on Participatory Forestry in Africa.* Arusha, Tanzania, February 18–22, 2002.

Forestry Outlook Study for Africa (FOSA). 2003. Forestry Outlook Study for Africa report. The African Forestry Commission and FAO.

Fries, J., and J. Heemans. 1992. Natural resource management in semi-arid Africa: Status and research needs. *Unasylva* 43:9–15.

Gillis, M. 1988. West Africa: Resource management policies and the tropical forest. In *Public policies and the misuse of forest resources*, eds. R. Repetto and M. Gillis. New York: Cambridge University Press.

Griffin, D. M. 1982. Questioning the relevance of graduate studies in forestry. *Unasylva* 34 (138).

Groombridge, B., ed. 1992. *Global biodiversity: Status of the earth's living resources.* Report of the World Conservation Monitoring Centre. New York: Chapman & Hall.

Hannah, L. 1992. *African people, African parks. An evaluation of development initiatives as a means of improving protected area conservation in Africa.* Washington, DC: Biodiversity Support Program and Conservation International.

Iddi, S. 2002. Community participation in forest management in the United Republic of Tanzania. Paper presented at the second workshop on participatory forestry in Africa. Arusha, Tanzania, February 18–22, 2002.

International Tropical Timber Organization (ITTO). 1998. *Criteria and indicators for sustainable management of natural tropical forest.* Yokohama: ITTO.

———. 2002. *Manual for the application of criteria and indicators for sustainable management of natural tropical forest.* Yokohama: ITTO.

Janzen, D. H. 1988. Tropical dry forests: The most endangered major tropical ecosystem. In *Biodiversity*, ed. E. O. Wilson. Washington, DC: National Academy Press.

Jones, B. T., and A. W. Mosimane. 2000. Empowering communities to manage natural resources. In *Empowering communities to manage natural resources. Case studies from Southern Africa*, eds. S. E. Shackelton and B. M. Campbell, 69–101. Lilongwe, Malawi: SADC Wildlife Sector—Natural Resources Management Programme.

Kajembe, G. C., G. C. Monela, and Z. S. K. Mvena. 2003. Making community man-agement work: A case study of Duru—Haitemba village forest reserve, Babati, Tanzania. In *Policies and governance structures in woodlands of southern Africa*, eds. G. Kowero, B. M. Campbell, and U. R. Sumalia. Bogor: CIFOR.

Kiss, A. 1990. Living with wildlife: Wildlife resource management with local participa-tion in Africa. World Bank Technical Paper no. 130, Africa Technical Department Series. Washington, DC: World Bank.

Kokou, K., K. Adjossou, and A. D. Kokutse. 2008. Considering sacred and riverside forests in criteria and indicators of forest management in low wood producing countries: The case of Togo. *Ecological Indicators* 8:158–69.

Lieberman, D. D. 1979. Dynamics of forest and thicket vegetation on the Accra Plains, Ghana. PhD thesis, University of Ghana, Legon.

MacKinnon, J., and K. MacKinnon. 1986. *Review of the protected areas system in the afro-tropical realm*. Gland, Switzerland: IUCN.

Mansur, E., and A. Cuco. 2002. Building a community forestry framework in Mozambigue. Local communities in sustainable forest management. Paper pre-sented at the second international workshop on participatory forestry in Africa, Arusha Tanzania, February 18–22, 2002.

McDonald, G. T., and M. B. Lane. 2004. Converging global indicators for sustainable forest management. *Forest Policy and Economics* 6:63–70.

McNeely, J. A. 1988. *Economics and biological diversity: Developing and using economic incentives to conserve biological resources*. Gland, Switzerland: IUCN.

Mogaka, H, S. Gacheke, J. Turpie, L. Emerton, and J. Karanja. 2001. Economic aspects of community involvement in sustainable forest management in eastern and southern Africa. IUCN Eastern Africa Programme. *Forest and Social Perspectives in Conservation* 8.

Monastersky, J. 1992. Warming Will Hurt Poor Nations Most. *Science News* 142:116.

Rametsteiner, E., 1999. Sustainable forest management certification: Framework con-ditions, system designs and impact assessment. Ministerial Conference on the Protection of Forest in Europe, Liaison Units, Vienna.

Reidar, P. 2003. *Assistance to forestry: Experiences and potential for improvement*. Jakarta, Indonesia: Center for International Forestry Research.

Renard, Y. 1992. Popular participation and community responsibility in natural resources management: A case from St. Lucia, West Indies. Draft. St. Lucia, West Indies: Caribbean Natural Resources Institute.

Sarrazin, K. 2002. Overview of the second international workshop on participatory forestry in Africa. Paper presented at the second workshop on participatory for-estry in Africa, Arusha Tanzania, February 18–22, 2002.

Shackleton, S. E., and B. M. Campbell. 2001. Devolution in natural resources manage-ment: International arrangements and power shifts: A synthesis of case stud-ies from southern Africa. USAID SADC NRM Project No. 690-0251. 12 WWF – SARPO, EU/CIFOR. Common property step project CSIR report.

Sonko, K., and K. Camara. 2000. Community forest implementation in Gambia: Its principles and prospects. Proceedings of the international workshop on com-munity forestry in Africa. Participatory forest management in Africa, Banjul, Gambia, April 26–30, 1999.

Stuart, S. N., R. J. Adams, and M. D. Jenkins. 1990. Biodiversity in sub-Saharan Africa and its islands: Conservation, management, and sustainable use. Occasional paper of the IUCN.

Talbott, K., and M. Furst. 1991. *Ensuring accountability: Monitoring and evaluating the preparation of national environmental action plans in Africa.* Washington, DC: World Resources Institute.

United Nations Environmental Programme (UNEP). 2000. The state of the environment. African biodiversity. Global Environment Outlook.

Wells, M., and K. Brandon. 1992. *People and parks. Linking protected area management with local communities.* Washington, DC: World Bank, World Wildlife Fund (WWF), and U.S. Agency for International Development (USAID).

Wily, L. A. 2002. Participatory forest management in the United Republic of Tanzania. Paper presented at the second international workshop on participatory forestry in Africa, Arusha, Tanzania, February 18–22, 2002.

World Rainforest Movement (WRM). 2007. Community-based forest management. *WRM Bulletin,* July 27.

World Resources Institute (WRI). 1993. Human interaction with the forests of central Africa. In *Central Africa global climate change and development: Technical reports.* Washington, DC: Biodiversity Support Program.

Wyatt-Smith, J. 1970. Training requirements for forestry in tropical Africa. *Unasylva* 96.

APPENDIX 13.1

Overview of the Applied EI in Forest Management

Most Applied EI in Forest Management	Use and Interpretation of Indicators
1. EI relative to the planning and use of the forest resources	
1.1. Forest health, ecological role, and biodiversity conservation	• Ecosystems diversity and land management characterization
	• Appropriation of different ways to protect components of the environment
	• Filling the function to preserve biodiversity
1.2. EI in Principles, Criteria, and Indicators (PCI) for forest management	• Restoration and improvement of social, economic, ecological, biological, and environmental functions of the forests and the possibility to appreciate the sense of changes
	• Clarification of the domains of intervention and the needs for human and material resources in sustainable management
1.3. Forest land area security and change over time	• Forest development, protected areas, and forest concessions management
	• Determination and localization of the FMU geographical boundaries, the yearly felling plan, the public and forest roads, the dwellings, and the hydrographic network
	• Control of timber production and community development in forest area

(continued)

APPENDIX 13.1 (continued)

Overview of the Applied EI in Forest Management

Most Applied EI in Forest Management	Use and Interpretation of Indicators
1.4. Forest zoning, conservation easements or restrictions, natural heritage sites, and forests enrolled in High Conservation Value (HCV)	• Spatial organization of FMU on the basis of socioeconomic, ecological, and environmental studies and forest stratification • FMU management in series of production, conservation, protection, communal development, and research forests and elaboration of management plans
1.5. Natural forests restoration	• Botanical and forest inventories • Creation of experimental stations for adequate silvicultural methods in different ecosystems • Convenient researches to save forest health • Appreciation of the capacity of the forests to provide timber and Non-timber Forest Products (NTFP) • Reliability of data on the potential availability of forest species in order to plan their sustainable use

2. EI relative to economic and fiscal policies

2.1. Forest products management	• Control of the wood market at every level to reduce pressure of forests • Increasing the economic role of the forests
2.2. Private sector in forest management	• Reduction of the public agencies' activities to favor private sector in forest management • Increasing the investment in forest management • Organization of forest in FMU endowed with a management plan • Development of private or communal forests
2.3. Forest certification and FLEGT process	• Participation involving the experience and aspiration of NGOs and the timber industry in forest management • Abolition of environmentally damaging subsidies and support for local industry development • Eradication of illegal loggings • Appropriate monitoring of timber harvesting and movement, which makes the need for an effective system to track the source of all timber an absolute imperative

3. EI relative to the institutional and social aspects

3.1. Stakeholders' involvement in forest management	• Decentralization and devolution strategies to create essential frameworks and instruments that allow communities access to forest resources and tenure rights • NGOs, donors, and government agencies' collaboration in forest management • Local communities' participation in the protection and management of natural resources • Population's responsibility for better ownership of forest development programs

APPENDIX 13.1 (continued)

Overview of the Applied EI in Forest Management

Most Applied EI in Forest Management	Use and Interpretation of Indicators
	• Emphasis on the role of women in rural areas to promote the forestry and the sustainable management of the natural resources
3.2. Education and training on forest management	• Availability to vocational workers, technicians, and professionals in forestry
	• Training putting emphasis on economic, social/cultural, and ecological/environmental issues that impact forestry
	• Regionalization of forest matter and development of regional initiatives and synergies

14

Using Ecological Indicators to Assess the Health of Marine Ecosystems: The North Atlantic

Villy Christensen and Philippe Cury

CONTENTS

14.1 Introduction

"Roll on, thou deep and dark blue ocean—roll! Ten thousand fleets sweep over thee in vain; Man marks the Earth with ruin—his control stops with the shore," Lord Byron wrote two hundred years ago. Much has happened since, and humans now impact the marine environment to an extent far greater than thought possible centuries or even decades ago.

The impact comes through a variety of channels and forcing factors. Eutrophication and pollution are examples, and while they may be important locally, they constitute less of a direct threat at the global scale. Habitat modification, especially of coastal and shelf systems, is of growing concern for marine ecosystems. Mangroves are being cleared at an alarming rate for aquaculture, removing essential habitat for juvenile fishes and invertebrates; coastal population density is exerting growing influence on coastal systems; and bottom trawls perform clear-cutting of marine habitats, drastically altering ecosystem form and functioning. The looming threat overall to the health of marine ecosystems, however, is the effect of overfishing,[1] and this is the focus of the present contribution.

In recent years we have witnessed a move from the perception that fisheries resources need to be developed by expanding the fishing fleet toward an understanding that the way we exploit the marine environment is bringing havoc to marine resources globally, endangering the very resources on which a large part of the human population relies for nutrition. Perhaps most alarming in this development is that the global fisheries production appears to have been declining steadily since 1990,[2] and the larger predatory fish stocks are being rapidly depleted,[3,4] while ecosystem structure and habitats are altered through intense fishing pressure.[5–8]

In order to evaluate how fisheries impact marine ecosystem health, we have to expand the toolbox traditionally applied by fisheries researchers. Fisheries management builds on assessments of fish populations. Over the years, a variety of tools for management have been developed, and a variety of population-level indicators have seen common use.[9] While such indicators serve and will continue to serve an important role for evaluating best practices for management of fish populations, the scope of fisheries research has widened. This is because of a growing understanding that where fish populations are exploited, their dynamics must be considered as integral components of ecosystem function, rather than as epiphenomena that operate independently of their environment. Internationally, there has been wide recognition of the need to move toward an Ecosystem Approach to Fisheries (EAF), a development strengthened by the Food and Agriculture Organization (FAO) through the Reykjavik Declaration of 2001,[10] and reinforced at the 2002 World Summit of Sustainable Development in Johannesburg, which requires nations to base policies for exploitation of marine resources on an EAF. Guidelines for how this can be implemented are developed through the FAO Code of Conduct for Responsible Fisheries.[11] The move is widely supported by regional and national institutions as well as academia, non-governmental organizations (NGOs), and the public-at-large, and it is mandated by the U.S. National Oceanic and Atmospheric Administration.[12]

Internationally, the first major initiative related to the use of ecosystem indicators for evaluating sustainable fisheries development was taken by the government of Australia in cooperation with the FAO, through a consultation in Sydney in January 1999, involving 26 experts from 13 countries. The consultation resulted in Technical Guidelines No. 8 for the FAO Code of Conduct for Responsible Fisheries: *Indicators for Sustainable Development of Marine Capture Fisheries*.[13] These guidelines were produced to support the implementation of the Code of Conduct, and deal mainly with the development of frameworks, setting the stage for using indicators as part of the management decision process.

The guidelines do not discuss properties of indicators, nor how they are used and tested in practice. This instead became the task of an international working group, established jointly by the Scientific Committee on Oceanic Research (SCOR) and the Intergovernmental Oceanographic Committee (IOC) of the United Nations Educational, Scientific, and Cultural Organization (UNESCO). SCOR/IOC Working Group 119 on Quantitative Ecosystem Indicators for Fisheries Management was established in 2001 with 32 members, drawn internationally. The objective of the working group was to support the scientific aspects of using indicators for an ecosystem approach to fisheries, to review existing knowledge in the field, to demonstrate the utility and perspectives for new indicators reflecting the exploitation and state of marine ecosystems, as well as to consider frameworks for their implementation.[14]

We see the key aspects of ecosystem health as a question of maintaining biodiversity and ecosystem integrity, in line with current definitions of the term. What actually constitutes a "healthy" ecosystem is a debatable topic. This debate includes the way we can promote reconciliation between conservation and exploitation interests. It also includes the recognition and understanding of system states to minimize the risk for loss of integrity when limits are exceeded.[15] From a practical perspective, we here assume that we can define appropriate indicators of ecosystem health and evaluate how far these are from a reference state considered representative of a healthy ecosystem. We will illustrate this by describing indicators in common use as well as the reference state they refer to.

14.2 Indicators

A vast array of indicators have been described and used for characterizing aspects of marine ecosystem health; a non-exhaustive review found upwards of two hundred related indicators.[16] With this background it is clear that the task at hand is not so much one of developing new indicators, but rather one of setting criteria for selecting indicators and evaluating the combination of indicators that may best be used to evaluate the health of marine ecosystems. Indeed, the key aspects of using indicators for management of ecosystems are centered on defining reference states and on developing indicator frameworks as discussed above.[17] Here we will, however, focus on a more practical aspect: What are the indicators that have actually been applied to evaluate the health status of marine ecosystems?

14.2.1 Environmental and Habitat Indicators

Human health is impacted by climate; many diseases break out during the colder winter months in higher latitudes or during the monsoon in the lower. We do not expect to see a similar, clear impact when discussing the marine environment, given that seasonal variability tends to be quite limited in the oceans. We do, however, see longer-term climate trends impacting ocean systems, typically over a time scale of decades, and often referred to as regime-shifts.[18,19] Climate changes especially become important when ecosystem indicators signal change—is a change caused by human impact through, e.g., fishing pressure, or are we merely observing the results of a change in, e.g., temperature? Understanding variability in environmental indicators is thus of fundamental importance for evaluating changes in the status of marine ecosystems. This conclusion is very appropriately supported by the first recommendation of the U.S. Ecosystem Principles Advisory Panel on developing a Fisheries Ecosystem Plan when stating, "[t]he first step in using an ecosystem approach to management must be to identify and bound the ecosystem. Hydrography, bathymetry, productivity and trophic structure must be considered; as well as how climate influences the physical, chemical and biological oceanography of the ecosystem; and how, in turn, the food web structure and dynamics are affected."[12]

A variety of environmental indicators are in common use, including atmospheric (wind, pressure, circulation), oceanographic (chemical composition, nutrients/eutrophication, temperature, and salinity), combined (upwelling, mixed layer depth), and indicators of the effect of environmental conditions for, e.g., primary productivity, plankton patterns, and fish distribution.[20]

Habitat impacts of fisheries have received increasing attention in recent years, focusing on biogenic habitats such as coral reefs, benthic structure, seagrass beds, and kelp forests, which are particularly vulnerable to mechanical damage from bottom trawl and dredging fisheries.[21] The trawling impact on marine habitats has been compared to forest clear-cutting and estimated to annually impact a major part of the oceans shelfs.[22] While habitat destruction has direct consequences for species that rely on benthic habitats for protection (as is the case for juveniles of many fish species),[23] it is less clear how even intensive trawling impacts benthic productivity.[24] One important study, though, found that the productivity of the benthic megafauna increased by an order of magnitude in study sites where trawling had ceased, compared to control sites with continued trawling.[25]

Habitat indicators for ecosystem health in other ecosystem types are often focused on describing communities and community change over time. As marine ecosystems generally are less accessible for direct studies, detailed surveys of habitats are still at a wanting stage. Indeed, for many ecosystems the only informative source may be charts, which traditionally include descriptions of bottom type as an aid to navigation. In recent years critical habitats have, however, received increased focus, and aided by improved capabilities for linking geo-positioning and underwater video or multibeam sonar surveys, habitat mapping projects are now becoming widespread, efficient activities, providing data material that in a foreseeable future will be of use to derive indicators of ecosystem health.[26]

As indicators for human impact on marine habitats, previous studies have used a variety of proxies such as, e.g., proportion of the seabed trawled annually, the ratio of bottom-dwelling and demersal fish abundance, and proportion of seabed area set aside for marine protected areas.

14.2.2 Species-Based Indicators

Indicators of the level of exploitation are central to management of fisheries, focusing on estimating population size and exploitation level of target species.[27] Such applications of indicators are, however, of limited use for describing fisheries impact on ecosystem health if they only consider target species. Instead, the aim for this is to identify species that may serve as indicators of ecosystem-level trends. The breeding success and feeding conditions of marine mammals and birds may, as an example, serve as indicators of ecosystem conditions.[28]

Another approach is to examine community-level effects of fishing, and indications are that indicators for which the direction of change brought about by fishing can be predicted may serve as useful indicators of ecosystem status.[29] Examples of potential indicators may be the average length of fishes or proportion of high-trophic-level species in the catch.

Most studies dealing with community aspects related to species in an ecosystem describe species diversity, be it as richness or evenness measures.[30] A variety of diversity indices have been proposed, with selection of appropriate indices very much related to the type of forcing function that is influencing ecosystem health. However, it is often a challenge when interpreting such indices to describe the reference states for "healthy" ecosystems.[31,32]

Using indicators to monitor individual species is of special interest where there are legal or other obligations, e.g., for threatened species. From an ecological perspective, special interest has focused on keystone species because of their capability to strengthen ecosystem resilience and thus positively impact ecosystem health.[33] Keystone species are defined as strongly interacting species that have a large impact on their ecosystems relative to their abundance. Who are they, and what are their roles in the ecosystem? The classic example from the marine realm is sea otters keeping a favorite prey, sea urchins, in check, allowing kelp forests to abound.[34] Eradication of sea otters has a cascading effect on sea urchins, which in turn deplete the kelp forests. Identification of keystone species is currently the focus of considerable research effort, reflecting that protection of such species is especially crucial for ecosystem health. Surprisingly, few examples of keystone species in marine systems have been published so far, but we are seeing new methods for identification of keystone species being developed.[35]

14.2.3 Size-Based Indicators

It was demonstrated more than 30 years ago that the size distribution of pelagic communities could be described as a linear, negative relationship between (log) abundance and size.[36] The intercept of the size distribution curve is a function of ecosystem productivity, while the slope is due to differential productivity with size. Forcing functions, such as fisheries, are expected to notably impact the slope of the size distribution curves, with increasing pressure associated with increased slopes as larger-sized organisms will be relatively scarce in an exploited system (Figure 14.1). The properties of size distribution curves and how they are impacted by fishing are well understood,[16,31,37] while there is some controversy around the possibility

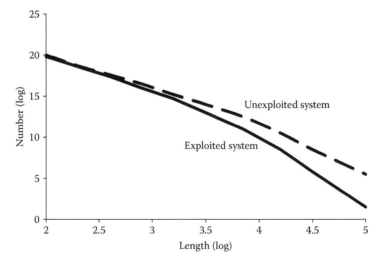

FIGURE 14.1
Particle size distribution curves for an ecosystem in unexploited and exploited states. Data are binned in size classes and logarithmic abundance (usually of numbers, occasionally of biomass) is presented. Exploitation is assumed to mainly reduce abundance of larger-sized organisms, while cascading may cause increase of intermediate sized (not shown here).

of detecting signals from changes in exploitation patterns based on empirical data sets.[38,39] Still, size distribution curves have been widely used to describe ecosystem effects of fishing, and studies have indeed shown promising results, as demonstrated in one of the main contributions to the 1999 International Symposium on Ecosystem Effects of Fishing.[40]

Fisheries impact fish populations by selectively removing larger individuals (see also the Fishing Down the Food Web section below), and thus by removing the faster-growing, large-size-reaching part of the populations.[41] It is widely assumed that if such phenotypic variability has a genetic basis, then exploitation will result in a selective loss in the gene pool with potentially drastic consequences.[42] There is, however, limited empirical evidence of such loss of genetic diversity and genetic drift, notably how to distinguish between genetic and environmental causes, but this may well be because of lack of effort and monitoring.[43]

14.2.4 Trophodynamic Indicators

Fish eat fish, and the main interaction between fish may well be through such means; indeed a large proportion of the world's catches are of piscivorous fishes.[44] There has, for this reason, been considerable attention for development of trophic models of marine ecosystems over the past decades,[45] and this has led to such modeling reaching a state of maturity where it is both widely applied and of use for ecosystem-based fisheries management.[46,47] When extracting and examining results from ecosystem models it becomes a key issue to select indicators to describe ecosystem status and health; we describe aspects of this in the next sections.

14.2.5 Network Analysis

One consequence of the current move toward ecosystem approaches to management of marine resources is that representations of key parameters and processes easily get really messy. When working with a single species it is fairly straightforward to present information in a simple fashion. But what do you do at the ecosystem level when dealing with a multitude of functional groups? One favored approach for addressing this question is network analysis, which has identification of ecosystem-level indicators at its root.

Network analysis is widely used in ecology, as discussed in several other contributions in this volume, and also in marine ecology.[48] In marine ecosystem applications, interest has focused on using network analysis to describe ecosystem development, notably through the work of R. E. Ulanowicz, centered around the concept of ecosystem ascendancy.[49,50] Related analyses have seen widespread application in fisheries-related ecosystem modeling where it is of interest to describe how humans impact the state of ecosystems.[51,52] Focus for many of the fisheries-related modeling has been on ranking ecosystems after maturity, *sensu* Odum.[53] The key aspect of these approaches is

linked to quantification of a selection of the 24 attributes of ecosystem maturity described by E. P. Odum, using rank correlation to derive an overall
measure of ecosystem maturity.[54]

14.2.6 Primary Production Required to Sustain Fisheries

How much do we impact marine ecosystems? This may be difficult to quantify, but the probable first global quantification that went beyond summing
up catches and incorporated an ecological perspective estimated that human
appropriation of primary production through fisheries around 1990 globally
amounted to around 6% of the total aquatic primary production, while the
appropriation where human impact was the biggest reached much higher
levels, for upwelling ecosystems 22%, for tropical shelves 20%, for non-
tropical shelves 26%, and for rivers and lakes 23%.[55] These coastal system-
levels are thus comparable to what has been estimated for terrestrial systems,
where humans appropriate 35% to 40% of the global primary production, be
it directly, indirectly, or foregone.[56]

   In order to estimate the primary production required (PPR) to sustain fisheries we use here an updated version of the approach used for the global
estimates reported above. Global, spatial estimates of fisheries catches are
now available for the period since 1950 along with estimates of trophic levels
for all catch categories (see www.fishbase.org and www.seaaroundus.org).[57]
We estimate the PPR for any catch category as follows:

$$PPR = C_y \cdot \left(\frac{1}{TE} \right)^{TL} \tag{14.1}$$

where C_y is the catch in year y for a given category with trophic level TL,
while TE is the trophic transfer efficiency for the ecosystem. We use a trophic
transfer efficiency of 10% per trophic level throughout based on a meta analysis,[55] and sum overall catch categories to obtain system-level PPR.

   We obtained estimates of total primary production from Nicolas Hoepffner,
the Institute for Environment and Sustainability, JRC, based on SeaWiFS
chlorophyll data for 1998 and the model of Platt and Sathyendranath,[58] as
available through www.seaaroundus.org.

14.2.7 Fishing Down the Food Web

Fishing tales form part of local folklore throughout the world. *I caught a big
fish.* What a big fish is, however, is a moving target as we all tend to judge
based on our own experience, making us part of a shifting-baseline syndrome.[59] As fishing impact intensifies, the largest species on top of the food
web become scarcer, and fishing will gradually shift toward more abundant,
smaller prey species. This forms part of a process termed "fishing down

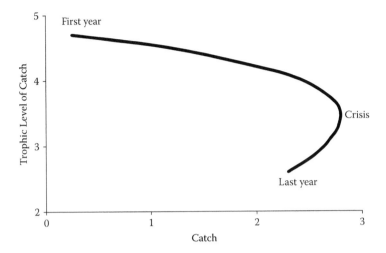

FIGURE 14.2

Illustration of "fishing down the food web" in which fisheries initially target high-trophic-level species with low catch rates. As fishing intensity increases (from first year toward last year) catches shift toward lower-trophic-level species. At high fishing intensity it has often been observed that catches will tend to decrease along with the trophic level of the catch (backward-bending part of curve, starting where "crisis" is indicted).

the food web"[7] in which successive depletion results in initially increasing catches as the fishery expands spatially and starts targeting low-trophic-level prey species rather than high-trophic-level predatory species, followed by a steady phase, and often a decreasing phase caused by overexploitation, possibly combined with shift in the ecological functioning of the ecosystems (see Figure 14.2).[7,44]

A series of publications based on detailed catch statistics and trophic level estimates typically from FishBase (www.fishbase.org) have demonstrated that fishing down the food web is a globally occurring phenomenon.[60–62] Indeed, there seems to be a general trend that the more detailed catch statistics that are available for the analysis, the more pronounced the phenomenon.[62]

As indicator for fishing down the food web we use the mean trophic level of the catch or, as it has been formally adopted by the Convention on Biological Diversity, the Marine Trophic Index. This index is easily estimated from the catches by species, combined with the trophic level of the species, as obtained from FishBase based on trophic models or isotope analysis.

14.2.8 Fishing-in-Balance

An important aspect of fishing down the food web is that we would expect to get higher catches of the more productive, lower-trophic-level catches of prey fishes in return for the loss of less productive, higher-trophic-level catches of predatory fishes. With average trophic transfer efficiencies of 10% between trophic level in marine systems,[55] we should indeed expect, at least theoretically, a 10-fold increase in catches if we could fully eliminate predatory species and replace them with catches of their prey species.

To quantify this aspect of fishing down the food web, an index termed Fishing-in-Balance (*FiB*) has been introduced.[63] The index is calculated based on the calculation of the PPR index, see Equation (14.1),

$$FiB = \log\left(\left[C_y \cdot \left(\frac{1}{TE}\right)^{TL_y}\right] \Bigg/ \left[C_1 \cdot \left(\frac{1}{TE}\right)^{TL_1}\right]\right) \qquad (14.2)$$

where, C_y and C_1 are the catches in year y and the first year of a time series, respectively, and TL_y and TL_1 are the corresponding trophic levels of the catches; TE is the trophic transfer efficiency (10%). The index will start at unity for the first year of a time series, and typically increase as fishing increases (due to a combination of spatial expansion and fishing down the food web), and then often show a stagnant phase followed by a decreasing trend. During the stagnant phase where the *FiB* index is constant, the effect of lower trophic level of catches will be balanced by a corresponding increase in catches. A decrease of 0.1 in the trophic level of the catches will, as an example, be balanced by a $10^{0.1}$ (25%) increase in catch level. There has so far been but few applications of the *FiB* index,[64] but indications are that the index has potential by virtue of being dimensionless, sensitive, and easy to interpret.

14.3 Application of Indicators

We illustrate the application of indicators by presenting accessible information for the North Atlantic Ocean, defined as comprising FAO Statistical Areas 21 and 27. The North Atlantic was the initial focus area for the Sea Around Us project (www.seaaroundus.org) through which information about ecosystem exploitation and resource status has been derived for the period since 1950.[3,65–67] During the second half of the twentieth century the catches increased from an already substantial level of seven million tons per year to reach double this level by the 1970s, then gradually declining (Figure 14.3). Catch composition changed over the period from being dominated by herring and large demersals to lower-trophic-level groups, with high landings of fish for fishmeal and oil. The biomass of higher-trophic-level fish in the North Atlantic has been estimated to have decreased by two-thirds over the half century.[3]

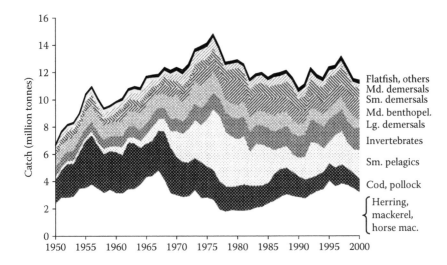

FIGURE 14.3
Total catches and catch composition for the North Atlantic (FAO Areas 21 and 27) estimated based on information from FAO, ICES, NAFO, and national sources. See www.seaaroundus .org for details.

14.3.1 Environmental and Habitat Indicators

There are indications, notably from the Continuous Plankton Recorder surveys of decadal changes, linked to the atmospheric North Atlantic Oscillation Index, and causing marked changes in productivity patterns as well as zooplankton composition.[68] Overall, the changes do not have consequences for ecosystem health, but they change the background at which to evaluate health, and as such should be considered.

Fishing pressure, notably by habitat-damaging bottom trawls, increased drastically during the second half of the twentieth century, where low-powered fleets of gill-netters, Danish seines, and other small-scale fisheries largely were replaced with larger-scale boats dominated by trawlers. The consequence of this has been widespread habitat damage, as illustrated by a large cold-water coral reef area south of Norway, where trawling was impossible until the 1990s when beam-trawlers had grown powerful enough to exploit and completely level the area within a few years.

It is unfortunately characteristic for fisheries science in the second half of the twentieth century that emphasis has been on fish population dynamics, and very little information is available about the effort exerted to exploit the resources, and of the consequences the exploitation has had on habitats. It is thus not possible at present to produce indices of habitat impact at the North-Atlantic scale (or of any larger part of the area for that matter).

14.3.2 Size-Based Indicators

Particle size distributions have been constructed for several areas of the North Atlantic illustrating how fisheries have reduced the abundance of larger fish.[60,69] We do, however, not yet have access to abundance information at the North Atlantic level that makes construction of particle size distributions possible at this scale. If we instead examine how the average of the maximum standard length of species caught in the North Atlantic has developed over the last 50 years we obtain the picture in Figure 14.4. The figure illustrates a gradual erosion of fish capable of reaching large sizes, with the average maximum size decreasing from 120 to 70 cm over the period. This finding links to what is presented below on trophodynamic indicators as size and tropic level are correlated measures.[70]

14.3.3 Trophodynamic Indicators

Network indicators covering the North Atlantic are not available as no model has been constructed for the area overall. There are a large number of models for various North Atlantic ecosystems, including some that cover the time period of interest here. We have, however, not been able to identify any network indicators that could be used to describe aspects of ecosystem health based on the available models. Instead we focus on other trophodynamic indicators that can be estimated from catch statistics as available through www.seaaroundus.org.

We estimate the primary production required (PPR) to sustain the North Atlantic fisheries to vary from 9% of the primary production in 1950, up to

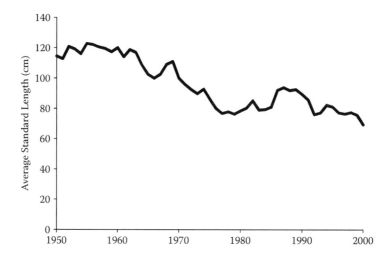

FIGURE 14.4
Average of maximum standard length for all catches of the North Atlantic obtained from information at www.seaaroundus.org.

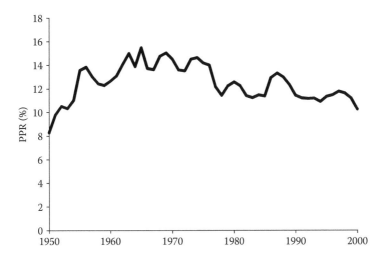

FIGURE 14.5
Primary production required to sustain the fisheries of the North Atlantic, expressed as percentage of the total primary production for the area.

nearly 16% in the late 1960s, then gradually decrease to 11% (Figure 14.5). The appropriation is thus in between the 6% and 26% estimated globally for open oceans and non-tropical shelves, respectively.[55] Since the vast majority of the North Atlantic area is oceanic, the PPR is relatively high compared with other areas. Examining the trend in PPR is by itself not very meaningful for drawing inferences about ecosystem status or health; it is more telling when including information about trends in trophic and catch levels in the considerations as demonstrated below.

The North Atlantic has been exploited for centuries, and has seen its fair share of devastation from the demise of Atlantic gray whales and on to more recent fisheries collapses throughout the area.[71] Reflective of the changes within the fish populations is the fishing down the food web index, which for the North Atlantic takes the shape presented in Figure 14.6. In the 1950s the average trophic level of the catches hovered around 3.5 to 3.55, then decreased sharply during the 1960s and 1970s, reaching a level around 3.3, where it has stayed since.

The decrease in trophic level that occurred during the 1960s and 1970s was, as expected, associated with an increase in catches (see Figure 14.7). The catches increased up to the mid-1960s without any impact on the average trophic level, indicating that the fisheries during this period were in a spatial expansion phase. Through the 1960s up to the mid-1970s the fisheries catches continued to increase but this was now associated with a marked decrease in trophic level of the catches. This in turn is indicative of a fishing down the food web effect, where higher-trophic-level species are replaced with more

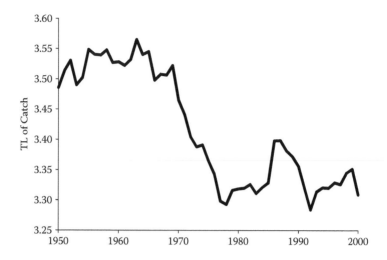

FIGURE 14.6
"Fishing down the food web" in the North Atlantic as demonstrated by the trend in the average trophic level of the catches during the second half of the twentieth century.

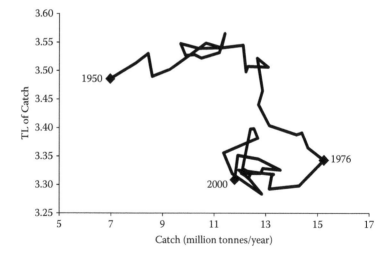

FIGURE 14.7
Phase plot of catches versus the average trophic levels of catches in the North Atlantic, 1950–2000.

productive lower-trophic-level species (Figure 14.8). From the mid-1970s the catches have been decreasing, while remaining at a low trophic level, and without any sign of a return to increased importance of high-trophic-level species. This backward-bending part of the catch—trophic level phase plot (Figure 14.7) seems to be a fairly common phenomenon, and may be

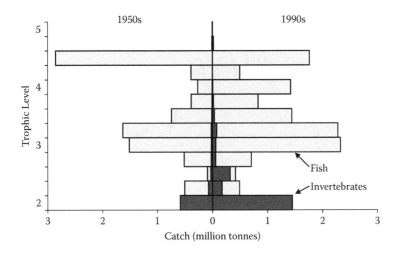

FIGURE 14.8
Catch composition of fish (*light-shaded bars*) and invertebrates (*dark-shaded bars*) by trophic level in the 1950s and the 1990s for the North Atlantic (FAO Areas 21 and 27) as obtained from FishBase (www.fishbase.org).

associated with a breakdown of ecosystem functioning and/or increased non-reported discarding.[7]

A closer examination of the catch composition for the North Atlantic in the 1950s compared with the 1990s shows that the more recent, lower trophic levels of the catches is indeed associated with lower catches of the highest-trophic-level species and higher catches of lower-trophic-level fish species as well as of invertebrates (Figure 14.8). The catch of the uppermost trophic-level category was nearly halved over the period.

As discussed, we would expect that a reduction in the trophic level of the catches should be associated with a corresponding increase in catches (as indeed observed in the 1960s), with the amount being a function of the trophic transfer efficiencies in the system. For the North Atlantic we estimate the corresponding Fishing-in-Balance (*FiB*) index as presented in Figure 14.9. As expected, the *FiB* index increased from its 1950 level up to the mid-1960 level, i.e., through the period characterized by spatial expansion and relatively low resource utilization. From the mid-1960s the index is stable for a decade, i.e., the fishing "was in balance." This was, however, followed with steady erosion from the mid-1970s through the century, where the index shows a clear decline, indicating that the reduction in the average trophic level of the catches is no longer compensated for by a corresponding increase in overall catch levels. The major conclusion that can be drawn from this is that the fisheries of the North Atlantic are unsustainable.

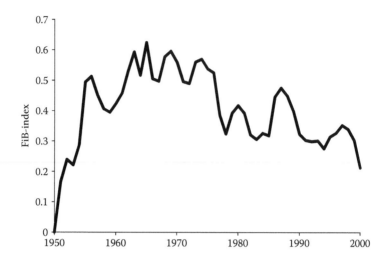

FIGURE 14.9
Fishing-in-Balance index for the North Atlantic, 1950–2000, estimated based on catches and the
average trophic level of the catches.

14.4 Conclusion

Ecosystem-based indicators have only recently become a central focus for the
scientific community working on marine ecosystems. However, there exists
a range of potential indicators that can provide useful information on ecological changes at the ecosystem level, and can help us move toward implementation of an Ecosystem Approach to Fisheries.[14]

We have used the North Atlantic Ocean as a case study to demonstrate the
use of indicators for describing aspects of ecosystem status and health. The
North Atlantic has been exploited for hundreds of years, for some species even
in a sustainable manner up to a few decades ago. Recent trends, however, are
far from encouraging, and the indicators we have selected largely indicate
that the fisheries of the North Atlantic are of a rather unsustainable nature.

If other aspects of the way we impact the North Atlantic are included it
does not improve the picture. This is clear from the detailed study of the fisheries and ecosystems of the North Atlantic presented by Pauly and Maclean,
who concluded by presenting a "report card" for the North Atlantic Ocean
where a "failing grade" was passed for its health status and the way we
exploit it (Table 14.1).[67]

There are no comparable report cards for other areas to facilitate drawing inferences at the global level; it is clear, however, that there are problems globally with the exploitation status of marine ecosystems. The North

TABLE 14.1

Report Card for the Health Status of the North Atlantic

NAME: North Atlantic Ocean

CLASS: Health Status

Subjects	Grade
Long-term productivity of fisheries	F
Economic efficiency of the fisheries	C–
Energy efficiency of the fisheries	D–
Ecosystem status	F
Effects of fisheries on marine mammals	D

Source: Pauly, D. and Maclean, J. L., *In a perfect ocean: The state of fisheries and ecosystems in the North Atlantic* (Washington, DC: Ocean Island Press, 2003).

Atlantic is no special case, indicating that the way the world's fisheries are being conducted is in general far from sustainable.[1]

There is, worldwide, much effort being directed toward improving the exploitation status for marine ecosystems as discussed earlier, and we need to consider how we track the success of such effort, should there be any. This question is very much related to how we assess ecosystem health, and we have here attempted to highlight some related, current research.

The indicators we have presented all relate to the composite ecosystem level, and we note that they all have maintenance of larger-sized, long-lived species as an integral component. We think that maintenance of such species in an ecosystem is important for ecosystem health status.[44] This is in accordance with E. P. Odum's maturity measures;[53] if large-size predators are depleted and marine ecosystems drastically altered through overfishing the risk drastically increases of radical changes in ecosystem status, e.g., through shifts from demersal to pelagic fish–dominated ecosystems or through outbreaks of jellies or red tide. At the decadal level ecosystems may experience alternate semistable states, with potential drastic consequences for food supply—the current problems with cod populations across the North Atlantic serving as a case in point. The safe approach for maintaining healthy, productive ecosystems involves maintaining reproductive stocks of marine organisms at all trophic levels.

Acknowledgments

We thank the members of SCOR/IOC Working Group 119 for a fruitful interaction, as well as SCOR, IOC, and other organizations for the support that made possible the working group's activities. VC also acknowledges support from the Sea Around Us project, initiated and funded by the Pew Charitable Trusts, Philadelphia, as well as from the Natural Sciences and Engineering Research Council of Canada. We thank Reg Watson and Dirk Zeller for making catch information and analysis available, and Daniel Pauly for discussion and suggestions that improved this contribution.

References

1. Pauly, D., Christensen, V., Guénette, S., Pitcher, T. J., Sumaila, U. R., Walters, C. J., Watson, R., and Zeller, D., Towards sustainability in world fisheries, *Nature* 418, 689–95, 2002.
2. Watson, R., and Pauly, D., Systematic distortions in world fisheries catch trends, *Nature* 414, 534–36, 2001.
3. Christensen, V., Guénette, S., Heymans, J. J., Walters, C. J., Watson, R., Zeller, D., and Pauly, D., Hundred-year decline of North Atlantic predatory fishes, *Fish and Fisheries* 4 (1), 1–24, 2003.
4. Myers, R. A., and Worm, B., Rapid worldwide depletion of predatory fish communities, *Nature* 423 (6937), 280–83, 2003.
5. Pauly, D., Alder, J., Bennett, E., Christensen, V., Tyedmers, P., and Watson, R., The future for fisheries, *Science* 302 (5649), 1359–61, 2003.
6. Hall, S. J., *The effects of fisheries on ecosystems and communities* (Oxford, UK: Blackwell, 1998).
7. Pauly, D., Christensen, V., Dalsgaard, J., Froese, R., and Torres, F., Jr., Fishing down marine food webs, *Science* 279 (5352), 860–63, 1998.
8. Kaiser, M., Clarke, K., Hinz, H., Austen, M., Somerfield, P., and Karakassis, I., Global analysis of response and recovery of benthic biota to fishing, *Marine Ecology Progress Series* 311, 1–14, 2006.
9. Collie, J. S., and Gislason, H., Biological reference points for fish stocks in a multispecies context, *Canadian Journal of Fisheries and Aquatic Sciences* 58 (11), 2167–76, 2001.
10. FAO, Report of the Reykjavik Conference on Responsible Fisheries in the Marine Ecosystem—Reykjavik, Iceland, 1-4 October 2001, FAO Fisheries Report No. 658, 2001.
11. FAO, The ecosystem approach to fisheries, FAO Technical Guidelines for Responsible Fisheries. No. 4, Suppl. 2 (Rome: FAO Fisheries Department, 2003), 112.
12. NMFS, Ecosystem-based fishery management: A report to Congress by the Ecosystems Principles Advisory Panel (Silver Spring, MD: National Marine Fisheries Service. U.S. Department of Commerce, 1999).
13. FAO, Indicators for sustainable development of marine capture fisheries: FAO Technical Guidelines for Responsible Fisheries No. 8 (Rome: FAO, 1999).
14. Cury, P., and Christensen, V., Quantitative ecosystem indicators for fisheries management, *ICES Journal of Marine Science* 62 (3), 307, 2005.
15. Fowler, C. W., and Hobbs, L., Limits to natural variation: Implications for systemic management, *Animal Biodiversity and Conservation* 25 (2), 7–45, 2002.

16. Rice, J. C., Evaluating fishery impacts using metrics of community structure, *ICES Journal of Marine Science* 57 (3), 682–88, 2000.

17. Rice, J. C., A generalization of the three-stage model for advice using the precautionary approach in fisheries, to apply broadly to ecosystem properties and pressures, *ICES J Mar Sci* 66 (3), 433–44, 2009.

18. Scheffer, M., Carpenter, S., Foley, J. A., Folke, C., and Walker, B., Catastrophic shifts in ecosystems, *Nature* 413 (6856), 591–96, 2001.

19. Contamin, R., and Ellison, A. M., Indicators of regime shifts in ecological systems: What do we need to know and when do we need to know it, *Ecological Applications* 19 (3), 799–816, 2009.

20. Brander, K., Cod recruitment is strongly affected by climate when stock biomass is low, *ICES Journal of Marine Science* 62, 339–343, 2005.

21. Moore, P. G., Fisheries exploitation and marine habitat conservation: A strategy for rational coexistence, *Aquatic Conservation—Marine and Freshwater Ecosystems* 9 (6), 585–91, 1999.

22. Watling, L., and Norse, E. A., Disturbance of the seabed by mobile fishing gear: A comparison to forest clearcutting, *Conservation Biology* 12 (6), 1180–97, 1998.

23. Sainsbury, K. J., Campbell, R. A., and Whitelaw, W. W., Effects of Trawling on the Marine Habitat on the North West Shelf of Australia and Implications for Sustainable Fisheries Management, in *Sustainable Fisheries through Sustainable Habitat*, ed. Hancock, D. A. (Canberra: Bureau of Rural Sciences Proceedings, AGPS, 1993), 137–45.

24. Steele, J., Roberts, S., Alverson, D., Auster, P., Collie, J., DeAlteris, J., Deegan, L., Briones, E., Hall, S., and Kruse, G., National Research Council study on the effects of trawling and dredging on seafloor habitat (Bethesda, MD: American Fisheries Society, 2005), 91.

25. Hermsen, J. M., Collie, J. S., and Valentine, P. C., Mobile fishing gear reduces benthic megafaunal production on Georges Bank, *Marine Ecology Progress Series* 260, 97–108, 2003.

26. Brown, C., and Blondel, P., Developments in the application of multibeam sonar backscatter for seafloor habitat mapping, *Applied Acoustics*, 2008.

27. Jennings, S., and Kaiser, M. J., The effects of fishing on marine ecosystems, *Advances in Marine Biology* 34, 201–351, 1998.

28. Iverson, S., Springer, A., and Kitaysky, A., Seabirds as indicators of food web structure and ecosystem variability: Qualitative and quantitative diet analyses using fatty acids, *Marine Ecology Progress Series* 352, 235–44, 2008.

29. Trenkel, V. M., and Rochet, M. J., Performance of indicators derived from abundance estimates for detecting the impact of fishing on a fish community, *Canadian Journal of Fisheries and Aquatic Sciences* 60 (1), 67–85, 2003.

30. Rice, J., Environmental health indicators, *Ocean & Coastal Management* 46 (3–4), 235–59, 2003.

31. Gislason, H., and Rice, J., Modelling the response of size and diversity spectra of fish assemblages to changes in exploitation, *ICES Journal of Marine Science* 55 (3), 362–70, 1998.

32. Rochet, M., and Trenkel, V., Why and How Could Indicators Be Used in an Ecosystem Approach to Fisheries Management?, in *The Future of Fisheries Science in North America*, eds. Beamish, R. J. and Rothschild, B. J. (Berlin: Springer, 2009), 209–26.

33. McClanahan, T., Polunin, N., and Done, T., Ecological states and the resilience of coral reefs, *Conservation Ecology* 6 (2), 18, 2002.
34. Kvitek, R. G., Oliver, J. S., Degange, A. R., and Anderson, B. S., Changes in Alaskan soft-bottom prey communities along a gradient in sea otter predation, *Ecology* 73 (2), 413–28, 1992.
35. Libralato, S., Christensen, V., and Pauly, D., A method for identifying keystone species in food web models, *Ecological Modelling* 195 (3–4), 153–71, 2006.
36. Sheldon, R. W., Prahask, A., and Sutcliffe Jr., W. H., The size distribution of particles in the ocean, *Limnology and Oceanography* 17, 327–40, 1972.
37. Shin, Y.-J., and Cury, P., Using an individual-based model of fish assemblages to study the response of size spectra to changes in fishing, *Canadian Journal of Fisheries and Aquatic Sciences* 61 (2), 414–31, 2004.
38. Rochet, M. J., and Trenkel, V. M., Which community indicators can measure the impact of fishing? A review and proposals, *Canadian Journal of Fisheries and Aquatic Sciences* 60 (1), 86–99, 2003.
39. Shin, Y., Rochet, M., Jennings, S., Field, J., and Gislason, H., Using size-based indicators to evaluate the ecosystem effects of fishing, *Ices Journal of Marine Science* 62 (3), 384, 2005.
40. Bianchi, G., Gislason, H., Graham, K., Hill, L., Jin, X., Koranteng, K., Manickchand-Heileman, S., Paya, I., Sainsbury, K., Sanchez, F., and Zwanenburg, K., Impact of fishing on size composition and diversity of demersal fish communities, *ICES Journal of Marine Science* 57 (3), 558–71, 2000.
41. Reiss, H., Hoarau, G., Dickey-Collas, M., and Wolff, W. J., Genetic population structure of marine fish: mismatch between biological and fisheries management units, *Fish and Fisheries* 10(4), 361–395, 2009.
42. Heino, M., and Godo, O. R., Fisheries-induced selection pressures in the context of sustainable fisheries, *Bulletin of Marine Science* 70 (2), 639–56, 2002.
43. Kuparinen, A., and Merilä, J., Detecting and managing fisheries-induced evolution, *Trends in Ecology & Evolution* 22 (12), 652–59, 2007.
44. Christensen, V., Managing fisheries involving predator and prey species, *Reviews in Fish Biology and Fisheries* 6 (4), 417–42, 1996.
45. Christensen, V., and Pauly, D., Ecopath II—A software for balancing steady-state ecosystem models and calculating network characteristics, *Ecological Modelling* 61 (3–4), 169–85, 1992.
46. Latour, R. J., Brush, M. J., and Bonzek, C. F., Toward ecosystem-based fisheries management: Strategies for multispecies modeling and associated data requirements, *Fisheries* 28 (9), 10–22, 2003.
47. Christensen, V., and Walters, C. J., Ecopath with Ecosim: Methods, capabilities and limitations, *Ecological Modelling* 172 (2–4), 109–39, 2004.
48. Wulff, F., Field, J. G., and Mann, K. H., *Network Analysis in Marine Ecology* (Berlin: Springer-Verlag, 1989).
49. Ulanowicz, R. E., *Growth and Development: Ecosystem Phenomenology* (New York: Springer-Verlag, 1986; reprinted by iUniverse, 2000).
50. Ulanowicz, R. E., *Ecology, the Ascendent Perspective* (New York: Columbia University Press, 1997).
51. Christensen, V., and Pauly, D., Trophic Models of Aquatic Ecosystems, *ICLARM Conference Proceedings* 26 (Manila, 1993), 390.
52. Christensen, V., and Pauly, D., Placing fisheries in their ecosystem context, an introduction, *Ecological Modelling* 172 (2–4), 103–107, 2004.

53. Odum, E. P., The strategy of ecosystem development, *Science* 104, 262–70, 1969.
54. Christensen, V., Ecosystem maturity—Towards quantification, *Ecological Modelling* 77 (1), 3–32, 1995.
55. Pauly, D., and Christensen, V., Primary production required to sustain global fisheries, *Nature* 374 (6519), 255–57 [Erratum in Nature, 376: 279], 1995.
56. Vitousek, P. M., Ehrlich, P. R., and Ehrlich, A. H., Human appropriation of the products of photosynthesis, *BioScience* 36, 368–73, 1986.
57. Pauly, D., The Sea Around Us Project: Documenting and communicating global fisheries impacts on marine ecosystems, *AMBIO: A Journal of the Human Environment* 36 (4), 290–95, 2007.
58. Platt, T., and Sathyendranath, S., Oceanic primary production: Estimation by remote sensing at local and regional scales, *Science* 241, 1613–20, 1988.
59. Pauly, D., Anecdotes and the shifting baseline syndrome of fisheries, *Trends in Ecology & Evolution* 10 (10), 430, 1995.
60. Jennings, S., Greenstreet, S. P. R., Hill, L., Piet, G. J., Pinnegar, J. K., and Warr, K. J., Long-term trends in the trophic structure of the North Sea fish community: Evidence from stable-isotope analysis, size-spectra and community metrics, *Marine Biology* 141 (6), 1085–97, 2002.
61. Heymans, J. J., Shannon, L. J., and Jarre, A., Changes in the northern Benguela ecosystem over three decades: 1970s, 1980s, and 1990s, *Ecological Modelling* 172 (2–4), 175–95, 2004.
62. Pauly, D., and Palomares, M. L., Fishing down marine food web: It is far more pervasive than we thought, *Bulletin of Marine Science* 76 (2), 197–211, 2005.
63. Pauly, D., Christensen, V., and Walters, C., Ecopath, Ecosim, and Ecospace as tools for evaluating ecosystem impact of fisheries, *ICES Journal of Marine Science* 57 (3), 697–706, 2000.
64. Christensen, V., Indicators for marine ecosystems affected by fisheries, *Marine and Freshwater Research* 51 (5), 447–50, 2000.
65. Guénette, S., Christensen, V., and Pauly, D., Fisheries Impacts on North Atlantic Ecosystems: Models and Analyses, *Fisheries Centre Research Reports* 9 (4), 344, 2001.
66. Zeller, D., Watson, R., and Pauly, D., Fisheries impact on North Atlantic marine ecosystems: Catch, effort and national and regional data sets, *Fisheries Centre Research Reports* 9 (3), 2001.
67. Pauly, D., and Maclean, J. L., *In a perfect ocean: The state of fisheries and ecosystems in the North Atlantic* (Washington, DC: Ocean Island Press, 2003).
68. Beaugrand, G., Reid, P. C., Ibanez, F., Lindley, J. A., and Edwards, M., Reorganization of North Atlantic Marine Copepod Biodiversity and Climate, *Science* 296 (5573), 1692–94, 2002.
69. Rice, J., and Gislason, H., Patterns of change in the size spectra of numbers and diversity of the North Sea fish assemblage, as reflected in surveys and models, *ICES Journal of Marine Science* 53 (6), 1214–25, 1996.
70. Jennings, S., Pinnegar, J. K., Polunin, N. V. C., and Warr, K. J., Linking size-based and trophic analyses of benthic community structure, *Marine Ecology Progress Series* 226, 77–85, 2002.
71. Mowat, F., *Sea of Slaughter* (Originally published: Boston: Atlantic Monthly Press, 1984).

15

Indicators for the Management of Coastal Lagoons: Sacca di Goro Case Study

J. M. Zaldívar, M. Austoni, M. Plus,
G. A. De Leo, G. Giordani and P. Viaroli

CONTENTS

15.1 Introduction

Coastal lagoons are subjected to strong anthropogenic pressures mainly from urban, agricultural, and industrial effluents and domestic sewage, but also due to the intensive shellfish farming (for a recent review, see Zaldívar, Cardoso, et al. 2008). For example, the Thau lagoon in southern France is an important site for oysters (*Crassostrea gigas*) and mussels (*Mytilus galloprovincialis*) farming (Bacher et al. 1995); the Adriatic lagoons in northern Italy—namely the Venice, Scardovari, and Sacca di Goro lagoons—on average attained a production of around 60,000 t yr⁻¹of the Manila clams, *Ruditapes philippinarum* (Solidoro et al. 2000). The combination of all these pressures calls for an integrated management that considers lagoon hydrodynamics, ecology, nutrient cycles, river runoff influence, shellfish farming, macroalgal blooms, and sediments, as well as the socioeconomical implications of different possible management strategies. Historically, coastal lagoons have undergone multiple and uncoordinated modifications undertaken with only limited sectorial objectives. For example, land-use modifications in the watershed affect the nutrient loadings to the lagoon, and modifications in lagoon bathymetry by dredging change the hydrodynamics. All these factors are responsible for community shifts, and trophic status changes up to eutrophic and dystrophic conditions are achieved, with macroalgal blooms (Viaroli et al. 2008), oxygen depletion, and sulphides production mainly in summer (Viaroli et al. 1995; Chapelle et al. 2001; Marinov et al. 2008; Giordani, Azzoni, et al. 2008).

Studies of lagoons under anthropogenic pressures require the support of hydrological, ecological, and biogeochemical models that integrate key variables and processes in the ecosystem structure and functioning and that link ecosystem responses to human activities. Such complex interactions are otherwise not possible to capture with more traditional statistical tools. Model results can be better handled with ecological indicators that ease comparison of ecosystem health conditions (Christian 2005) and are useful tools for scenario analysis. Models, indicators, and scenarios can be definitely assembled in decision support system (DSS) tools allowing multicriteria analysis, which also incorporates economical evaluations and accountability (Mocenni et al. 2008). Ecosystem health is usually assessed with chemical or biological indicators, e.g., phosphorus and nitrogen (Vollenweider 1992), oxygen (Viaroli and Christian 2004; Best et al. 2007), macrophytes (Orfanidis et al. 2001), and benthic macrofauna (Borja et al. 2000). Chemical and biological indicators are handled with different metrics that provide an assessment of water quality, trophic status, and ecological status. However, such indices are generally not broad enough to reflect the complexity of ecosystems. It is therefore necessary for the indicators to include structural and functional properties of the ecosystem. To cope with these aspects, new indices have been developed (see Chapter 2), among which eco-exergy and related values, i.e., structural eco-exergy, specific eco-exergy, have been recently used to assess ecosystem health in freshwater and marine ecosystems (Xu et al. 1999; Jørgensen 2000).

In this chapter, a short review of indicators capable of accurately reflecting the ecological status and quality of coastal lagoons is first presented. The management of macroalgae harvesting in the Sacca di Goro lagoon is then assessed, providing a bio-economical evaluation for macroalgae removal in relation with clam farming. This study was performed with previously developed models (Zaldívar, Cattaneo, et al. 2003; Zaldívar, Plus, et al. 2003), using specific eco-exergy (Jørgensen 1997) and costs/benefits analysis (De Leo et al. 2002; Cellina et al. 2003). Here, costs are associated with the macroalgal biomass harvesting and disposal, whereas benefits result from both the increase of shellfish productivity and the decrease of their mortality due to anoxic crises. For analyzing the ecosystem health we used specific eco-exergy, calculated in terms of biomass of the different biological components of the aquatic ecosystems and its information content. The comparison between both approaches has allowed us to develop a management strategy that aims at improving the ecosystem health and at reducing the economic losses associated with clam mortality during anoxic crises.

15.2 Short Review of Indicators for Coastal Lagoons

The objective of this section is to review the main class of indicators that have been recently proposed to assess the ecological status in coastal lagoons and the definition of class boundaries proposed for each methodology, in accordance with the Water Framework Directive (WFD; European Commission 2000). Because coastal lagoons are naturally stressed, the use of indicators developed for other ecosystems is not applicable (Dauvin 2007; Blanchet et al. 2008).

The WFD has developed the concept of Ecological Quality Status, which is based on the evaluation of the main biological components of given aquatic ecosystems, which for transitional waters (coastal lagoons) are phytoplankton, macrophytes, benthic macrofauna, and fishes. In addition, transparency, thermal conditions, salinity, oxygen, and nutrient concentrations are requested as supporting chemical and physicochemical variables.

15.2.1 Biological Indicators

Coastal lagoons show great internal patchiness and heterogeneity, which can either amplify or bias the stressor effects when considering indicators that have large spatial and temporal variability. Variability and meaning of certain variables are also constrained by lagoon depth and morphology. For example, in shallow coastal lagoons meroplanktonic rather than planktonic communities can establish, due to tight benthos-plankton coupling, which is caused by sediment resuspension. Nonetheless, species composition and richness of phytoplankton community has been used for assessing water quality in the lagoon of Venice (Bianchi et al. 2003). Recently, Mouillot et al. (2006) proposed an index, based on body size or size spectra of a given species. However, although there is evidence of relationships between phytoplankton size structure and environmental conditions, the standardization of this method is rather difficult, which biases its application in the ecological status classification.

Due to the shallow depth, benthic phanerogams and macroalgae represent the main components within the primary producer community. A clear succession from phanerogam to macroalgae species was also depicted as a result of the increased nutrient loading and inherent eutrophication (Viaroli et al. 2008). Assuming that benthic phanerogams are less tolerant to adverse environmental conditions than opportunistic macroalgae, they have been used to assess the ecological status, e.g., with the Ecological Evaluation Index (EEI; Orfanidis et al. 2001). The EEI ranges from the pristine state with late-successional species (Ecological State Group I [ESG I]) to the degraded state with opportunistic species (Ecological State Group II [ESG II]). The first group comprises taxa with a thick or calcareous thallus, low growth rates, and long life cycles (perennials), whereas the second group includes filamentous genera with high growth rates and short life cycles (annuals). Seagrasses were included in the first group, whereas Cyanophyceae and species with a coarsely branched thallus were included in the second group. The evaluation of ecological status into five categories from high to bad includes a cross comparison in a matrix of the ESG and a numerical scoring system.

Benthic fauna components are sedentary and have relatively long life spans; thus they can respond to local environmental conditions and give integrated responses to water and sediment quality variations (Bilyard 1987; Dauer 1993; Weisberg et al. 1997; Paul et al. 2001). Benthic fauna respond to anthropogenic and natural stressors and they are sensitive to different kinds of pollutants (Pearson and Rosenberg 1978; Magni 2003; Magni et al. 2009). Benthic macroinvertebrates are characterized by different trophic roles and different stress tolerance. For these reasons benthic macrofauna is the most used indicator alone or with different index metrics (Elliot 1994). Among others, in Europe three main indices are now applied: AMBI and M-AMBI (Borja et al. 2000; Borja et al. 2004; Muxica et al. 2006), and BENTIX (Simboura and Zenetos 2002). However, there are still some problems with these indicators since it is not clear to what extent they are able to distinguish between natural and anthropogenic stressors (Elliot and Quintino 2007). Other indices that use taxonomic-free attributes such as body-size, abundance distribution among functional groups, functional diversity and productivity (Mouillot et al. 2006), and biomass size structure index (Reizopoulou and Nicolaidou 2007) could prove to be more effective and relevant for these ecosystems. Recently, benthic communities in coastal lagoons have been assessed with synoptic information on benthic faunal condition, e.g., measures of community composition, and controlling natural abiotic factors, e.g., sediment organic matter (Magni 2003).

Unfortunately, there is little information available concerning fish and fish indicators developed for transitional water systems. Several possible indicators are discussed in Mouillot et al. (2006). Functional diversity, defined as the value and range of functional traits of the organisms present in a given ecosystem (Diaz and Cabido 2001), is one promising approach. The idea is that when environmental constraints increase, functional redundancy or similarity between fish assemblages should increase and, therefore, functional biodiversity would decrease (Mouillot et al. 2006). A measure of fish biomass also suggested seems too problematic to produce any relevant indicator of ecological quality. Nonetheless, the authors propose experiments based on cages to measure mortality and growth of juveniles. Even so, it seems that there is still a considerable amount of work to be performed before producing a relevant indicator.

15.2.2 Water Quality and Trophic Status Indicators

Chemical and physicochemical supporting elements in the WFD are selected from cause-effect relationships (Devlin et al. 2007). Also in this case, the high spatial and temporal variability in transitional waters is considered as detrimental, mainly for those variables that undergo significant day-night and seasonal changes, like dissolved oxygen (Viaroli and Christian 2004; Icely et al. 2007).

There is a very large amount of literature concerning methodologies for assessing the eutrophication in aquatic systems (Vollenweider et al. 1992; Nixon 1995; Cloern 2001), which can be broadly divided into three categories: screening methods, model-based, and mixed approaches (Nobre et al. 2005; Ferreira et al. 2007). Screening methods have been created to provide an assessment of eutrophication status based on few diagnostic physical and biogeochemical variables. Typical examples are the OSPAR common procedure on eutrophication assessment (OSPAR 2003) and the U.S. National Estuarine Eutrophication Assessment (NEEA; see Bricker et al. 1999). Concerning transitional water systems, several screening methodologies have been proposed (Zaldívar, Cardoso, et al. 2008).

The main trophic status indicators have been developed with phosphorus and phytoplankton chlorophyll-a, basically referring to the trophic reference system proposed by Vollenweider et al. (1992) and Nixon (1995). Furthermore, Vollenweider et al. (1998) proposed a trophic index (TRIX) that integrates chlorophyll-a, oxygen saturation, total nitrogen, and total phosphorus to characterize the trophic state of coastal marine waters, which is nowadays largely applied to coastal lagoons. TRIX is based on the assumption that eutrophication processes are mainly reflected by changes in the phytoplankton community, which is certainly not the case for shallow coastal lagoons and estuaries (excluding deep estuaries) where both microphytobenthos and benthic vegetation are the main components of the primary producer community. Most often, the issues to be analyzed are complex and cannot be resolved by considering only simple variables and linear relationships (De Wit et. al. 2001). Nevertheless, one can identify a set of basic variables that are indicative of ecosystem properties and functions, which can be easily measured and can be applied for classification and assessment of sensitivity to external stressors (Viaroli and Christian 2004). Among these the authors considered morphometric parameters, hydrological variables, sediment characteristics, and biological elements. Overall, most of these descriptors have been implemented for deep to relatively deep aquatic systems; therefore, they have to be further calibrated and validated for shallow coastal lagoons and estuaries using the "weight-of-evidence" approach. To bring together information from multiple indicators, metrics that allow integration or combination of multiple variables will also greatly improve the capacity of representing ecological status or sensitivity to a given stressor (Viaroli et al. 2004).

Recently, water budgets and nutrient loadings have been widely used for assessing the net ecosystem metabolism (NEM) of coastal lagoons with a wide array of primary producer communities (Giordani et al. 2005). Basically, NEM is calculated with the LOICZ biogeochemical model from P loadings; thus NEM gives a measure of the trophic status and of its dependency on nutrient delivery from watershed. A version that also considers the sediment component in the LOICZ models has been recently developed (Giordani, Zaldívar, et al. 2008).

A metric to assess production-to-respiration ratios/relationships in shallow aquatic environments has been presented by Rizzo and Christian (1996), the Benthic Trophic Status Index (BTSI), and Viaroli and Christian (2004), the Trophic Oxygen Status Index (TOSI). Both are designed to provide classification of benthic systems relative to their potential for heterotrophic and photoautotrophic activities as measured through hourly oxygen uptake in the dark (community respiration) and production or uptake at light saturation (maximum net community production).

Sedimentary variables can be integrated with water quality using simple metrics. Recently, an index (Transitional Water Quality Index [TWQI]) for assessing trophic status and water quality in transitional aquatic ecosystems of Southern Europe was developed (Giordani et al. 2009). The index was implemented from the water quality index of the U.S. National Sanitation Foundation and integrates the main causal factors (inorganic nutrients), key biological elements (primary producers), and indicator of effects (oxygen) of eutrophication. Six main variables were used: relative coverage of benthic phanerogams and opportunistic macroalgae species, and concentrations of dissolved oxygen, phytoplankton chlorophyll a, dissolved inorganic nitrogen, and phosphorus. Nonlinear functions were used to transform each measured variable into its quality value. Each quality value was then multiplied by a weighting factor, to take into account the relative contribution of each variable to the overall water quality. Finally, the index value was calculated as the sum of the weighted quality values, ranging from 0 (poorest state) to 100 (best condition). The index was tested and validated in six transitional water ecosystems that differed for anthropogenic pressures and eutrophication levels and was compared with the IFREMER classification scheme (France) and TRIX (Italy). Overall, the TWQI and the IFREMER evaluations (Souchu et al. 2000) correlated significantly, while TWQI was in disagreement with the TRIX responses. We suggested that indices based on phytoplankton only (e.g., TRIX) in shallow coastal transitional waters, where benthic vegetation controls primary productivity, are unsuitable.

With the increase of computer power, model-based methodologies have been developed. Normally, they are based on a hydrodynamic model that incorporates a biogeochemical model that considers the dynamics of organic and inorganic nutrients (Marinov et al. 2008). Normally, such models are site specific; therefore, they are not generally applicable. However, they are useful tools to analyze the environmental responses to changes in pressures as well as to provide environmental managers with an approximate idea of the time constants of their system. In addition, they may be used for scenario analysis of environmental options (Carafa et al. 2006; Plus et al. 2006; Marinov et al. 2007; Carafa et al. 2009) as well as, when coupled with socioeconomic models, with decision support systems (Loubersac et al. 2007; Mocenni et al. 2008).

Hybrid or mixed approaches try to combine the screening methods with simplified model-based approaches in order to develop general tools that have the advantages of both approaches in terms of applicability and predictive power. For TW several approaches of this type have been developed. Nobre et al. (2005) combined the ASSETS (Assessment of Estuarine Trophic Status) screening model with an ecological model to analyze the results from several scenarios or for defining homogeneous water bodies in estuaries (Ferreira et al. 2006). A similar approach was followed by Giordani et al. (2009) for Sacca di Goro, where a modified version of the Water Quality Index was coupled with a biogeochemical model and interfaced with a decision support system (Mocenni et al. 2008).

Concerning the development of integrated indicators, in a first step, Austoni et al. (2007) explored the application of specific eco-exergy on macrophytes. For this reason, they extended the calculation of β values (information content) for 244 seaweed and seagrass species that are common in Mediterranean coastal lagoons. A good agreement was found between the estimated β values and the macrophyte-based indicators EEI (Orfanidis et al. 2001) and the IFREMER classification scheme (Souchu et al. 2000). Furthermore, the specific eco-exergy was calculated for 71 sites in coastal lagoons of Southern France and compared with classification schemes that consider macrophytes, finding a good agreement between them. Therefore specific eco-exergy may be used as an integrated index that is able to synthesize and complement existing approaches.

15.2.3 Ecosystem Thresholds and Quality Class Boundaries

An important concept that has been recognized during the last decade (Scheffer et al. 2001) is the existence of thresholds (a critical value of a pressure beyond which a state indicator shifts to a different regime). Even though the concept of thresholds has been embedded in ecological risk assessment for a long time (Suter 1993), starting from the dose-response curves to a contaminant, only in the last years has the concept that "a gradual change in pressure would provoke a gradual change in the ecosystem" been modified by the realization that this gradual change may be interrupted by a sudden and drastic effect in the ecosystem. It has been suggested that the existence of such thresholds be used as a conceptual framework for the development of strategies for sustainable management of natural resources (Mudarian 2001; Huggett 2005).

When analyzing the response of an ecosystem to changes in an external factor (control variable), it is possible to distinguish between two different frameworks: a continuous approach provided by bifurcation theory (e.g., Dingjun et al. 1997), or when recording an abrupt change in a time series of an indicator of ecosystem status (Andersen et al. 2009). In the first case, bifurcation theory was developed to classify possible quantitative changes that a dynamical system may experience when one or several parameters are changed. In the second case time series analysis developed a set of detection methods in different fields, e.g., econometrics, ecotoxicology, oceanography, statistics, etc. These methods may be divided between statistical methods, parametric and non-parametric, e.g., Qian et al. (2003) and Zaldívar, Strozzi, et al. (2008); model based, e.g., Cox (1987) and Klepper and Bedaux (1997); or time series analysis, e.g., detection of abrupt changes in some characteristic property of the series, e.g., Basseville and Nikiforov (1993) and Zeileis et al. (2003), among others. For a complete review on the statistical methods for identification of thresholds, the reader is referred to Andersen et al. (2009).

Typical examples of regime shifts in transitional waters are the shift in macrophyte community due to nutrient increase (Viaroli et al. 2008; Zaldívar et al. 2009), in benthic communities with hypoxia due to increase of organic matter (Conley et al. 2007), in caged fish farms in terms of organic matter inputs (Holmer et al. 2007).

The WFD defines five categories of water quality: High, Good, Moderate, Poor, and Bad. However, the most important boundary is between moderate and good. In this case restoration measures have to be taken into account, which implies economic considerations. Assuming that our system responds in a nonlinear fashion, then assessing the five categories should take into account that when a threshold point is reached enormous restoration measures would be necessary, or in some cases no remedial actions would be possible. This assumes that the boundaries should be fixed at a value of pressure, and ecosystem status measured by a relevant biological metric that gives enough guarantee that the ecosystem is not close to the threshold value, which in the case of strong nonlinear response could imply passing from High to Bad as, for example, in the case of seagrass extinction. In these circumstances, the common practice of dividing the system's indicator in equal segments probably does not hold. In addition, in the case of a sharp threshold, the intermediate categories Moderate and Poor could be difficult to distinguish and probably a temporal dimension should be added during the evaluation to establish the line between Moderate and Poor. Furthermore, the point of no return should be assessed, also taking into account socioeconomic considerations.

15.3 Sacca di Goro Case Study

The Sacca di Goro (see Figure 15.1) is a shallow-water embayment of the Po River Delta (44° 47′ – 44° 50′ N and 12° 15′ – 12° 20′ E). The surface area is 26 km² and the total water volume is approximately 40×10^6 m³. Numerical models (Marinov et al. 2006, 2008) have demonstrated a clear zonation of the lagoon with the low-energy eastern area separated from two higher-energy zones, including both the western area influenced by freshwater inflow from the Po di Volano and the central area influenced by the Adriatic Sea. The eastern zone, called Valle di Gorino, is very shallow (with a maximum depth of 1 m) and accounts for one-half of the total surface area and for one-fourth of the water volume of the lagoon.

The bottom is flat and the sediment is composed of typical alluvial mud with a high clay and silt content in the northern and central zones, while

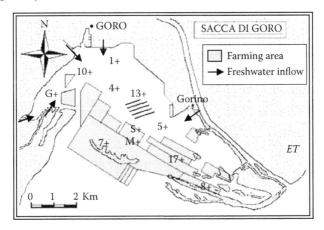

FIGURE 15.1
General layout of Sacca di Goro lagoon with the main farming areas indicated in gray and freshwater inflows by arrows.

sand is more abundant near the southern shoreline, and sandy mud predominates in the eastern area.

The watershed, Burana-Volano, is a lowland, flat basin located in the Po Delta and covering an area of about 3000 km². On the northern and eastern side it is bordered by a branch of the Po River entering the Adriatic Sea. A large part of the catchment area is below sea level with an average elevation of 0 m, a maximum elevation being 24 m and a minimum of –4 m. About 80% of the watershed is dedicated to agriculture. All the land is drained (irrigated) through an integrated channel network and various pumping stations. Point and non-point pollution sources discharge a considerable amount of nutrients in the lagoon from small tributaries and drainage channels (Po di Volano and Canal Bianco).

The catchment is heavily exploited for agriculture, while the lagoon is one of the most important aquacultural systems in Italy. About 10 km² of the aquatic surface are exploited for the Manila clam (*R. philippinarum*) farming, with an annual production of about 8000 tons (Figure 15.2). Fish and shellfish production provides work, directly or indirectly, for 5,000 people. The economic annual revenue has been varying during the last few years around €100 million.

Water quality is a major problem due to (1) the large supply of nutrients, organic matter, and sediments that arrive from the freshwater inflows; (2) the limited water circulation due to little water exchange with the sea (total water

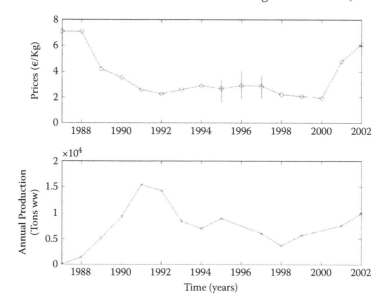

FIGURE 15.2
Averaged prices for *Ruditapes philippinarum* in Northern Adriatic (Bencivelli, personal communication, and Solidoro et al. 2000) and time evolution of estimated clams annual production in Sacca di Goro (Bencivelli, personal communication).

exchange time between two and four days (Marinov et al. 2006); and (3) the intensive shellfish production. In fact, from 1987 to 1992 the Sacca di Goro experienced an abnormal proliferation of macroalgae (*Ulva* sp.), which gradually replaced phytoplankton populations (Viaroli et al. 1992) (see Figure 15.3). This was a clear symptom of the rapid degradation of environmental conditions and of an increase in the eutrophication of this ecosystem.

The decomposition of *Ulva* in summer (at temperatures of 25°C to 30°C) produces the depletion of oxygen (Figure 15.4), which can lead to anoxia in the water column. In the beginning of August 1992, after a particularly severe anoxic event that resulted in a high mortality of farmed populations of mussels and clams, a 300- to 400-m wide, 2-m deep channel was cut through the sand bank to allow an increase in the seawater inflow and the water renewal in the Valle di Gorino. This measure temporarily solved the situation— during the following years a reduction of the *Ulva* cover (Viaroli et al. 1995) and a clear increase in phytoplankton biomass values were observed (Sei et al. 1996). However, in 1997 another anoxic event took place when an estimated *Ulva* biomass of 100,000 to 150,000 metric tons (enough to cover half of the lagoon) started to decompose. The economic losses due to mortality of the farmed clam populations were estimated around €7.5 million to €10 million (Bencivelli 1998).

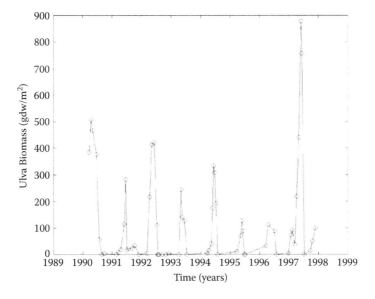

FIGURE 15.3
Measured annual trends of *Ulva* biomasses in the water column in the sheltered zone of the Sacca di Goro lagoon (Viaroli et al. 2006).

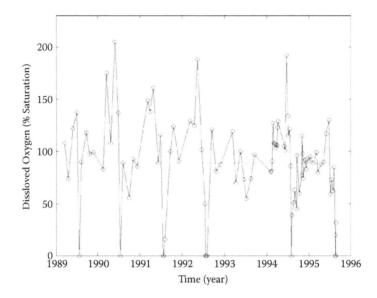

FIGURE 15.4

Experimental annual trends of dissolved oxygen saturation concentrations in the water column in the sheltered zone of the Sacca di Goro lagoon, from Viaroli et al. (2001).

15.3.1 Simulation Models

15.3.1.1 Biogeochemical Model

A model of the Sacca di Goro ecosystem has been developed and partially validated with field data from 1989 to 1998 (Zaldívar, Cattaneo, et al. 2003). The model considers the nutrient cycles in the water column and in the sediments as schematically shown in Figure 15.5. Nitrogen (nitrates plus nitrites and ammonium) and phosphorous have been included in the model since these two nutrients are involved in phytoplankton growth in coastal areas. Silicate has been introduced to distinguish between diatoms and flagellates, whereas consideration of the dissolved oxygen was necessary to study the evolution of hypoxia and the anoxic events that have occurred in Sacca di Goro during the past few years.

With regard to the biology, the model considers two types of phytoplankton and zooplankton communities. The phytoplankton model is based on Lancelot et al. (2002), and it explicitly distinguishes between photosynthesis (directly dependent on irradiance and temperature) and phytoplankton growth (dependent on both nutrients and energy availability). The microbial loop includes the release of dissolved and particulate organic matter with two different classes of biodegradability into the water column (Lancelot et al. 2002).

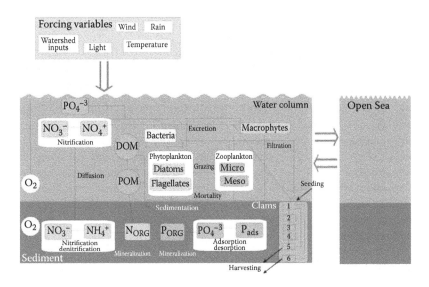

FIGURE 15.5
General schema of the biogeochemical model for Sacca di Goro.

Detrital particulate organic matter undergoes sedimentation. Furthermore, the evolution of bacteria biomass is explicitly taken into account.

In shallow lagoons, sediments play an important role in biogeochemical cycles (Chapelle et al. 2000). The sediments have several roles: they act as sinks of organic detritus material through sedimentation, and they consume oxygen and supply nutrients through bacterial mineralization, nitrification, and benthic fauna respiration. Indeed, depending on the dissolved oxygen concentration, nitrification or de-nitrification takes place in sediments, and for the phosphorous the sediments usually act as a buffer through adsorption and desorption processes. For all these reasons, the model considers the dynamics of nutrients in the sediment compartment.

Ulva sp. has become an important component of the ecosystem in Sacca di Goro. The massive presence of this macroalgae has heavily affected the lagoon ecosystem and has prompted several interventions aimed at removing its biomass in order to avoid anoxic crises, especially during the summer. In this case, *Ulva* biomass and the nitrogen concentration in macroalgae tissues are considered as other state variables (Solidoro et al. 1997).

The state space of dynamical variables considered is summarized in Table 15.1. We consider 38 state variables: there are five for nutrients in the water column and five in the sediments; organic matter is represented by 15 state variables in the water column and two in the sediments; 11 state variables represent the biological variables: six for phytoplankton, two for zooplankton, one for bacteria, and another two for *Ulva*.

TABLE 15.1

State Variables Used and Units in the Biogeochemical Model

Variable Name	Unit	Variable name	Unit
Inorganic nutrients, water column		*Biological variables, water column*	
Nitrate	mmol m^{-3}	Micro-phytopk (20–200 µm):	
Ammonium	mmol m^{-3}	Diatoms	mg C m^{-3}
Reactive phosphorous	mmol m^{-3}	Flagellates	mg C m^{-3}
Silicate	mmol m^{-3}	Micro-zoopk (40–200 µm)	mg C m^{-3}
Dissolved oxygen	g O$_2$ m^{-3}	Meso-zoopk (>200 µm)	mg C m^{-3}
Organic matter (OM), water column		Bacteria	mg C m^{-3}
Monomeric dissolved OM (C)	mg m^{-3}	Ulva	g dw l^{-1}
Monomeric dissolved OM (N)	mmol m^{-3}	Nitrogen in *Ulva* tissue	mg gdw^{-1}
Detrital biogenic silica	mmol m^{-3}	*Sediments*	
High biodegradability:		Ammonium (i.w.)	mmol m^{-3}
Dissolved polymers (C)	mg m^{-3}	Nitrate (i.w.)	mmol m^{-3}
Dissolved polymers (N, P)	mmol m^{-3}	Phosphorous (i.w.)	mmol m^{-3}
Particulate OM (C)	mg m^{-3}	Inorganic adsorbed phosphor.	µg P g^{-1} PS
Particulate OM (N, P)	mmol m^{-3}	Dissolved oxygen (i.w.)	g O$_2$/m^3
Low biodegradability:		Organic particulate phosphor.	µg g^{-1} PS
Dissolved polymers (C)	mg m^{-3}	Organic particulate nitrogen	µg g^{-1} PS
Dissolved polymers (N, P)	mmol m^{-3}		
Particulate OM (C)	mg m^{-3}		
Particulate OM (N, P)	mmol m^{-3}		

C, carbon; gdw, gram-dry-weight; i.w., interstitial waters; N, nitrogen; P, phosphorous; PS, particulate sediment, i.e., dry sediment.

15.3.1.2 Discrete Stage-Based Model of R. philippinarum

Knowing the importance of *R. philippinarum* in the Sacca di Goro ecosystem, it is clear that a trophic model that takes into account the effects of shellfish farming activities in the lagoon is necessary. For this reason a discrete stage-based model has been developed (Zaldívar, Plus, et al. 2003). The model considers six stage-based classes (see Figure 15.5). The first one corresponds to typical seeding sizes, whereas the last two correspond to the marketable sizes "medium" (37 mm) and "large" (40 mm) according to Solidoro et al. (2000). The growth of *R. philippinarum* is based on the continuous growth model from Solidoro et al. (2000), which depends on the temperature and phytoplankton in the water column. This model has been transformed into a variable stage duration for each class in the discrete stage-based model. Furthermore, the effects of harvesting as well as the mortality due to anoxic crisis are taken into account by appropriate functions, as well as the evolution of cultivable area and the seeding and harvesting strategies in use in Sacca di Goro. Recently, the model was implemented for analyzing the spatial distribution of clam productivity into the lagoon and to assess several environmental scenarios (Marinov et al. 2007).

15.3.1.3 *The Harvesting Model of* Ulva

During macroalgal blooms, mechanical harvesting of *Ulva* biomass is usually performed with vessels. To model the *Ulva*'s biomass harvested by one vessel per unit of time, we followed the model developed by De Leo et al. (2002), assuming that the vessel harvesting capacity, q, is $1.3 . 10^{-5}$ g dry weight per (gdw l^{-1}) per hour, which corresponds approximately to 100 metric tons of wet weight of *Ulva* per day. Therefore, we have incorporated into the *Ulva*'s model a term that takes into account this:

$$H(U,E) = \begin{cases} q \cdot E \cdot R(U) & \text{if } U(t) \geq U_{th} \\ 0 & \text{if } U(t) < U_{th} \end{cases} \qquad (15.1)$$

where E is the number of vessels, U is the *Ulva* biomass (gdw l^{-1}), and U_{th} is the threshold density above which the vessels start to operate. R is a function developed by Cellina et al. (2003) to take into account that the harvesting efficiency of vessels decreases when algal density is low. R was defined as:

$$R(U) = \frac{U^2}{U^2 + \delta} \qquad (15.2)$$

where δ is the semisaturation constant set to $2.014 . 10^{-4}$ (gdw l^{-2}) according to Cellina et al. (2003).

The function $H(U,E)$ acts as another mortality factor in the *Ulva*'s equation, with the difference that the resulting organic matter is not pumped into the microbial loop but is removed from the lagoon. The removal of this organic matter decreases the severity and number of anoxic crises in the lagoon and, hence, reduces the mortality of the clam population.

15.3.1.4 *Cost/Benefits Model*

The direct costs of *Ulva* harvesting have been evaluated to be €1000 per vessel per day including fuel, wages, and insurance, whereas the costs of biomass disposal are in the range of €150 per metric ton of *Ulva* wet weight (De Leo et al. 2002). Damages from shellfish production caused by *Ulva* are due to oxygen depletion and the subsequent mortality increase in the clam population. To take into account this factor we have evaluated the total benefits obtained from simulating the biomass increase using the averaged prices for *R. philippinarum* in the northern Adriatic (Figure 15.2). Therefore, an increase in clams biomass harvested from the lagoon will result in an increase in benefits. The total value obtained (CB = *Costs-Benefits*) is the difference between the costs associated with the operation of the vessels as well as the disposal of the harvested *Ulva* biomass minus the profits obtained by selling the shellfish biomass harvested in Sacca di Goro.

15.3.2 Eco-Exergy Calculation

The definitions and calculations of eco-exergy and structural eco-exergy (or specific eco-exergy) are discussed in Chapter 2.

The Sacca di Goro model considers several state variables for which the exergy should be computed. These are organic matter (detritus), phytoplankton (diatoms and flagellates), zooplankton (micro- and meso-), bacteria, macroalgae (*Ulva* sp.), and shellfish (*R. philippinarum*). The exergy was calculated using the data from Table 15.2 on genetic information content, and all biomasses were reduced to gdw l^{-1} using the parameters in Table 15.3.

15.3.3 Results and Discussion

15.3.3.1 The Existing Situation

Sacca di Goro has been suffering from anoxic crises during the warm season. Such crises are responsible for considerable damage to the aquaculture industry and to the ecosystem functioning. To individuate the most effective way to avoid such crises, it is important to understand the processes leading to anoxia in the lagoon. Figure 15.6 (top) shows the experimental and simulated *Ulva* biomasses. The model is able to predict the *Ulva* peaks and for some years their magnitude. For comparing experimental and simulated results

TABLE 15.2

Parameters Used to Evaluate the Genetic Information Content

Ecosystem Component	Number of Information Genes	Conversion Factor (Wi)
Detritus	0	1
Bacteria	600	2.7 (2)
Flagellates	850	3.4 (25)
Diatoms	850	3.4
Micro-zooplankton	10000	29.0
Meso-zooplankton	15000	43.0
Ulva sp.	2000[a]	6.6
Shellfish (bivalves)	—	287[b]

Source: From Jørgensen, S. E. 2000. Application of exergy and specific exergy as ecological indicators of coastal areas. *Aquatic Ecosystem Health and Management* 3:419–30.

[a] Coffaro et al. (1997).

[b] Marques et al. (1997) and Fonseca et al. (2000).

TABLE 15.3

Parameters Used for the Calculation of the Eco-Exergy for the Sacca di Goro Lagoon Model

Ecosystem Component	C:dw (gC gdw⁻¹)	-ln P_i
Detritus	—	7.5×10^5
Bacteria	0.4	12.6×10^5
Flagellates	0.22	17.8×10^5
Diatoms	0.22	17.8×10^5
Micro-zooplankton	0.45	209.7×10^5
Meso-zooplankton	0.45	314.6×10^5
Ulva sp.	—	41.9×10^5
Shellfish (bivalves)	—	2145×10^5

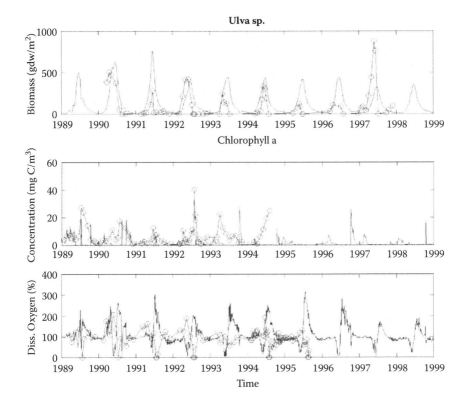

FIGURE 15.6

Experimental and simulated *Ulva* biomasses; Chlorophyll a and oxygen concentrations in Sacca di Goro.

we have assumed a constant area in the lagoon of 16.5 km². As observed by Viaroli et al. (2001), the rapid growth of *Ulva* sp. in spring is followed by a decomposition process, usually starting from mid-June. This decomposition stimulates microbial growth. The combination of organic matter decomposition and microbial respiration produces anoxia in the water column, mostly in the bottom water. This is followed by a peak of soluble reactive phosphorous that is liberated from the sediments (Giordani et al. 2008).

Oxygen evolution in the water column is highly influenced by the *Ulva* dynamics. In fact, high concentrations are simulated in correspondence with high algal biomass growth rates. Furthermore, when *Ulva* biomass starts to decompose the oxygen starts to deplete (Marinov et al. 2008). Experimental and simulated data are shown in Figure 15.6 (bottom). As can be seen, anoxic crises have occurred practically every year in the lagoon.

Figure 15.7 shows the comparison between the estimated and simulated total clam biomass in Sacca di Goro. It can be observed that there is a good agreement between estimated and simulated values. Oxygen also has a strong influence on *R. philippinarum* dynamics since anoxic crises are responsible for high mortality in the simulated total population (see Figure 15.8). Furthermore, population dynamics in the first stages is controlled by the seeding strategy performed in the lagoon. According to Castaldelli (private communication) there are two one-month seeding periods. The first begins in March, the second from mid-October to mid-November. The dynamics in Classes 5 and 6, which correspond to marketable sizes, are controlled by harvesting, since in the model they are harvested all year with an efficiency of 90% and 40%, respectively.

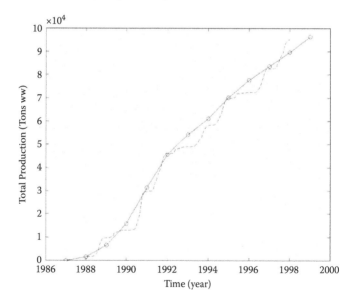

FIGURE 15.7
Estimated (Bencivelli, personal communication; *continuous line*) and simulated (*discontinuous line*) total production of *Ruditapes philippinarum*.

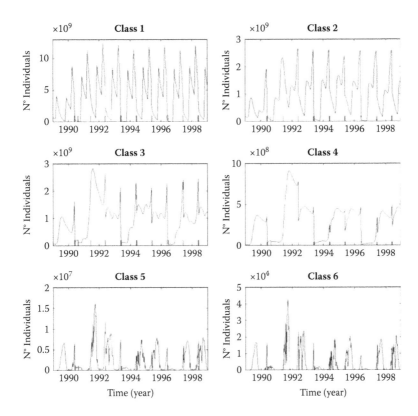

FIGURE 15.8
Simulated *Ruditapes philippinarum* population dynamics in Sacca di Goro. The simulated anoxic periods (oxygen concentration below 2 mg l⁻¹) are indicated by small bars.

Figure 15.9 shows the values calculated for the eco-exergy and specific eco-exergy. It can be seen that the calculations do not show the annual cycles one should expect in the lagoon, with low eco-exergy during the winter and autumn accompanied by an increase during spring and summer. This is due to the fact that the eco-exergy is practically controlled by shellfish biomass. This can be seen in Figure 15.10, where the contribution to the eco-exergy of each ecosystem compartment is plotted as a percentage. Concerning specific eco-exergy there is less variation. The changes are due to the effects of anoxic crises that affect the biomass distribution. As can be seen in Figure 15.10 there are localized peaks of *Ulva* in correspondence with the decrease in *R. philippinarum* biomass due to an increase in mortality during anoxic episodes.

15.3.3.2 Harvesting Ulva Biomass

A measure that has been taken in Sacca di Goro to control macroalgal blooms consists of harvesting vessels that remove the *Ulva* in zones where clam

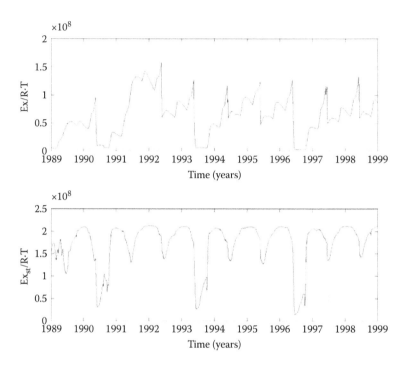

FIGURE 15.9
Computed eco-exergy (g l⁻¹) and specific eco-exergy for the Sacca di Goro model, from 1989 to 1999. Parameters used for the calculation of the genetic information content are given in Tables 15.2 and 15.3.

fishery is located. However, it was not clear how the vessels should operate to reduce their costs and obtain the maximum benefit for the shellfish industry. In a series of recent studies De Leo et al. (2002) and Cellina et al. (2003) developed a stochastic model that allowed the assessment of harvesting policies in terms of cost-effectiveness, that is, the number of vessels and the *Ulva* biomass threshold at which the harvesting should start.

In this work, we inserted their costs model in the coupled continuous biogeochemical model and discrete stage-based *R. philippinarum* population models. Furthermore, no specific functions for evaluating the effects of anoxic crises on *Ulva* and clams dynamics have been introduced. Benefits are calculated as a function of the number of harvested clams in the lagoon and their selling price (see Figure 15.2).

Several hundreds of simulation runs from 1989 to 1994, using the same initial conditions and forcing functions, have been carried out to estimate the optimum solution in terms of costs and benefits, number of operational vessels (from 0 to 20 vessels), and ecosystem (specific eco-exergy) improvement at different *Ulva* biomass thresholds (0.01 gwd l⁻¹ to 0.16 gdw l⁻¹, which corresponds approximately to 20 gdw m⁻² and 380 gdw m⁻², respectively). The results are

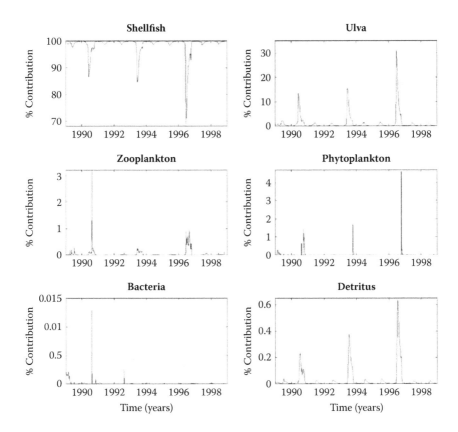

FIGURE 15.10
Contributions of the different models' compartments to the total eco-exergy of the system.

summarized in Figures 15.11 and 15.12, which show the relative estimated costs and benefits, $(CB_i - CB_0)/CB_0$, and specific eco-exergy improvement, Ex_{st}^i / Ex_{st}^0, where 0 refers to the existing situation and i to the specific number of vessels and different *Ulva* biomass thresholds. The optimum solution would be the one with lower costs and higher specific eco-exergy improvement.

As it can be seen in Figure 15.11, there is an optimal solution concerning the costs and benefits: work at low *Ulva* biomass thresholds 0.02 to 0.03 gdw l^{-1} (50–70 gdw m^{-2}) with 10 to 12 vessels—that is 0.6 to 0.7 vessels km^{-2} operating in the lagoon. These values are in agreement with previous studies. De Leo et al. (2002) obtained around 0.5 vessels km^{-2} and *Ulva* thresholds between 70 and 90 gdw m^{-2}, whereas Cellina et al. (2003) found values between 50 and 75 gdw m^{-2} for 6 to 10 vessels operating in the lagoon.

For the case of relative specific eco-exergy (see Figure 15.12), there is not a global maximum since relative specific eco-exergy continues to increase as we increase the number of vessels operating in the lagoon at low *Ulva* biomass thresholds. However, the optimal solution from the cost-benefit

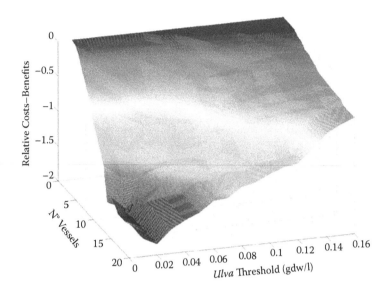

FIGURE 15.11
Simulated results in terms of relative costs and benefits in Sacca di Goro by changing the number of vessels and the *Ulva* biomass threshold at which they start to operate.

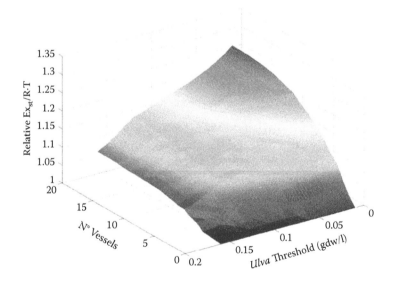

FIGURE 15.12
Simulated results in terms of relative specific eco-exergy improvement in Sacca di Goro by
changing the number of vessels and the *Ulva* biomass threshold at which they start to operate.

analysis would improve the specific eco-exergy by approximately 21% in
comparison with the "do nothing" strategy. The maximum improvement
calculated is around 25%.

15.3.3.4 Reduction in Nutrient Inputs

Another possible measure to improve the ecosystem functioning would be to reduce the nutrient loads in Sacca di Goro. For this study, we established a scenario that considers the reduction in nutrient loads arriving from Po di Volano, Canale Bianco, and Po di Goro compared with the maximum values established by National Italian legislation (based on EU Nitrate Directive) for Case III (poor quality, polluted: $NH_4^+ < 0.78$ mg N l^{-1}, $NO_3^- < 5.64$ mg N l^{-1}, $PO_4^{3-} < 0.17$ mg P l^{-1}). Furthermore, we have not considered the improvement that the Adriatic Sea should experience if reduction in nutrient loads is accomplished in the Po River. To take into account these effects a 3-D simulation of the North Adriatic Sea considering the nutrients load reduction scenarios should be carried out to properly account for these effects in our model.

Figures 15.13 and 15.14 present the evolution of eco-exergy and specific eco-exergy under the two proposed scenarios—*Ulva* removal and nutrient load reduction—in comparison with the "do nothing" alternative. As can be seen, the eco-exergy and specific eco-exergy of both scenarios increase. This is due to the fact that in our model both functions are dominated by clam biomass. Therefore, it implies that the biomass of *R. philippinarum* in Sacca di Goro would have been increased whatever the scenario used. This can be seen in Figure 15.15, where the optimal solution in terms of operating vessels would have multiplied approximately by a factor of three the harvested *R. philippinarum* biomasses with the subsequent economic benefits.

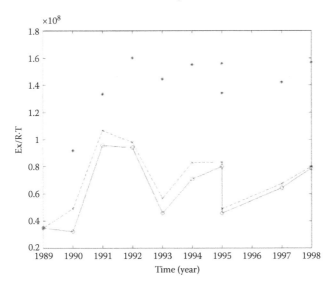

FIGURE 15.13
Eco-exergy mean annual values: present scenario (*continuous line*); removal of *Ulva*, optimal strategy from cost-benefit point of view (*dotted line*); nutrient load reduction from watershed (*dashed line*).

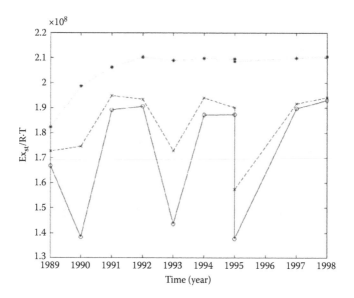

FIGURE 15.14
Specific eco-exergy mean annual values: present scenario (*continuous line*); removal of *Ulva*, optimal strategy from cost-benefit point of view (*dotted line*); nutrient load reduction from watershed (*dashed line*).

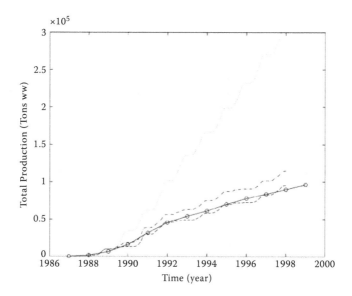

FIGURE 15.15
Estimated (Bencivelli, personal communication, *continuous line*) and simulated: total production of *Ruditapes philippinarum* present scenario (*dashed line*); removal of *Ulva*, optimal strategy from cost-benefit point of view (*dotted line*); nutrients load reduction from watershed (*dashed/ dotted line*).

An evaluation of the costs associated with a reduction in nutrient loads is beyond the scope of this chapter. However, this evaluation should be carried out when the WFD (European Commission 2000) enters in force, but taking into account the dimensions and importance of the Po River the costs will certainly be higher than the removal of *Ulva* by vessels.

15.4 Conclusions

The assessment of the health of an ecosystem is not an easy task and it may be necessary to apply several indicators simultaneously to obtain a proper estimation. Several researchers have proposed different indicators that cover different aspects of the ecosystem health, but it seems clear that only a coherent application of them would lead us to have a correct indication of the analyzed ecosystem. Between these indicators, eco-exergy expresses the biomass of the system and the genetic information that this biomass is carrying and specific eco-exergy can tell us how rich on information the system is. These indicators are able to cover a considerable amount of ecosystem characteristics, and it has been shown that they are correlated with several important parameters such as respiration, biomass, etc. However, it has been found (Jørgensen 2000) that eco-exergy is not related to biodiversity and, for example, a very eutrophic system often has a low biodiversity but high eco-exergy.

It also seems clear that both values would give a considerable amount of information when analyzing the ecological status of inland and marine waters as requested by the Water Framework Directive (European Commission 2000). However, there is still work to be done in several areas. The first area consists of standardizing the genetic information content for the species occurring in EU waters and, hence, allowing a uniform calculation of exergy, which will permit a useful comparison between studied sites. A step in this direction has been carried out in Austoni et al. (2007) by calculating the genetic information content of more than 200 seaweed and seagrass species that are common in Mediterranean coastal lagoons, and in Austoni (2007) where the genetic information content was calculated for more than 500 macrobenthic species. The calculated specific eco-exergy results were then compared with developed indices (Orfanidis et al. 2001; Souchu et al. 2000; Borja et al. 2000; and Simboura and Zenetos 2002) finding a good relationship between information content and the ecological role of broad functional groups, e.g., for the case of macrophytes: opportunistic drifting macroalgae, perennial macroalgae, and seagrass. A similar approach was carried out using FAO fisheries data (Kernegger et al. 2008); however, in this case, due to the absence of good genetic data, it was not possible to calculate genetic information content at the species level. The second area where research is still needed consists of developing a methodology that would allow the calculation of eco-exergy from monitoring data, already considered in Annex V of the Water Framework Directive (Murray et al. 2002; Quevauviller et al. 2005). Unfortunately, ecological data in terms of biomasses of important elements in an ecosystem are not normally available, and therefore an important aspect would be to study how to use the physico-chemical parameters (normally the values for which most historical data are currently available) for the estimation of the eco-exergy and specific eco-exergy of a system.

Finally, to transform the concept of eco-exergy on an operational tool as an ecological indicator on inland and marine waters, it would be necessary to develop a methodology for its calculation when models are not available. It is always possible to calculate the eco-exergy, if one has enough data on the ecosystem composition and biomasses. However, the data normally available consist mainly of nutrient concentrations and phytoplankton in the form of chlorophyll a, which cannot directly provide a good estimation of the eco-exergy of an ecosystem. It is clear that both aspects are related, nutrients allow the growth and development of the ecosystem and their change has a direct effect on the eco-exergy values of our system (see Sacca di Goro case study), but how do we convert these monitoring parameters into a formulation that allows the calculation of eco-exergy? However, when biological monitoring is established, following WFD, this approach could become operational, allowing managers with a global parameter that could allow the inter-comparison between ecosystems.

The results of the model are, in general, in good agreement with the stochastic models developed by De Leo et al. (2002) and Cellina et al. (2003). All these results point toward starting macroalgal removal earlier, when *Ulva* biomasses are relatively low. At higher biomasses, due to the high growth rates of *Ulva* and the nutrient availability in Sacca di Goro, it is more difficult to prevent the anoxic crises. From the point of view of improving specific exergy in the Goro lagoon the best approach would consist of using the maximum amount of vessels operating at thresholds as low as possible. However, the optimal result from the cost-benefit analysis will considerably improve the ecological status of the lagoon in terms of specific exergy. The nutrient reduction scenario considers a small reduction. However, more realistic scenarios could be implemented when data on the river basin management plans from Burana-Volano and Po River watersheds become available. A step in this direction has been carried out in Marinov et al. (2007), where different scenarios considering clam productivity were identified and analyzed using a coupled watershed and 3-D biogeochemical model with and without macroalgal blooms. Model simulations indicated that macroalgal blooms have an important negative impact on clam productivity due to the risk of anoxia and subsequent clam mortality. Furthermore, simulation results evidenced that meteorological conditions also affect clam productivity, especially in a dry year due to a shortage of food supply.

For these reasons, the use of biogeochemical modeling, ecological indicators, and cost-benefit analysis seems an adequate combination for developing integrated tools able to build up strategies for sustainable ecosystem management, including ecosystem restoration or rehabilitation.

Acknowledgments

This research was partially supported by the EU-funded project DITTY (Development of Information Technology Tools for the management of European southern lagoons under the influence of river-basin runoff, EVK3-CT-2002-00084) in the Energy, Environment, and Sustainable Development program of the European Commission and by the Exploratory Research Project "Application of exergy as an ecological indicator in Mediterranean coastal lagoons" at the Joint Research Centre, Institute for Environment and Sustainability.

References

Andersen, T., J. Carstensen, E. Hernandez-Garcia, and C. M. Duarte. 2009. Ecological thresholds and regime shifts: Approaches to identification. *Trends in Ecology and Evolution* 24:49–57.

Austoni, M. 2007. Development and characterization of bio-indicators for Mediterranean coastal lagoons: Towards the application of specific exergy as an integrated indicator. PhD thesis, Parma University, Italy.

Austoni, M., G. Giordani, P. Viaroli, and J. M. Zaldívar. 2007. Application of specific exergy to macrophytes as an integrated index of environmental quality for coastal lagoons. *Ecol Indicators* 7:229–38.

Bacher, C., H. Bioteau, and A. Chapelle. 1995. Modelling the impact of a cultivated oyster population on the nitrogen dynamics: The Thau lagoon case (France). *Ophelia* 42:29–54.

Basseville, M., and I. V. Nikiforov. 1993. *Detection of abrupt changes: Theory and application*. Englewood Cliffs, NJ: Prentice-Hall.

Bencivelli, S. 1998. La Sacca di Goro: La situazzione di emergenza dell'estate 1997. In *Lo stato dell'ambiente nella provincia di Ferrara. Anno 1997*. Amministrazione Provinciale di Ferrara. Servizio Ambiente, 61–66.

Best, M. A., A. W. Wither, and S. Coates. 2007. Dissolved oxygen as a physico-chemical supporting element in the Water Framework Directive. *Marine Pollution Bulletin* 55:53–64.

Bianchi, F., F. Acri, F. Bernardi Aubry, A. Berton, A. Boldrin, E. Camatti, D. Cassin, and A. Comaschi. 2003. Can plankton communities be considered as bio-indicators of water quality in the Lagoon of Venice? *Mar Poll Bull* 46:964–71.

Bilyard, G. R. 1987. The value of benthic fauna in marine pollution monitoring studies. *Mar Poll Bull* 18:581–85.

Blanchet, H., N. Lavesque, T. Ruellet, J. C. Dauvin, P. G. Sauriau, N. Desroy, C. Desclaux, M. Leconte, G. Bachelet, A. L. Janson, C. Bessineton, S. Duhamel, J. Jourde, S. Mayot, S. Simon, and X. de Montaudouin. 2008. Use of biotic indices in semi-enclosed coastal ecosystems and transitional waters habitats—Implications for the implementation of the European Water Framework Directive. *Ecol Ind* 8:360–72.

Borja, A., J. Franco, and I. Muxika. 2004. The biotic indices and the water framework directive: The required consensus in the new benthic monitoring tools. *Mar Poll Bull* 48:405–408.

Borja, A., J. Franco, and V. Perez. 2000. A marine biotic index to establish the ecological quality of soft-bottom benthos within European estuarine and coastal environments. *Mar Poll Bull* 40:1100–14.

Bricker, S. B., C. G. Clement, D. E. Pirhalla, S. P. Orlando, and D. R. G. Farrow. 1999. National estuarine eutrophication assessment. Effects of nutrient enrichment in the nation's estuaries. NOAA-NOS Special Projects Office.

Carafa, R., D. Marinov, S. Dueri, J. Wollgast, G. Giordani, P. Viaroli, and J. M. Zaldívar. 2009. A bioaccumulation model for herbicides in *Ulva rigida* and *Tapes philippinarum* in Sacca di Goro lagoon (Northern Adriatic). *Chemosphere* 74:1044–52.

Carafa, R., D. Marinov, S. Dueri, J. Wollgast, J. Ligthart, E. Canuti, P. Viaroli, and J. M. Zaldívar. 2006. A 3D hydrodynamic fate and transport model for herbicides in Sacca di Goro coastal lagoon (Northern Adriatic). *Mar Poll Bull* 52:1231–248.

Cellina, F., G. A. De Leo, A. E. Rizzoli, P. Viaroli, and M. Bartoli. Economic modelling as a tool to support macroalgal bloom management: A case study (Sacca di Goro, Po river delta). *Oceanologica Acta* 26:139–47.

Chapelle, A., P. Lazuer, and P. Souchu. 2001. Modelisation numerique des crises anoxiques (malaiguës) dans la lagune de Thau (France). *Oceanologica Acta* 24:S99–12.

Chapelle, A., A. Ménesguen, J. M. Deslous-Paoli, P. Souchu, N. Mazouni, A. Vaquer, and B. Millet. 2000. Modelling nitrogen, primary production and oxygen in a Mediterranean lagoon. Impact of oysters farming and inputs from the watershed. *Ecol Model* 127:161–81.

Christian, R. R. 2005. Beyond the Mediterranean to global observations of coastal lagoons. *Hydrobiologia* 550:1–8.

Cloern, J. E. 2001. Our evolving conceptual model of the coastal eutrophication problem. *Mar Ecol Prog Ser* 210:223–53.

Coffaro, G., M. Bocci, and G. Bendoricchio. 1997. Application of structural dynamical approach to estimate space variability of primary producers in shallow marine waters. *Ecol Model* 102:97–114.

Conley, D. J., J. Carstensen, G. Ærtebjerg, P. B. Christensen, T. Dalsgaard, J. L. S. Hansen, and A. B. Josefson. 2007. Long-term changes and impacts of hypoxia in Danish coastal waters. *Ecol Appl* 17:S165–84.

Cox, C. 1987. Threshold dose-response models in toxicology. *Biometrics* 43:511–23.

Dauer, D. M. 1993. Biological criteria, environmental health and estuarine macrobenthic community structure. *Mar Poll Bull* 26:249–57.

Dauvin, J. C. 2007. Paradox of estuarine quality: Benthic indicators and indices, consensus or debate for the future. *Mar Poll Bull* 55:271–81.

De Leo, G., M. Bartoli, M. Naldi, and P. Viaroli. A first generation stochastic bioeconomic analysis of algal bloom control in a coastal lagoon (Sacca di Goro, Po river Delta). *Mar Ecol* 410:92–100.

De Wit, R., L. J. Stal, B. A. Lomstein, R. A. Herbert, H. Van Gemerden, P. Viaroli, V.-U. Cecherelli, F. Rodríguez-Valera, B. Bartoli, G. Giordani, R. Azzoni, B. Schaub, D. T. Welsh, A. Donelly, A. Cifuentes, J. Antón, K. Finster, L. B. Nielsen, A.-G. Underlien Pedersen, A. Turi Neubeurer, M. A. Colangelo, and S. K. Heijs. 2001. "ROBUST: The ROle of BUffering capacities in STabilising coastal lagoon ecosystems. *Continental Shelf Research* 21:2021–41.

Devlin, M., S. Painting, and M. Best. 2007. Setting nutrient thresholds to support an ecological assessment based on nutrient enrichment, potential primary production and undesirable disturbance. *Mar Poll Bull* 55:65–73.

Diaz, S., and M. Cabido. 2001. Vive la difference: Plant functional diversity matters to ecosystem processes. *Trends in Ecology and Evolution* 16:646–55.

Dingjun, L., W. Xian, Z. Deming, and H. Maoan. 1997. Bifurcation theory and methods of dynamical systems. Singapore: World Scientific.

Elliot, M. 1994. The analysis of macrobenthic community data. *Mar Poll Bull* 28:62–64.

Elliot, M., and V. Quintino. 2007. The estuarine quality paradox, environmental homeostasis and the difficulty of detecting anthropogenic stress in naturally stressed areas. *Mar Poll Bull* 54:640–45.

European Commission. 2000. Directive 2000/60/EC of the European Parliament and of the Council of 23 October 2000 establishing a framework for community action in the field of water policy. *OJ Eur Communities* 43:1–72.

Ferreira, J. G., S. Bricker, and T. C. Simas. 2007. Application and sensitivity testing of a eutrophication assessment method on coastal systems in the United States and European Union. *J Environ Management* 82:433–45.

Ferreira, J. G., A. M. Nobre, T. C. Simas, M. C. Silva, A. Newton, S. B. Bricker, W. J. Wolff, P. E. Stacey, and A. Sequeira. 2006. A methodology for defining homogeneous water bodies in estuaries—Application to the transitional systems of the EU Water Framework Directive. *Estuarine, Coastal and Shelf Science* 66:468–82.

Fonseca, J. C., J. C. Marques, A. A. Paiva, A. M. Freitas, V. M. C. Madeira, and S. E. Jørgensen. 2000. Nuclear DNA in the determination of weighing factors to estimate exergy from organisms biomass. *Ecol Model.* 126:179–89.

Giordani, G., R. Azzoni, and P. Viaroli. 2008. A rapid assessment of the sedimentary buffering capacity towards free sulphides. *Hydrobiologia* 611:55–66.

Giordani, G., P. Viaroli, D. P. Swaney, C. N. Murray, J. M. Zaldívar, and J. I. Marshall Crossland, eds. 2005. *Nutrient fluxes in transitional zones of the Italian coast.* LOICZ Reports and Studies No. 28. Texel, the Netherlands: LOICZ.

Giordani, G., J. M. Zaldívar, D. P. Swaney, and P. Viaroli. 2008. Modelling ecosystem functions and properties at different time and spatial scales in shallow coastal lagoons: An application of the LOICZ biogeochemical model. *Estuarine Coastal & Shelf Science* 77:264–77.

Giordani, G., J. M. Zaldívar, and P. Viaroli. 2009. Simple tools for assessing water quality and trophic status in transitional water ecosystems. *Ecological Indicators* 9:982–91.

Holmer, M., N. Marbà, E. Diaz-Almela, C. M. Duarte, M. Tsapakis, and R. Danovaro. 2007. Sedimentation of organic matter from fish farms in oligotrophic Mediterranean assessed through bulk and stable isotope (δ13C and δ15N) analyses. *Aquaculture* 262:268–80.

Huggett, A. J. 2005. The concept and utility of 'ecological thresholds' in biodiversity conservation. *Biological Conservation* 124:301–10.

Icely, J., A. Newton, S. Mudge, and P. Oliveira. 2007. Episodic hypoxia: A problem for lagoon management. Proceedings of EcoSummit 2007, Beijing, China.

Jørgensen, S. E. 1997. *Integration of ecosystem theories: A pattern*, 2nd ed. Dordrecht, The Netherlands: Kluwer.

———. 2000. Application of exergy and specific exergy as ecological indicators of coastal areas. *Aquatic Ecosystem Health and Management* 3:419–30.

Kernegger, L., J. Carstensen, and J. M. Zaldívar. 2008. Application of specific eco-exergy to FAO fisheries data. *The Open Fish Science Journal* 1:11–18.

Klepper, O., and J. J. M. Bedaux. 1997. Nonlinear parameter estimation for toxicological thresholds model. *Ecol Model* 102:315–24.

Lancelot, C., J. Staneva, D. Van Eeckhout, J. M. Beckers, and E. Stanev. 2002. Modelling the Danube-influenced north-western continental shelf of the Black Sea. II. Ecosystem response to changes in nutrient delivery by the Danube river after its damming in 1972. *Est Coast Shelf Sci* 54:473–99.

Loubersac, L., T. Do Chi, A. Fiandrino, M. Jouan, V. Derolez, A. Lemsanni, H. Rey-Valette, S. Mathe, S. Pages, C. Mocenni, M. Casini, S. Paoletti, M. Pranzo, F. Valette, O. Serais, T. Laugier, N. Mazouni, C. Vincent, P. Got, M. Troussellier, and C. Aliaume. 2007. Microbial contamination and management scenarios in a Mediterranean coastal lagoon (Etang de Thau, France): Application of a Decision Support System within the Integrated Coastal Zone Management context. *Transitional Waters Monographs* 1:107–27.

Magni, P. 2003. Biological benthic tools as indicators of coastal marine ecosystem health. *Chemistry and Ecology* 19:363–72.

Magni, P., D. Tagliapietra, C. Lardicci, A. Castelli, S. Como, G. Frangipane, G. Giordani, J. Hyland, F. Maltagliati, G. Pessa, A. Rispondo, M. Tataranni, P. Tomassetti, and P. Viaroli. 2009. Animal-sediment relationships: Evaluating the Pearson-Rosenberg paradigm in Mediterranean Coastal lagoons. *Mar Poll Bull* 54:478–86.

Marinov, D., L. Galbiati, G. Giordani, P. Viaroli, A. Norro, S. Bencivelli, and J. M. Zaldívar. 2007. An integrated modelling approach for the management of clam farming in coastal lagoons. *Aquaculture* 269:306–20.

Marinov, D., A. Norro, and J. M. Zaldívar. 2006. Application of COHERENS model for hydrodynamic investigation of Sacca di Goro coastal lagoon (Italian Adriatic Sea shore). *Ecol Model* 193:52–68.

Marinov, D., J. M. Zaldívar, A. Norro, G. Giordani, and P. Viaroli. 2008. Integrated modelling in coastal lagoons: Sacca di Goro case study. *Hydrobiologia* 611:147–65.

Marques, J. C., M. A. Pardal, S. N. Nielsen, and S. E. Jørgensen. 1997. Analysis of the properties of exergy and biodiversity along an estuarine gradient of eutrophication. *Ecol Model* 102:155–67.

Mocenni, C., M. Casini, S. Paoletti, G. Giordani, P. Viaroli, and J. M. Zaldívar. 2008. A Decision Support System for the management of the Sacca di Goro (Italy). In *Decision Support Systems for risk-based management of contaminated sites*, eds. A. Marcomini, G. W. Sutter II, and A. Critto, 399–422 (Ch. 19). New York: Springer.

Mouillot, D., S. Spatharis, S. Reizopoulo, T. Laugier, L. Sabetta, A. Basset, and T. Do Chi. 2006. Alternatives to taxonomic-based approaches to assess changes in transitional water communities. *Aquatic Conservation* 16:469–82.

Mudarian, R. 2001. Ecological thresholds: A survey. *Ecol Econ* 38:7–24.

Murray, C. N., G. Bidoglio, J. M. Zaldívar, and F. Bouraoui. 2002. The Water Framework Directive: The challenges of implementation for river basin-coastal research. *Fresenius Environmental Bulletin* 11:530–41.

Muxica, I., A. Borja, and J. Bald. 2006. Using historical data, expert judgement and multivariate analysis in assessing reference conditions and benthic ecological status, according to the European Water Framework Directive. *Mar Poll Bull* 55:16–29.

Nixon, S. W. 1995. Coastal marine eutrophication: A definition, social causes, and future concerns. *Ophelia* 41:199–219.

Nobre, A. M., J. G. Ferreira, A. Newton, T. Simas, J. D. Icely, and R. Neves. 2005. Management of coastal eutrophication: Integration of field data, ecosystem-scale simulations and screening models. *J Mar Systems* 56:375–90.

Orfanidis, S., P. Panayotidis, and N. Stamatis. Ecological evaluation of transitional and coastal waters: A marine benthic macrophytes-based model. *Mediterranean Marine Science* 2:45–65.

OSPAR. 2003. Draft starting point for the further development of the common procedure for the identification of the eutrophication status of the OSPAR Maritime Area. Meeting of the Eutrophication Committee (EUC) London, December 8–12, 2003.

Paul, J. F., K. J. Scott, D. E. Campbell, J. H. Gentile, C. S. Strobel, R. M. Valente, S. B. Weisberg, A. F. Holland, and J. A. Ranasinghe. 2001. Developing and applying a benthic index of estuarine condition for the Virginian Biogeographic Province. *Ecol Ind* 1:83–99.

Pearson, T. H., and R. Rosenberg. 1978. Macrobenthic succession in relation to organic enrichment and pollution of the marine environment. *Oceanography and Marine Biology: An Annual Review* 16:229–311.

Plus, M., I. La Jeunesse, F. Bouraoui, J. M. Zaldívar, A. Chapelle, and P. Lazure. 2006. Modelling water discharges and nutrient inputs into a Mediterranean lagoon. Impact on the primary production. *Ecol Model* 193:69–89.

Qian, S. S., R. S. King, and C. J. Richardson. 2003. Two statistical methods for the detection on environmental thresholds. *Ecol Model* 166:87–97.

Quevauviller, P., P. Balananis, C. Fragakis, M. Weydert, M. Oliver, A. Kaschl, G. Arnold, A. Kroll, L. Galbiati, J. M. Zaldívar, and G. Bidoglio. 2005. Science-policy integration needs in support of the implementation of the EU Water Framework Directive. *Environmental Science and Policy* 8:203–11.

Reizopoulou, S., and A. Nicolaidou. 2007. Index of Size Distribution (ISD): A method of quality assessment for coastal lagoons. *Hydrobiologia* 577:141–49.

Rizzo, W. M., and R. R. Christian. 1996. Significance of subtidal sediments to heterotrophically-mediated oxygen and nutrient dynamics in a temperate estuary. *Estuaries* 19:475–87.

Scheffer, M., S. R. Carpenter, J. A. Foley, C. Folke, and B. Walker. 2001. Catastrophic shifts in ecosystems. *Nature* 413:591–96.

Sei, S., G. Rossetti, F. Villa, and I. Ferrari. 1996. Zooplankton variability related to environmental changes in a eutrophic coastal lagoon in the Po Delta. *Hydrobiologia* 329:45–55.

Simboura, N., and A. Zenetos. 2002. Benthic indicators to use in ecological quality classification of Mediterranean soft bottoms marine ecosystems, including a new biotic index. *Mediterranean Marine Science* 3:77–111.

Solidoro, C., V. E. Brando, C. Dejak, D. Franco, R. Pastres, and G. Pecenik. 1997. Long term simulations of population dynamics of *Ulva r.* in the lagoon of Venice. *Ecol Model* 102:259–72.

Solidoro, C., R. Pastres, D. Melaku Canu, M. Pellizzato, and R. Rossi. 2000. Modelling the growth of *Tapes philippinarum* in Northern Adriatic lagoons. *Mar Ecol Prog Ser* 199:137–48.

Souchu, P., M. C. Ximenes, M. Lauret, A. Vaquer, and E. Dutrieux. 2000. Mise à jour d'indicateurs du niveau d'eutrophisation des milieux lagunaires méditerranéens, août 2000. Ifremer-Créocean-Université Montpellier II, 412.

Suter, G. W. 1993. *Ecological risk assessment*. Chelsea, MI: Lewis Publishers.

Viaroli, P., R. Azzoni, M. Bartoli, G. Giordani, and L. Tajè. 2001. Evolution of the trophic conditions and dystrophic outbreaks in the Sacca di Goro lagoon (Northern Adriatic Sea). In *Structures and processes in the Mediterranean ecosystems*, eds. F. M. Faranda, L. Guglielmo, and G. Spezie, 59, 467–75. Milan: Springer-Verlag.

Viaroli, P., M. Bartoli, C. Bondavalli, and M. Naldi. 1995. Oxygen fluxes and dystrophy in a coastal lagoon colonized by *Ulva rigida* (Sacca di Goro, Po River Delta, Northern Italy). *Fresenius Envir Bull* 4:381–86.

Viaroli, P., M. Bartoli, G. Giordani, P. Magni, and D. T. Welsh. 2004. Biogeochemical indicators as tools for assessing sediment quality/vulnerability in transitional aquatic ecosystems. *Aquatic Conservation: Marine and Freshwater Ecosystems* 14:S19–29.

Viaroli, P., M. Bartoli, G. Giordani, M. Naldi, S. Orfanidis, and J. M. Zaldívar. 2008. Community shifts, alternative stable states, biogeochemical controls and feedbacks in eutrophic coastal lagoons: A brief overview. *Aquatic Conservation Freshwater and Marine Ecosystems* 18:S105–17.

Viaroli, P., and R. R. Christian. 2004. Description of trophic status, hyperautotrophy and dystrophy of a coastal lagoon through a potential oxygen production and consumption index—TOSI: Trophic Oxygen Status Index. *Ecol Ind* 3:237–50.

Viaroli, P., G. Giordani, M. Bartoli, M. Naldi, R. Azzoni, D. Nizzoli, I. Ferrari, J. M. Zaldívar, S. Bencivelli, G. Castaldelli, and E. A. Fano. 2006. The Sacca di Goro and an arm of the Po river. In *The handbook of environmental chemistry, Volume 5. Water pollution: Estuaries*, ed-in-chief O. Hutzinger, volume ed. P. J. Wangersky, 197–232. Berlin: Springer-Verlag.

Viaroli, P., A. Pugnetti, and I. Ferrari. 1992. *Ulva rigida* growth and decomposition processes and related effects on nitrogen and phosphorus cycles in a coastal lagoon (Sacca di Goro, Po River Delta). In *Marine eutrophication and population dynamics*, eds. G. Colombo, I. Ferrari, V. U. Ceccherelli, and R. Rossi, 77–84. Fredensborg, Denmark: Olsen & Olsen.

Vollenweider, R. A., F. Giovanardi, G. Montanari, and A. Rinaldi. 1998. Characterization of the trophic conditions of marine coastal waters with special reference to the NW Adriatic Sea: Proposal for a trophic scale, turbidity and generalized water quality index. *Environmetrics* 9:329–57.

Vollenweider, R. A., A. Rinaldi, and G. Montanari. 1992. Eutrophication, structure and dynamics of a marine coastal system: Results of a ten year monitoring along the Emilia-Romagna coast (Northwest Adriatic Sea). In *Marine coastal eutrophication*, eds. R. A. Vollenweider, R. Marchetti, and R. Viviani, 63–106. Proceedings of an International Conference, Bologna, Italy, March 21–24, 1990.

Weisberg, S. B., J. A. Ranasinghe, D. M. Dauer, L. C. Schaffner, R. J. Diaz, and J. B. Frithsen. 1997. An estuarine benthic index of biotic integrity (B-IBI) for Chesapeake Bay. *Estuaries* 20:149–58.

Xu, F.-L., S. E. Jørgensen, and S. Tao. 1999. Ecological indicators for assessing freshwater ecosystem health. *Ecol Model* 116:77–106.

Zaldívar, J. M., F. Bacelar, S. Dueri, D. Marinov, P. Viaroli, and E. Hernandez. 2009. Modeling approach to regime shifts of primary production in shallow coastal ecosystems: Competition between seagrass and macroalgae. *Ecol Model* 220:3100–3110.

Zaldívar, J. M., A. C. Cardoso, P. Viaroli, A. Newton, R. de Wit, C. Ibañez, S. Reizopoulou, F. Somma, A. Razinkovas, A. Basset, M. Holmer, and C. N. Murray. 2008. Eutrophication in transitional waters: An overview. *Transitional Waters Monographs* 2:1–78.

Zaldívar, J. M., E. Cattaneo, M. Plus, C. N. Murray, G. Giordani, and P. Viaroli. 2003. Long-term simulation of main biogeochemical events in a coastal lagoon: Sacca di Goro (Northern Adriatic Coast, Italy). *Continental Shelf Research* 23:1847–75.

Zaldívar, J. M., M. Plus, C. N. Murray, G. Giordani, and P. Viaroli. 2003. Modelling the impact of clams in the biogeochemical cycles of a Mediterranean Lagoon. In *Proceedings of the Sixth International Conference on the Mediterranean Coastal Environment. MEDCOAST 03*, ed. E. Ozhan, 1291–1302. Ravenna, Italy, October 7–11.

Zaldívar, J. M., F. Strozzi, S. Dueri, D. Marinov, and J. P. Zbilut. 2008. Recurrence quantification analysis as a method for the detection of environmental thresholds. *Ecol Model* 210:58–70.

Zeileis, A., C. Keliber, W. Kramer, and K. Hormik. 2003. Testing and dating structural changes in practice. *Computational Statistics & Data Analysis* 44:109–23.

16

Ecosystem Indicators for the Integrated Management of Landscape Health and Integrity

Felix Müller and Benjamin Burkhard

CONTENTS

16.1 Introduction

Throughout the last decades, ecosystem approaches seem to have grown out of puberty: For a rising number of ecologists the high complexity of ecological systems has not only become an accepted fact, but also an interesting object of investigation. In parallel, the successful reductionistic methodology has been accomplished steadily by holistic concepts that stress systems approaches and syntheses, and elucidate the linkages between the multiple compartments of ecological and human-environmental systems within structural, functional, and organizational entities. For instance, in Germany five Ecosystem Research Centers have been installed and supported within the last decades (see, e.g., Fränzle 1998; Fritz 1999; Gollan and Heindl 1998; Hantschel et al. 1998; Widey 1998; Wiggering 2001), and additional research projects have been carried out in national parks (e.g., Kerner et al. 1991), biosphere reservations (e.g., Schönthaler et al. 2001), and coastal districts (e.g., Dittmann et al. 1998; Kellermann et al. 1998). With these initiatives the comprehension and the acceptance of ecosystem approaches have made a big step forward (for an overview, see Schönthaler et al. 2003). The listed approaches have been accomplished by several Long-Term Ecological Research Program (LTER) initiatives and several projects that are based on the UN Commission on Biodiversity (CBD) ecosystem approach (see http://www.ecology.uni-kiel .de/salzau2006/).

Also in environmental practice, ecosystemic attitudes are becoming more and more favorable: While in the past, environmental activities were restricted to specific ecological resorts, today—in the age of the sustainability principle and the ecosystem services concept (see http://www.ecology.uni-kiel.de/ salzau2008/)—we can find environmental politics that try to integrate individual resorts. Instead of a concentration on environmental sectors, ecosystems are becoming focal objects, and interdisciplinary cooperation is increasing continuously, also in environmental practice (see Schönthaler et al. 2003).

The major problem of these modern approaches is to cope with the enormous complexity of environmental systems that arises from the various elements, subsystems, and interrelations that ecosystems provide. Hence, scientific approaches to reduce this complexity with a valid and theory-based methodology have become basic requirements for a high qualitative development of systemic approaches in science, technology, and practice (see Müller and Li 2004). One concept to reduce the complexity of ecological and human-environmental systems is a representation of the most significant parameters of an observer-defined system by indicators, which are quantified variables that provide information on a certain phenomenon with a synoptic distinctness (Radermacher et al. 1998). Often indicators are used if the indicandum—the focal object of the demanded information—is too complex to be measured directly or if its features are not accessible with the available methodologies.

There are certain acknowledged requirements for indicators. For instance, they should be easily measurable, they should be able to be aggregated, and they should depict the investigated relationships in an understandable manner. The indicandum should be clearly and unambiguously represented by the indicators, and these variables should comprise an optimal sensitivity, include normative loadings in a defined extent only, and provide a high utility for early warning purposes. As Table 16.1 shows, there are many further needs for the quality of indicator sets, which often can only hardly be met if complex interrelations have to be represented.

TABLE 16.1

Criteria and Requirements for Ecological Indicators

Political relevance	High level of aggregation
Political independence	Target-based orientation
Spatial comparability	Usable measuring requirements
Temporal comparability	Usable requirements for quantification
Sensitivity concerning the indicandum	Unequivocal assignment of effects
Capability of being verified	Capability of being reproduced
Validity	Spatiotemporal representativeness
Capability of being aggregated	Methodological transparency
Transparency for users	Comprehensibility

Note: The listed items should be realized to an optimum degree to produce an applicable indicator system. According to Müller and Wiggering (2004).

Concerning these requirements, the existing holistic indicator sets comprise different potentials, advances, and limitations. For example, with respect to indicator complexity, on the one hand we can find very complex indicator sets with a very high number of proposed variables (e.g., Schönthaler et al. 2001; Statistisches Bundesamt et al. 2002), and on the other there are approaches that include a reduction up to one parameter only (e.g., Jørgensen 2000; Ulanowicz 2000; Odum et al. 2000). Between these indicator systems there is a broad wingspan regarding the necessary database, the demanded measuring efforts, the complexity of the aggregation methodology, and the comprehensibility of the results as well as the cognitive transparency for the users.

Within this polarization, we have tried to find a representative holistic indicator set on the basis of the concepts, results, and theoretical background of the R&D project, Ecosystem Research in the Bornhöved Lakes District (Fränzle 1998, 2000; Fränzle et al. 2008). Secondary investigations have been executed in the R&D project, Macro Indicators to Represent the State of the Environment for the National Environmental-Economic Accounting System of Germany (Statistisches Bundesamt et al. 2002). The respective investigations have led to a set of eight ecosystem variables that are suitable to represent the focal elements of the Pressure-State-Response and the Drivers-Pressure-State-Impact-Response indicator approaches (Rapport and Singh 2006; Burkhard and Müller 2008a), the state of ecosystems on an integrative level. The indicators are proposed for use as representatives for the capacity of self-organization in ecological systems, which is the selected indicandum to depict the degree of integrity or health in ecological entities (Burkhard et al. 2008).

This chapter tries to demonstrate the derivation and application of the aggregated ecosystem indicator set. In the beginning, the basic principles and the specific requirements for the indicator selection are described. These resulting conceptual forcing functions come from ecosystem analysis, ecosystem theory, and the normative principles of ecosystem integrity. The respective framework for indicator selection is clarified, and thereafter the indicators are presented together with information on the used methodologies for their quantifications on different scales. On this basis, case studies are presented. That sequence starts with a comparison of different ecosystems, and it is continued by a description of applications on the landscape scale. In Section 16.2, the potentials of the indicator set for monitoring schemes are discussed, and finally an application in sustainable landscape management is described. The chapter ends with a discussion and a prospect to future developments.

16.2 Basic Principles for Indicator Derivation

Besides the requirements summarized in Table 16.1, three pillar principles have been considered as basic conceptual points of departure for the indicator derivation. The first guideline, which guarantees a high applicability and a general correctness, has origins in fundamental ideas from ecosystem theory: Ecosystems are regarded as self-organizing entities, and the degree of self-organizing processes and their effects have been chosen as an aggregated measure to represent the systems' actual states. The basic theoretical principles of this approach stem from the thermodynamic fundamentals of self-organization and from the orientor principle, which is also used by many other concepts published in this book. A second pillar consists of the methodologies of ecosystem analysis: to depict ecological entities in a holistic manner, structures as well as functions have to be taken into account, the latter representing the performance of the ecosystems. Finally, for utilization in environmental management, the basic approaches that emerge from these principles have to be reflected on a normative level. As the factual evaluation of the concrete indicator values is a societal (not an ecological) task, a useful indicator set has to be based on political concepts and targets. In this case, the preconditions for environmental decision making are formulated by a specific definition of ecological integrity (Barkmann et al. 2001), which includes several items that are valid for the ecosystem health approach as well.

16.2.1 Ecosystem Theory—Conceptual Background

To reach an optimal applicability of scientific methodology, theoretical considerations seem to be a good starting point, even if applicable indicators for practical purposes have to be developed. In ecosystem theory there are many different approaches (see Jørgensen 1996; Müller 1997; Jørgensen et al. 2007) that can easily be condensed and aggregated within the theory of self-organization. This approach not only provides a unifying concept of ecosystem dynamics, it also depicts a high agreement with basic ideas from the ecosystem health concept (see Table 16.2) that

TABLE 16.2

Axioms of Ecosystem Health, after Costanza et al. (1993)

Dynamism: nature is a set of processes, more than a composition of structures
Relatedness: nature is a network of interactions
Hierarchy: nature is built up by complex hierarchies of spatio-temporal scales
Creativity: nature consists of self-organizing systems
Different fragilities: nature includes various sets of different resiliences

Note: The listed parameters reflect the basic system-related fundamentals of the health approach, which are also valid for the concept of ecological integrity.

stresses the creativity of nature, which is nothing else than the potential for self-organization.

In a generalized outline of the selected theoretical concept, the order of ecological systems emerges from spontaneous processes that operate without consciously regulating influences from the system's environment. Actually, these processes are constrained by human activities (see Müller 2005; Müller et al. 1997a, 1997b; Müller and Nielsen 2000), but although such constraints can reduce the degrees of freedom for ecosystem development, the self-organized processes cannot be set aside. The consequences of these processes have been condensed within the orientor approach (Bossel 1998; Müller and Leupelt 1998), a systems-based theory about ecosystem development that is founded on the general ideas of non-equilibrium thermodynamics (Jørgensen 1996, 2000; Schneider and Kay 1994; Kay 2000) and network development (Fath and Pattten 1998, 2000) on the one hand and succession theory on the other (e.g., Odum 1969; Dierssen 2000).

Self-organized systems are capable of creating structures and gradients, if they receive a flow through of exergy (usable energy, or the energy fraction of a system that can be transferred into mechanical work; see Jørgensen 2000). The typical exergy input path into ecosystems is solar radiation. This "high quality" energy fraction is transformed within metabolic reactions (e.g., respiration, heat export), producing non-convertible energy fractions (entropy) that are exported to the environment of the system. As a result of these energy conversion processes, under certain circumstances (Ebeling 1989) gradients (structures) are built up, and maintained (Müller et al. 2008). There are two extremal thermodynamic principles that take these conditions into account and postulate an optimizing behavior of open, biological systems: Jørgensen (2000) states that self-organized ecological systems tend to move away from thermodynamic equilibrium, that is, build up ordered structures and store the imported exergy within biomass, detritus, and information (e.g., genetic information) that can be indicated by structural diversities. In addition, Schneider and Kay (1994) state that the degradation of the applied gradients is an emerging function of self-organized systems.

As a consequence of these physical principles, throughout the undisturbed complexifying development of ecosystems—between Holling's exploitation and conservation stages (Holling 1986; Gunderson and Holling 2003)—there are certain characteristics that are increasing steadily and slowly. These features are developing toward an attractor state restricted by specific site conditions and prevailing ecological functions. As the development seems to be regularly oriented toward that attractor basin, the respective state variables are called orientors (Bossel 2000).

Using these ecosystem features as indicators, the naturalness of an ecosystem's development can be depicted. Figure 16.1 shows some of these orientors. In general it can be postulated that throughout an undisturbed development, the complexity of the ecosystems will be increasing asymptotically up to the state of maturity (Odum 1969). Within this development, exergy storage will be rising, on a materialistic level as well as on a structural basis: more and more gradients are built up. With this increasing structural diversity, the diversity of flows and the system's ascendancy (Ulanowicz 2000) will also grow, as well as certain network features (Fath and Patten 2000), and therefore also the energy necessary for the maintenance of the developing system will increase. Therefore, exergy storage as well as exergy degradation are typical orientors, and their dynamics can be explained in a contemporary manner. These basic thermodynamic principles have many consequences on other ecosystem features. For instance, the food web will become more and more complex; heterogeneity, species richness, and connectedness will rise; and many other attributes, as shown in Figure 16.1, will follow a similar long-term trajectory.

This orientation is a theoretical principle that can hardly be found in reality due to the continuous effects of disturbances. Especially in the case of high external inputs, the orientor values might decrease rapidly, proceeding into a retrogressive direction. In the following sequence, an adaptive or resilient system will find the optimization trajectory again, while a heavily disturbed ecosystem might no longer be able to improve the values of the orientors. Therefore, the robustness of ecosystems can be indicated by the orientors as well. Consequently, their values are also suitable to represent the ecological risk that is correlated to external inputs or changes of the prevailing boundary conditions. Yet we have to be aware of the fact that high orientor values do not guarantee high stability or high buffer capacity. Following Holling's ideas on ecosystem resilience and development, at the mature stage complex ecosystems become "brittle," their adaptivity decreases because of the high internal connectedness and respective interdependencies. Thus, the dynamics of external variables can force the mature system to break down and start with another developmental sequence.

An indication for ecosystem self-organization has been proposed in a small number of case studies only. Most of them refer to the concepts of ecosystem health (e.g., Rapport 1989; Haskell et al. 1993; Rapport and Moll 2000) or ecological integrity (e.g., Karr 1981; Woodley et al. 1993). Besides multivariate approaches (e.g., Schneider and Kay 1994; Kay 1993, 2000) and aggregated approaches (e.g., Costanza 1993) some authors propose to use highly integrated

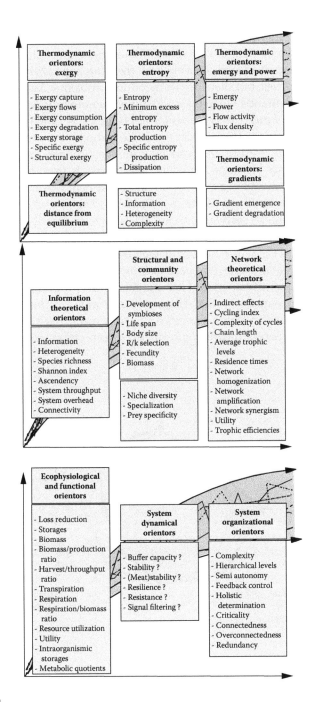

FIGURE 16.1
Ecological orientors from different theoretical origins. The listed ecosystem properties regularly show an optimizing behavior during the long-term development in undisturbed situations, according to Jørgensen and Müller (2000a).

variables like exergy (Jørgensen 2000), emergy (Odum et al. 2000; Ulgiati et al. 2003) or ascendancy (Ulanowicz 2000). These bright concepts are very original, they are discussed very actively, and they can cope with the concept of emergent properties, but there are tremendous problems, data requirements, and modeling demands when trying to apply them in practice.

One example of multivariate orientor applications is shown in Figure 16.2. Two different German stream ecosystems are compared on the basis of emergent ecosystem properties that can take on the function as orientors. The depicted values are based on intensive measurements from Mejer (1992) in a Black Forest stream and in a lowland stream ecosystem within the Bornhöved Lakes District in Northern Germany (Pöpperl 1996). These data have been used to run the model software ECOPATH 3.0, which describes the food web structures quantifying the standing stock, production, and consumption of the elements and the whole system as well as the flow of matter between the ecosystem compartments (average annual rates per square meter). Additionally, the model can quantify a series of holistic ecosystem properties.

The diagram elucidates that there are enormous differences between the investigated ecosystems. Specifically, concerning the primary production-based parameters (primary production, respiration, total system flow

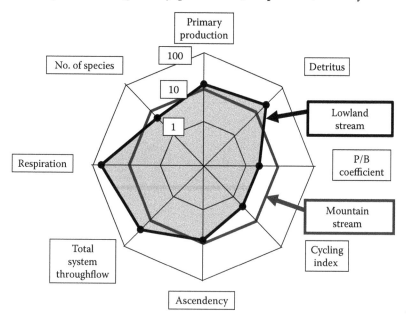

FIGURE 16.2
Amoeba diagram depicting the relative indicator values for a mountain stream and a lowland stream on the basis of a throphic ECOPATH model that has been applied to data sets from Mejer (1992) and Pöpperl (1996). The model was calibrated and run by R. Pöpperl and S. Opitz. The mountain stream values represent 100% in the graphics, and the comparison depicts the consequences of eutrophication for some orientor values of the Northern German lowland stream.

through), the lowland stream provides typical values for a strongly eutrophicated ecosystem. On the other hand, the more complex structure (no. of species), the relative diversity of flows and related parameters (cycling index, p/b coefficient) show that the mountain stream represents a much higher degree of ecosystem integrity.

16.3 Ecosystem Analysis—Empirical Background

Besides the theoretical considerations, there are other good reasons to use an ecosystem approach for environmental assessments. In Table 16.3, some of these motivations are listed. Various case studies from forest dieback research, ecotoxicology, and eutrophication research have documented that indirect effects, chronic effects, and de-localized effects are much more significant than direct interactions (see also Patten 1992). Furthermore, many disturbances do not affect only one environmental sector, but the whole ensemble of ecological compartments via webs of interactions and consequences. Last but not least, the ecosystem approach makes it possible to include phenomena like self-organization, emergent properties, and ecological complexity (Fränzle 2000). Therefore, the conceptual combination of structural and functional approaches into an organizational concept is a fine starting point to fulfill the empirical requirements for health or integrity indication (Costanza et al. 2000; Golley 2000; Müller and Windhorst 2000).

TABLE 16.3

Arguments Stressing the Methodological Significance of Ecosystem Approaches in Environmental Management, as They Can Provide a Better Consideration of the Following Items

Indirect effects (e.g., webs of reactions concerning forest dieback)

Chronic effects (e.g., accumulation of toxic substances)

De-localized effects (e.g., forest effects of ammonia from slurry)

Integration of ecological processes and relations into planning procedures

Representation of ecological complexity

Consideration of features of self-organization

Aggregation of structures and functions

Integration of different ecological media (e.g., soil-vegetation-atmosphere)

Integration of different environmental sectors (e.g., immission and erosion)

Utilization of improved extents and resolutions
- In terms of time (multiple interacting temporal scales)
- In terms of space (multiple interacting spatial scales)
- In terms of content and disciplines (multiple scientific approaches)
- In terms of analytical depth (multiple levels of aggregation and reduction)

The respective scientific approaches focus on "models of networks consisting of biotic and abiotic interactions in a certain area" (Jørgensen and Müller 2000b; Müller and Breckling 1997). Schönthaler et al. (2003) have defined ecosystem research as a "media spanning research of element and energy cycling, of structures and dynamics, of control mechanisms and of criteria for ecosystem resilience with the aim to learn how to understand the steering and feedback processes in ecological entities." Kaiser et al. (2002) have accomplished this description as follows: "Ecosystem research analyses the interactions of biological ecosystem components with each other, with their inanimate environment and with man. It delivers basic knowledge on structures, dynamics, element and energy flows, ecosystem stability and resilience."

Besides structural aspects (e.g., items of abiotic and biotic heterogeneity and their dynamics), ecosystem research investigates the imports, exports, and storages, and the internal flows of energy, water, and nutrients (e.g., carbon, nitrogen, potassium, calcium, sodium, magnesium) through the compartments of ecological entities (e.g., soil horizons; the unsaturated zone; the groundwater layer; plants on different structural levels and in different layers, but also with different internal functional subunits; animals with different positions in the food webs, or micro-organisms that can be found in different spatial compartments; and the atmospheric compartment) including the derivation of efficiency and cycling attributes (e.g., different ratios of biomass, respiration, production, or water movement, cycling index).

As there are many variables that can be taken into account to measure these items, and as they are linked within very complex webs of interactions, it is hard to select a small number of indicators that are capable of representing the whole variety of aspects describing the state of ecological systems. To proceed with this task, a combination has to be made reflecting the theoretical items, the empirical requirements, and the normative targets of the indicator set.

16.4 Ecosystem Health and Ecological Integrity— Normative Background

As the aspired indicators have to be used as information sources in environmental decision making, societal and normative arguments are also important prerequisites of their selection. The indicators have to refer to the leading concept of environmental management, which actually is the global political principle of a sustainable development. It has been discussed in various papers and political statements (e.g., Hauff 1987; World Commission on Environment and Development [WCED] 1987; Daily 1997; Costanza 2000),

TABLE 16.4

Basic Features and Requirements of Sustainable Landscape Management Strategies, According to Müller and Li (2004)

Long-term strategies...	...think in generations
Multiscale strategies...	...compare human vs. ecological time scales
Interdisciplinary strategies...	...realize that ecology is only one part
Holistic strategies...	...consider structures *and* functions
Realistic strategies...	...include uncertainties
Nature-oriented strategies...	...take nature as a model
Theory-based strategies...	...ensure correctness
Hierarchical strategies...	...realize constraints and scales
Goal-oriented strategies...	...joint definition of the targets

and in essence we are asked to use natural resources in a way that enables future generations access to these resources at least in a similar mode as applied today. The main conceptual innovations of the sustainability principle are the interdisciplinary linkage of social and natural items and the large spatiotemporal scales that have to be taken into account (Allen and Holling 2008). Thus, specific requirements arise from this principle. They are summarized in Table 16.4.

An important outcome of the described self-organized processes in the ecosphere is the potential of using the outputs of ecosystems' performances by man; ecosystem structures and functions provide certain environmental services that are the benefits people obtain from ecosystem organization, thus being basic requirements for human life (see Costanza et al. 2000; Millennium Assessment Board 2003). One potential classification of these services is based on the works of de Groot (1992): From his point-of-view, the performance of ecosystems can be classified as follows:

- *General provisions (carrier services)*: Ecosystem structures are providing space and suitable substrates for human activities.
- *Products*: Ecosystem development provides natural resources for human use.
- *Information*: Ecosystems are providing cultural attributes.
- *Regulations*: Ecosystem functions are regulating the availability of basic demands for human life. All ecological processes can be assigned to this category as they buffer external influences in a way that enables man to continue life in an environment with suitable climatic, chemical, and physical conditions.

Taking into account the terms and concepts mentioned in the last chapter, it is possible to use an alternative formulation for the ecological components

		Ecosystem Services			
		Supporting services	Provisioning services	Regulating services	Cultural services
Components of ecological integrity	Exergy capture	X	X	X	
	Exergy dissipation	X		X	
	Biotic waterflows	X	X	X	
	Metabolic efficiency	X		X	
	Nutrient loss	X	X	X	
	Storage capacity	X	X	X	
	Biotic diversity	X	X	X	X
	Organization	X	X	X	X

FIGURE 16.3
Interrelations between the proposed integrity indicators and ecosystem services.

of sustainable development: "Meet the needs of future generations" in this context means "keep available the ecosystem services on a long-term, intergenerational, and broad scale, intragenerational level." The potential direct contributions of the integrity variables can be seen in Figure 16.3, where the interrelations to the ecosystem service classes from the Millennium Assessment are depicted (Millennium Assessment Board 2003; Müller and Burkhard 2007). Obviously, the integrity indicators show extreme similarities with the supporting services of the Millennium Assessment.

From a synoptic viewpoint at these service categories, one fact becomes obvious: all ecosystem services are strongly dependent on the performance of the regulation functions. The correlated processes not only influence production rates, but in the long run they also determine the potentials of ecosystems to provide carrier functions and cultural services. And if we finally link all argumentations of this chapter, it becomes clear that the respective benefits are strictly dependent on the degrees and the potentials of the fundamental self-organizing processes. To maintain these services, the ability for future self-organizing processes within the respective system has to be preserved (Kay 1993). This demand is considered as a focal point of modern environmental management models, such as ecosystem health or ecological integrity. In a recent paper, Barkmann et al. (2001) defined ecological integrity as a political target for the preservation against nonspecific ecological risks, which are general disturbances of the self-organizing capacity of ecological systems. Thus, the goal should be a support and preservation of those processes and structures that are essential prerequisites of the ecological ability for self-organization.

16.4.1 The Selected Indicator Set

The three basic pillars for the presented indicator selection result in a set of variables that are able to depict the state of ecosystems on the basis of their features concerning the degree of self-organization and the potential to proceed in this way. Referring to the orientors presented in Figure 16.1, it becomes obvious that many of them cannot be easily measured or even modeled under usual circumstances. Some orientors can only be calculated on the basis of very comprehensive data sets that are measured on a very small number of sites. Other orientors can only be quantified by model applications. Therefore, the selected orientors have to be represented by variables that are accessible by traditional methods of ecosystem quantification. Consequently, the next step of indicator derivation is a "translation" of the thermodynamic, organizational, network, and information theoretical items into ecosystem analytical variables. Within this step, it has to be reflected that the number of indicators should be reduced as far as possible (see Table 16.1). Thus, many of the ecosystem variables depicted in Figure 16.1 cannot be taken into account. Instead, a small set consisting of the most important items that can be calculated or measured in many local instances is what we have to look for. This set should furthermore be based on the focal variables that are usually investigated in ecosystem research and that can be made accessible in comprehensive monitoring networks (Müller et al. 2000). The general subsystems that should be taken into account to represent ecosystem organization are listed below as elements of ecosystem orientation:

- Ecosystem structures: While ecosystems are evolving, the number of integrated species is regularly increasing steadily and also the abiotic features are becoming more and more complex. This development is accompanied by a rising degree of information, heterogeneity, and complexity. Also, specific life forms (e.g., symbiosis) and specific types of organisms (r/k strategists, organisms with rising life spans and body masses) become predominant throughout the orienting development (see Jørgensen et al. 2007).

- Ecosystem functions: Due to the increasing number of structural elements, the translocation processes of energy, water, and matter are becoming more and more complex, the significance of biological storages is growing as well as the degree of storage in general, and consequently the residence times of the input fractions are increasing. These processes influence the budgets of the respective fractions that can be measured by input-output analysis. Due to the high degree of mutual adaptation throughout the long developmental time, the efficiencies of the single transfer reactions are rising, cycling is optimized, and thus losses of matter are reduced. The respective ecosystem functions are usually investigated within three classes of processes that are interrelated to a very high degree:

- Ecosystem energy balance: Exergy capture (uptake of usable energy) is rising during undisturbed development, the total system throughput is growing (maximum power principle; see Odum et al. 2000) as well as the articulation of the flows (ascendancy, see Ulanowicz 2000). Due to the high number of processors and the growing amount of biomass, the energetic demand for maintenance processes and respiration is growing as well (entropy production; see Svirezhev and Steinborn 2001; Steinborn 2001).

- Ecosystem water balance: Throughout the undisturbed development of ecosystems and landscapes, more and more elements have to be provided with water. This means that specificially the water flows through the vegetation compartments show a typical orientor behavior (Kutsch et al. 1998). These fluxes provide another high significance, because they demonstrate an important prerequisite for all cycling activities in terrestrial ecosystems: the water uptake by plants regulated by the degree of transpiration.

- Ecosystem matter balance: Imported nutrients are transferred within the biotic community with a growing partition throughout undisturbed ecosystem development. Therefore, the biological nutrient fractions are rising as well as the abiotic carbon and nutrient storages, the cycling rate is growing, and the efficiencies are being improved. As a result, the loss of nutrients is reduced.

Based on these features, a general indicator set to describe the ecosystem or landscape state in terrestrial environments has been derived. It is shown in Table 16.5. The basic hypothesis concerning this set is that a holistic representation of the degree and the capacity for complexifying ecological processes on the basis of an accessible number of indicators can be fulfilled by these variables. They also represent the basic trends of ecosystem development; thus they show the developmental stage of an ecosystem or a landscape. As a whole this variable set represents the degree of self-organization in the investigated system. Hence, it can be postulated that (with the exception of mature stages that are in fact very seldom in our cultural landscape) the potential for future self-organization can also be depicted with this indicator set.

Of course this parameter set cannot provide a complete indication of sustainability, because the social and economic subsystems are not taken into account (e.g., driving force or response indicators). Also, external inputs and other pressures are not represented. But the focal ecological branch of

TABLE 16.5

Proposed Indicators to Represent the Organizational State of Ecosystems and Landscapes.

Orientor Group	Indicator	Potential Key Variable(s)
Biotic structures	Biodiversity	Number of species
Biotic structures	Biotope heterogeneity	Index of heterogeneity
Energy balance	Exergy capture	Gross or net primary production
	Entropy production	Entropy production after Aoki
		Entropy production after Svirezhev and Steinborn (2001)
		Output by evapotranspiration and respiration
	Metabolic efficiency	Respiration per biomass
Water balance	Biotic water flows	Transpiration per evapotranspiration
Matter balance	Nutrient loss	Nitrate leaching
	Storage capacity	Intrabiotic nitrogen
		Soil organic carbon

Note: The nominated key variables can be regarded as an optimal indicator set. If these parameters are not available other variables may be chosen to reflect the respective indicandum. Doing this, the observer must realize that the quality of the indicator-indicandum relations may be sinking.

sustainability can be described on the basis of the orientor state indication. In spite of this strategic restriction, the integrity indication provides potential linkages to the human-based indicators of the Driver–Pressure–State–Impact–Response (DPSIR) scheme. A comparison with the basic ecosystems services after de Groot (1992) shows that regulation services are well represented in this indicator set and that there are high interrelations with the production services while carrier and information services are not represented in a satisfactory manner.

16.5 Case Studies and Applications

16.5.1 Indicating Health and Integrity on the Ecosystem Scale

This indicator set has been applied within several case studies on different scales, whereby the linkages between data sources, model outputs, and indicator demands have been an object of methodological optimization throughout the last years. In the following paragraphs one example will be shown from the ecosystem research project in the Bornhöved Lakes, which was conducted between 1988 and 2001 in northern Germany. Within the main research area Altekoppel comparative empirical ecosystem studies were carried out in agro ecosystems and forests (Hörmann et al. 1992). A precise description of the methodologies used for the indicator quantification can be found in Schimming and von Stamm (1993), Baumann (2001), and Barkmann (2001). The respective measurements were conducted by numerous colleagues from the Bornhöved Lakes Project (see also http://www.ecology .uni-kiel.de) whose investigations are summarized, for instance, in Hörmann et al. (1992), Breckling and Asshoff (1996), or Fränzle et al. (2008).

In the following case study, results from a 100-year-old beech forest and a directly neighboring arable land ecosystem are demonstrated. Both ecosystems had a similar agricultural use before the forest was planted. Thus the question is which ecosystem features and which ranges of the self-organization capacity have been modified by the different land use schemes (see also Kutsch et al. 2001; Kutsch et al. 1998; Windhorst et al. 2004).

Figure 16.4 shows the differences between the two ecosystems with respect to their biocenotic structures. This variable represents the biotic complexity of ecosystems, and it reflects the amount of exergy stored in information. Nearly all investigated organism groups show higher numbers of species in the forest ecosystem. One exception is the group of small mammals, who can find very good food conditions in the arable land and who are well adapted to this ecosystem type. The second structural indicator is the abiotic heterogeneity, which was calculated with a geographic information system (GIS)-based neighborhood method after Reiche (Baumann 2001). While the index of the forest ecosystem is 0.56 referring to the soil organic matter, the maize field has a value of only 0.08. Also corresponding to the soil chemical constituents H^+, Ca^{2+}, Mg^{2+}, K^+, and phosphate, the forest soil heterogeneity is higher than the respective value on the arable land (Reiche et al. 2001). Therefore, we can constitute very high differences concerning the structural patterns of these ecosystems.

Investigating the storage capacities of the two ecosystems, the biomass and the intrabiotic nutrients were used as indicators. They are capable of representing the ecosystem pools as another compartment of exergy storage, the

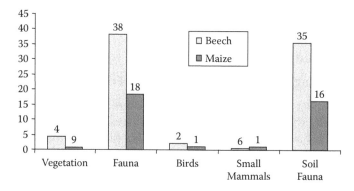

FIGURE 16.4
Comparison of the species numbers in some community groups of the investigated ecosystems; data were compiled from Hörmann et al. (1992).

chemical buffer capacities, and the availability of nutrients for the further development of the system. The depicted data are based on direct measurements, yield analyses, and modeling results (see Baumann 2001). The living biomass varied from 131 t C/ha in the beech forest to 6.5 t C/ha in the arable land, and the relations for the soil organic carbon is 80 t C/ha vs. 56 t C/ha, respectively. The correlated ecosystem comparison concerning the intrabiotic nutrients is sketched in Figure 16.5. It shows that the higher values can be found in the forest ecosystem for both nitrogen and phosphorus compounds.

Another important functional parameter used is the loss of nutrients. It shows the irreversible export of chemical compounds as well as the efficiency of the recycling regime in the ecosystem. Data from Figure 16.6 are based on

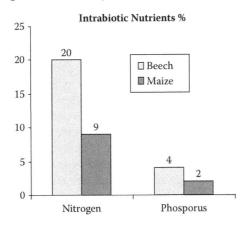

FIGURE 16.5
Comparison of the intrabiotic nutrient contents of the investigated ecosystems; data from Kutsch et al. (1998).

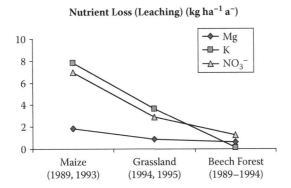

FIGURE 16.6
Comparison of the nutrient loss from the investigated ecosystems; data from H. Wetzel and C. G. Schimming (unpublished). The figure demonstrates that there are significant differences between the ecosystems concerning the leaching quantities of the three chemical compounds.

chemical analyses of the soil solution and model applications concerning the balances and the output path into the atmosphere. The figure shows that there are enormous differences between the two systems. Of course, this is a consequence of the different import and export regimes. But besides these extreme examples, the loss of nutrients seems to be a very general effect resulting from ecological disturbances. This may be caused by opening the food webs and cycles, which usually become more and more closed in undisturbed developmental phases. Hence, nutrient loss is a suitable candidate for a key indicator of ecosystem health.

Similar results were obtained concerning the biotic water flows, which represent a biological efficiency measure and symbolize the basic prerequisites for all cycling processes. The data are based on hydrological and microclimatological measurements and transpiration modeling with a two-layer Soil–Water–Atmosphere–Transfer (SWAT) model (Herbst et al. 1999). The percentage of transpiration from the total evapotranspiration loss was 63% in the case of the forest ecosystem and 34% concerning the field. This signalizes the distinct significance of biological flows in the site budgets of water. This item could also be understood as an ecosystemic water use efficiency, because it is strongly correlated with the capacity of nutrient cycling, and because transpiration is a very important factor of the temperature regulation of ecosystems.

Also, the metabolic efficiency (respiration/biomass) of the forest was much higher than the efficiency of the arable land ecosystem. This elucidates the different degrees of flow organization and the energetic demand to maintain the existing structures. The entropy production was calculated with a methodology after Aoki (1998) and on the basis of the exergy radiation balance (Steinborn 2001). While the first method does not produce a satisfying sensitivity, the radiation balance approach can discriminate both ecosystems very well (see Baumann 2001).

A synopsis of the indicator values is presented in Figure 16.7. Looking at the whole figure, it is obvious that all values of the forest ecosystem are higher than the respective numbers of the arable land system with one exception: exergy capture. This indicandum has been represented by the gross primary production. The high value of the arable land ecosystem demonstrates that the farmer has been successful in optimizing the production of his site. The consequences of this economic orientation can be seen in all other variables: summarizing, they show that the degree of self-organization—and with this the ecological integrity—of the forest is much higher than it is in the field. In the case of new external disturbances this system bears a much higher risk of retrogressive changes than the forest, which represents a higher state of self-organizing capacity.

16.5.2 Indicating Landscape Health

While the case study stated before is totally based on small-scaled measured data, additional approaches have been developed on the landscape

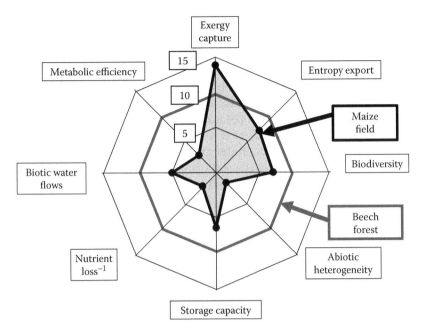

FIGURE 16.7
Synopsis of the indicator values for the two compared ecosystems. The beech values were taken as reference values (100%).

scale where the demands for empirical measurements are much smaller. To extend the indicator system to the landscape level, it was linked with the GIS-coupled modeling system "Dilamo" (digital landscape analysis and modeling; Reiche 1996). Using this instrument, many of the explained indicators can be calculated on the landscape scale. The integrated models "Wasmod" and "Stomod" have been enabled stepwise to calculate the parameters described in Table 16.5 in a validated and reliable manner. This methodology has been applied in different areas: M. Meyer (2000) used the modeling procedure to foresee the outcome of three land use scenarios for the whole Bornhöved Lakes District. He could show that especially the nutrient budget indicators show high differences due to distinct land use strategies. The municipality of Plön in Northern Germany was analyzed by Barkmann (2001) to show the dynamics of the integrity indicators in different years, using the same methodology. He also could underline that particularly the loss of nutrients seems to be very sensitive to changes in ecosystem structures and functions. U. Meyer (2001) has conducted a similar study for two catchments in the Biosphere Reservation "Rhön" in Central Germany (Schönthaler et al. 2001).

Taking a similar approach, Schrautzer et al. (2007) have derived landscape balances for the different ecosystem types of the Bornhöved Lakes District, including water, matter, and energy budgets for the whole watershed. With this contribution, the methodological linkage between the modeling systems, the GIS, and the proposed indicator set has been transferred into a highly applicable form. This case study is based on an ecosystem classification that was conducted for all terrestrial ecosystems in the catchment of Lake Belau (447 ha). This watershed includes a high proportion of wetland ecosystems. The ecosystem classification (conducted by U. Heinrich, J. Schrautzer, and H. P. Blume; see Fränzle et al. 2008) takes into account the vegetation types, soil criteria, and dominant land use structures. The resulting ecosystem types have been calibrated with data on the groundwater table, the C/N ratios, the pH values, and the S values of the soil compartments. The result was a map of ecosystem types that was elaborated with a GIS.

Based on that classification the resulting ecosystem types were analyzed with the computer-based "digital landscape analysis system" (Reiche 1996). The four information layers of soils, topography, linear landscape elements, and land use were used to produce more detailed digital maps, which were joined with the classification maps. In the next step the modeling system Wasmod-Stomod (Reiche 1996) was used to simulate the dynamics of water budgets, nutrients, and carbon fluxes based on a 30-year series of daily data about meteorological and hydrological forcing functions. The model outputs were validated by measured data in some of the systems (Schrautzer 2002). Furthermore, the model outputs were extended to include data sets concerning the ecosystem indicators by the following variables:

- Exergy capture: net primary production (NPP)
- Entropy production: microbial soil respiration
- Storage capacity: nitrogen balance, carbon balance
- Ecosystem efficiency: evapotranspiration/transpiration, NPP/soil respiration
- Nutrient loss: N net mineralization, N leaching, denitrification
- Ecosystem structures: number of plant species (measured values)

In the following example, these indicators were used to investigate different stages of a retrogressive wetland succession. The wet grasslands of the Bornhöved Lakes District are managed in a way that includes the following measures: drainage, fertilization, grazing, and mowing in a steep gradient of ecosystem disturbances. The systems are classified due to these external input regimes, and in Figure 16.8 the consequences can be seen in a synoptic manner: While the farmer's target (improving the production and the yield of the systems), indicated by the NPP, is growing by a factor of 10, the structural indicator is decreasing enormously throughout the retrogression. Also, the efficiency measures (NPP/soil respiration) are going down, and the biotic water flows get smaller. On the other hand, the development of the N and C balances demonstrates that the system is turning from a sink function into

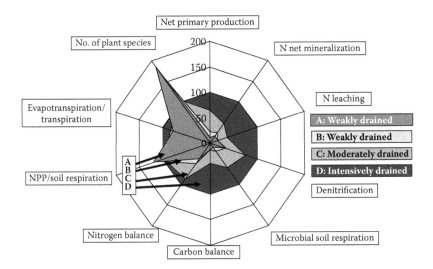

FIGURE 16.8
Synoptic illustration of the development of ecosystem indicators in a retrogressive succession of wet grasslands in the watershed of Lake Belau, according to Schrautzer et al. (2007). The most disturbed system type (the eutrophicated and strongly drained grasslands) is used as a reference state (100%).

a source, the storage capacity is being reduced, and the loss of carbon and nitrogen compounds (all indicators on the right side of the figure) is rising enormously. With these figures we can state an enormous decrease of ecosystem health, and because many of the processes are irreversible, the capacity for future self-organization is reduced up to a very small degree.

In a subsequent study (Müller et al. 2006; Schrautzer et al. 2007) several retrogression stages of wetland degradation in Northern Germany were assigned to the developmental scheme of Figure 16.9. Data were compiled by literature studies, measurements, and modeling exercises. For the comprehension of the case study, each stage is illustrated by one amoeba diagram. The consequences of land use intensification can be followed in the upper row, leading from the left (mesotrophic alder carr) to the right side (wet pasture). Due to the arrangement of the indicators the degree of integrity is symbolized by the position of the gray areas in the diagrams: The more they are situated on the right side, the stronger is the effect of the land-use-based disturbance, the smaller is their overall degree of integrity.

The results of an abandonment of these sites can be discovered in the lower rows; in all cases they lead to wetland sites with dominating alder trees. In some cases species protection measures can provoke a resilient development between two stages, then the abandoned systems can re-achieve the integrity features of their initial conditions. If the state of a wet pasture has been reached, only very intensive measures, e.g., landscape rewetting, can provoke a development toward a more healthy system. Observing such pathways only

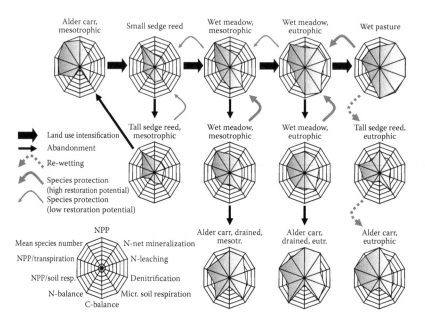

FIGURE 16.9
Distinction between multiple stages of wetland retrogression. The amoeba diagrams illustrate the typical values of ecosystem integrity variables and their development after land use intensification. The rounded arrows symbolize possible results of different restoration measures. Small dark arrows depict the results of abandonment.

from a structural viewpoint, the result might be satisfying: alder trees will be found again—even though this process may take some decades. But if we look into the details of those alder systems, it becomes obvious that only in one case can the original quality be attained again (small sedge reed—tall sedge reed alder carr); in all the other cases several indicators provide distinct values; there is no functional resilience, as soon as the systems have been changed into wet meadows.

16.5.3 Application in Sustainable Landscape Management

The following case study was taken from a European project about strategies for a sustainable reindeer herd in Northern Fenno-Scandinavia (RENMAN; see http://www.urova.fi/home/renman/). In this case, the indication of ecosystem health was accomplished by social and economic data to build a science-based fundamental for the outstanding land use decision-making processes in the region. Besides big ethnic problems between the Sámi native inhabitants and the Fenno-Scandinavian population, the key problems of reindeer herding can be put down to the fact that in the past decades there was an immense loss of grazing land for the big reindeer herds. Causes are growing demands for electricity (hydropower plants with huge artificial lakes), an increasing demand for tourist areas, a nonsustainable, clear cut–based, intensive forestry, and a fence system that reduces the original mobility of the reindeer herds to an extreme degree (see Burkhard and Müller 2008b; Burkhard et al. 2003; Vihervaara et al., in press). As a result, now there is a relatively high number of animals within a smaller area. Additionally, the traditional differentiation of summer and winter pastures (which were situated at very distant areas) is no more realized. Consequently, the reindeer herds today are in danger of destroying their traditional winter fodder during the summer grazing periods: ground lichen (*Cetraria nivalis* and *Cladina* sp.) are a specific fodder that becomes the focal food during the winter season. During the summer seasons, lichen can dry very rapidly, getting brittle and easily disturbed. If the herds are using the winter grazing grounds during a dry summer period, the ground lichen can be easily destroyed. Besides this problem, the abundances of the arboreal tree lichen (*Alectoria* spp., *Bryoria* spp., *Usnea* spp.), which are an alternative winter food, has decreased enormously due to the intensive forestry practices.

In this conflict field between different land use strategies, we carried out a systems analysis concerning land use structures, ecological items, and social and economic problems. The methodology was based on landscape mapping, measurements of ecological variables, and modeling with the Wasmod-Stomod system (Reiche 1996), which was briefly introduced before. Three scenarios were carried out, referring to A: a business as usual strategy; B: an intensification; and C: a reduction of reindeer herding. While the ecological data were measured or calculated, the other items were investigated on the basis of expert interviews. The experts were asked to foresee the consequences of the scenario conditions within a time span of 25 years by estimating the development on a scale from –5 (high decrease of the indicator values) to +5 (high increase), using the indicators depicted in Figure 16.10. This is one example for the scenario outcomes, referring to the scenario (C), "Reduction of reindeer herding."

The land use amoeba demonstrates an impressive decrease of reindeer herding, while before all forestry will be intensified, new artificial lakes are expected to be built, and tourism as well as mining will have a higher significance in the land use structure. Concerning the consequences for the indigenous Sámi population, a very high economic risk has been postulated: all values of the economic amoeba will decrease, the employment situation might become fatal, and the autonomy of the Lapland region will rapidly go down. Furthermore, the social amoeba demonstrates that the experts are afraid of a high demographic loss of population, that the Sámi ethnic identity will be diminished, and social security and health will be confronted with huge problems. The last part of the figure concerns the ecological outcome. Here the model and field measurement results show rather small alterations: The reduced number of reindeer causes less trampling, which leads to a decrease in abiotic diversity. All other indicators show a slight increase, if reindeer herding is decreased. This can be interpreted as a first sign of successional

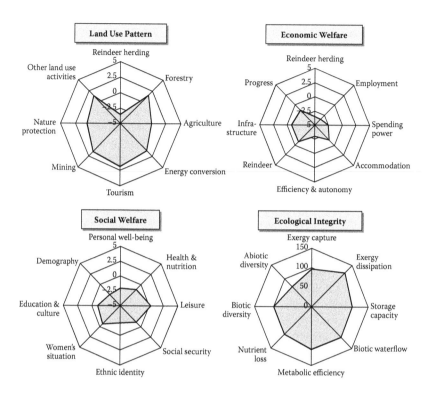

FIGURE 16.10
An overview of the indicator values concerning the land use scenario C (reduction of reindeer herding). The values for the indicator groups' land use, and economic and social welfare
were taken from the expert interviews in Lapland. The amoeba bodies show the results of the
experts' estimations as derivations (between +5 and –5) from the reference value 0 (situation
in 2003). The ecological values are model outcomes or field measurements on differently managed pasture areas (Burkhard and Müller 2008b).

development following the disturbance related to intensive reindeer herding. But all these changes are minor transformations. Summarizing, the ecological consequences of different herding regimes are not very significant.
Thus, the focal argumentation for the finding of a sustainable management
of Lapland's landscapes is an economic and social question. To solve this
problem, our indicators hopefully can be helpful tools.

16.5.4 Indicating Dynamics in Marine Ecosystems

In one of our latest case studies, the concepts of ecological integrity and
corresponding indicators were transferred to the marine ecosystem of the
German North Sea (as part of the project *Zukunft Küste—Coastal Futures*,
www.coastal-futures.org; Burkhard et al. 2009). In the German exclusive
economic zone of the North Sea, a pattern of different human uses can be

found: shipping, fishery, military areas, raw material exploitation, nature protection, and tourism. Recently, new plans for the establishment of large offshore wind farms were made, providing a new form of use to the North Sea (www.bsh.de). To assess the impacts that the installation of thousands of huge wind turbines, including their foundations and scour protections, will have on marine ecosystems, a combined model–GIS approach was used to quantify integrity indicators. The main question was whether the installation of wind farms has the potential to cause significant ecosystem dynamics, ranging from systems' degradations to the development of highly productive and diverse artificial reef systems. The construction of wind farms was simulated with the ecosystem box model ERSEM (Lenhart et al. 2006) by increasing the suspended particulate matter (SPM) concentrations in the water column, simulating disturbances during ramming of wind turbine piles and cable scavenging. ERSEM outcomes gave information about net primary production (integrity indicator exergy capture), winter turnover rate of nutrients (nutrient cycling), and transport loss of nutrients (nutrient loss). Furthermore, ERSEM output data were used as input information for marine food web simulations with Ecopath (www.ecopath.org; Christensen and Pauly 1992a, 1992b), indicating entropy production (C per year from respiration), storage capacity (C stored in biomass), and ecosystem organization (ascendancy). Abiotic heterogeneity was indicated by sediment dynamics and alterations in water currents, which were modeled with MIKE21 (www .dhigroup.com; Danish Hydraulic Institute [DHI] 1999). GIS data on resting seabirds were used to indicate impacts of offshore wind turbines on (above water) biotic diversity (evenness) (Dierschke and Garthe 2006). The results shown in Figure 16.11 illustrate that some ecosystem components are sensitive to impacts caused during the construction of offshore wind parks: the increase in SPM during the construction phase causes a light limitation in water, which leads to a decreasing primary production, respectively exergy capture. Nutrient cycling was reduced as well, whereas nutrient loss increased slightly. Biotic diversity decreased during the construction of wind turbines and did not return to the reference state, indicating that resting seabird communities are disturbed in the long term. During the operation of offshore wind parks, the integrity parameters, besides biotic diversity, returned to the reference state as early as one year after construction, indicating a resilient behavior. It has to be pointed out that the modeling of the operation phase of offshore wind farms, following their construction, covered a short time period only. To simulate long-term effects and possible emergence of artificial reef systems, further modeling must be carried out in the project at the moment. A more detailed elaboration of methods and results can be found in Burkhard et al. (2009).

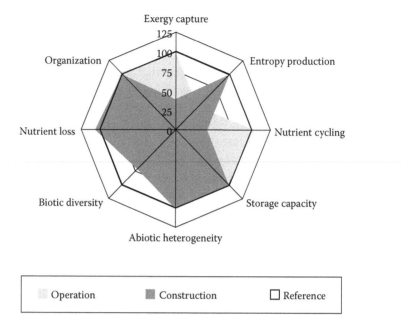

FIGURE 16.11
Impacts of offshore wind farm installations and operation on the German North Sea ecosystem. The reference values (100%) refer to the situation today, without any offshore wind farm (Burkhard et al. 2009).

16.5 Discussion and Conclusions

In this chapter an approach for a holistic indication of ecosystem health features with a medium number indicator set was described. It is based on ecosystem theory, empirical ecosystem research, and a self-organization–based aspect of ecological integrity. Some applications were briefly demonstrated on different scale levels, demonstrating a methodological shift from comprehensive ecosystem measurements to model applications on the landscape level. Finally, case studies were used to show potential applications in environmental management as a part of the search for sustainable landscape management strategies.

The indicator system was used to show the consequences of different land use systems throughout a developmental duration of 100 years in which the optimization of agricultural production led to a loss of many other important functional and structural features. It could also be shown that the economically oriented management of wet grasslands modifies the selected ecological attributes enormously, i.e., it provokes the development of a source function of these ecosystems, which subsequently produce a high fraction of entropic flows (carbon loss, nitrogen loss) into their environment. The reindeer management example demonstrates that structural changes must not always be accomplished by big functional modifications and that in some cases the human dimensions of sustainability are much more significant than the ecological ones. Finally, the offshore wind power study showed that the indicator set and appropriate models can be used to make predictive analyses of potential systems dynamics.

Recently, we tried to develop another application of the indicator set in environmental monitoring. There are many advantages to changing the sectoral monitoring approaches into an ecosystem-based concept. Attempts are made in monitoring networks of the German Counties (Bundesländer), in biosphere reservations and national parks, and the indicator set has also been implemented into a new concept of environmental-economic accounting in Germany. Regrettably, these applications are still far from being realized, because the change to a holistic attitude in official networks is hampered by many psychological and administrative restrictions.

In parallel to these applications, the variable set has to remain in development. If we look back at Table 16.1, it becomes obvious that some of the requirements mentioned have not been fulfilled up to now. For instance, the comprehensibility and the methodological transparency have to be improved, the indicators provide different sensitivities for specific environmental developments, and due to the theoretical and complex background, the high political relevance of the indicators has not become obvious for politicians up to now. Consequently, besides the scientific tasks of improving the indication, the conviction of the potential users to change their concepts toward a higher consideration of ecosystem attributes, and toward a fruitful application of the health or integrity concepts, will be a main task of future activities.

References

Allen, C., and C. S. Holling. 2008. *Discontinuities in ecosystems and other complex systems*. New York: Columbia University Press.

Aoki, I. 1998. Entropy and exergy in the development of living systems: A case study of lake ecosystems. *Journal of the Physical Society of Japan* 67:2132–39

Barkmann, J. 2001. Modellierung und I\indikation nachhaltiger landschaftsentwicklung—Beiträge zu den grundlagen angewandter ökosystemforschung. Diss., University of Kiel.

Barkmann, J., R. Baumann, U. Mejer, F. Müller, and W. Windhorst. 2001. Ökologische integrität: Risikovorsorge im nachhaltigen landschaftsmanagement. *Gaia* 10 (2): 97–108.

Baumann, R. 2001. Konzept zur indikation der selbstorganisationsfähigkeit terrestrischer ökosysteme anhand von daten des ökosystemforschungsprojekts Bornhöveder Seenkette. Diss., University of Kiel.

Bossel, H. 1998. Ecological orientors: Emergence of basic orientors in evolutionary self-organization. In *Eco targets, goal functions and orientors*, eds. F. Müller and M. Leupelt, 19–33. Berlin: Springer.

———. 2000. Sustainability: Application of systems theoretical aspects to societal development. In *Handbook of ecosystem theories and management*, eds. S. E. Jørgensen and F. Müller, 519–36. Boca Raton, FL: CRC Press.

Breckling, B., and M. Asshoff. 1996. Modellbildung und simulation im Projektzentrum Ökosystemforschung. *Ecosys Bd* 5:342P.

Burkhard, B., T. Kumpula, and F. Müller. 2003. Renman—An integrative study in Northern Scandinavia. *Ecosys* 10:116–24.

Burkhard, B., and F. Müller. 2008a. Drivers-pressure-state-impact-response. In *Ecological indicators*, vol. 2 of Encyclopedia of ecology (5 vols.), eds. S. E. Jørgensen and B. D. Fath, 967–70. Oxford, UK: Elsevier.

———. 2008b. Indicating human-environmental system properties: Case study Northern Fenno-Scandinavian reindeer herding. *Ecological Indicators* 8:828–40.

Burkhard, B., F. Müller, and A. Lill. 2008. Ecosystem health indicators: Overview. In *Ecological indicators*, vol. 2 of Encyclopedia of ecology (5 vols.), eds. S. E. Jørgensen and B. D. Fath, 1132–38. Oxford, UK: Elsevier.

Burkhard, B., S. Opitz, H.-J. Lenhart, K. Ahrendt, S. Garthe, B. Mendel, and W. Windhorst. 2009. Ecosystem-based modeling and indication of ecological integrity in the German North Sea—Case study offshore wind farms. *Ecological Indicators*. (in press).

Christensen, V., and D. Pauly. 1992a. A guide to the EcoPath II software system (version 2.1). *ICLARM Software* 6:1–72.

———. 1992b. EcoPath II—A software for balancing steady-state models and calculation of network charactersistics. *Ecol Modeling* 61:169–85.

Costanza, R. 1993. Towards an operational definition of ecosystem health. In *Ecosystem health*, eds. R. Costanza, B. G. Norton, and B. D. Haskell, 239–56. Washington, DC: Island Press.

———. 2000. Societal goals and the valuation of ecosystem services. *Ecosystems* 3:4–10.

Costanza, R., C. Cleveland, and C. Perrings. 2000. Ecosystem and economic theories in ecological economics. In *Handbook of ecosystem theories and management*, eds. S. E. Jørgensen and F. Müller, 547–60. Boca Raton, FL: CRC Press.

Costanza, R., B. G. Norton, and B. D. Haskell. 1993. *Ecosystem health*. Washington, DC: Island Press.

Daily, G. C. 1997. *Nature's services: Societal dependence on natural systems*. Washington, DC: Island Press.

Danish Hydraulic Institute (DHI). 1999. Horns Rev Wind Power Plant—Environmental impact assessment of hydrography. Baggrundsrapport nr. 8. Danish Hydraulic Institute report 50396-01.

de Groot, R. S. 1992. *Functions of nature*. Netherlands: Wolters-Noorhoff.

Dierschke, V., and S. Garthe. 2006. Literature review of offshore wind farms with regards to seabirds. *BfN-Skripten* 186:131–98.

Dierssen, K. 2000. Ecosystems as states of ecological successions. In *Handbook of ecosystem theories and management*, eds. S. E. Jørgensen and F. Müller, 427. Boca Raton, FL: CRC Press.

Dittmann, S., U. Schleier, S. P. Günther, M. Villbrandt, V. Niesel, A. Hild, V. Grimm, H. Bietz, and C. Dohn. 1998. Elastizität des Ökosystems Wattenmeer. Projektsynthese. Forschungsbericht für den BMBF, Bonn.

Ebeling, W. 1989. *Chaos—Ordnung—Information*. Frankfurt: Verlag Harri Deutsch.

Fath, B., and B. C. Patten. 1998. Network orientors: A utility goal function based on network synergism. In *Eco targets, goal functions and orientors*, eds. F. Müller and M. Leupelt, 161–76. Berlin: Springer.

———. 2000. Ecosystem theory: Network environ analysis. In *Handbook of ecosystem theories and management*, eds. S. E. Jørgensen and F. Müller, 345–60. Boca Raton, FL: CRC Press.

Fränzle, O. 1998. Ökosystemforschung im bereich der Bornhöveder Seenkette. In *Handbuch der ökosystemforschung*, eds. O. Fränzle, F. Müller, and W. Schröder (Hrsg.), Ch. V-4.3. Landsberg: Ecomed-Verlag.

———. 2000. Ecosystem research. In *Handbook of ecosystem theories and management*, eds. S. E. Jørgensen and F. Müller, 89–102. Boca Raton, FL: CRC Press.

Fränzle, O., L. Kappen, H. P. Blume, and K. Dierssen, eds. 2008. *Ecosystem organization in a complex landscape*. Ecological Studies vol. 202. Berlin: Springer.

Fritz, P. 1999. UFZ-Forschungszentrum Leipzig-Halle. In *Handbuch der ökosystemforschung*, eds. O. Fränzle, F. Müller, and W. Schröder (Hrsg.), Ch. V-4.5. Landsberg: Ecomed-Verlag.

Gollan, T., and B. Heindl. 1998. Bayreuther Institut für terrestrische ökosystemforschung. In *Handbuch der ökosystemforschung*, eds. O. Fränzle, F. Müller, and W. Schröder (Hrsg.), Ch. V-4.6. Landsberg: Ecomed-Verlag.

Golley, F. 2000. Ecosystem structure. In *Handbook of ecosystem theories and management*, eds. S. E. Jørgensen and F. Müller, 21–32. Boca Raton, FL: CRC Press.

Gundersson, L. H., and C. S. Holling, eds. 2003. *Panarchy*. Washington, DC: Island Press.

Hantschel, R., M. Kainz, and J. Filser. 1998. Forschungsverbund Agrarökosysteme München. In *Handbuch der ökosystemforschung*, eds. O. Fränzle, F. Müller, and W. Schröder (Hrsg.), Ch. V-4.7. Lansberg: Ecomed-Verlag.

Haskell, B. D., B. G. Norton, and R. Costanza. 1993. Introduction: What is ecosystem health and why should we worry about it? In *Ecosystem health*, eds. R. Costanza, B. G. Norton, and B. D. Haskell, 3–22. Washington, DC: Island Press.

Hauff, V. (Hrsg.). 1987. Unsere Gemeinsame Zukunft. Der Brundlandt-Bericht der Weltkommission für Umwelt und Entwicklung. Greven.

Herbst, M., C. Eschenbach, and L. Kappen. 1999. Water use in neighbouring stands of beech and black alder. *Ann For Sci* 56:107–20.

Holling, C. S. 1986. The resilience of terrestrial ecosystems: Local surprise and global change. In *Sustainable development of the biosphere*, eds. W. M. Clark and R. E. Munn, 292–320. Oxford, UK: IBP reports.

Hörmann, G., U. Irmler, F. Müller, J. Piotrowski, R. Pöpperl, E. W. Reiche, G. Schernewski, C. G. Schimming, J. Schrautzer, and W. Windhorst. 1992. Ökosystemforschung im bereich der Bornhöveder Seenkette. Arbeitsbericht 1988–1991. *Ecosys* 1:338.

Jørgensen, S. E. 1996. *Integration of ecosystem theories—A pattern*. Dordrecht, The Netherlands: Kluwer.

———. 2000. The tentative fourths law of thermodynamics. In *Handbook of ecosystem theories and management*, eds. S. E. Jørgensen and F. Müller, 161–76. Boca Raton, FL: CRC Press.

Jørgensen, S. E., B. Fath, S. Bastianoni, J. Marquez, F. Müller, S. N. Nielsen, B. Patten, E. Tiezzi, and R. Ulanowicz. 2007. *A new ecology—The systems perspective.* Amsterdam: Elsevier Publishers.

Jørgensen, S. E., and F. Müller. 2000a. Ecological orientors—A path to environmental applications of ecosystem theories. In *Handbook of ecosystem theories and management*, eds. S. E. Jørgensen and F. Müller, 561–76. Boca Raton, FL: CRC Press.

———. 2000b. Ecosystems as complex systems. In *Handbook of ecosystem theories and management*, eds. S. E. Jørgensen and F. Müller, 5–20. Boca Raton, FL: CRC Press.

Kaiser, M., T. Mages-Delle, and R. Oeschger. 2002. Gesamtsynthese ökosystemfor schung wattenmeer. Berlin: UBA-Texte, 45/02

Karr, J. R. 1981. Assessment of biotic integrity using fish communities. *Fisheries* 6:21–27.

Kay, J. J. 1993. On the nature of ecological integrity: Some closing comments. In *Ecological integrity and the management of ecosystems*, eds. S. Woodley, J. Kay, and G. Francis. Ottawa: University of Waterloo and Canadian Park Service.

———. 2000. Ecosystems as self-organized holarchic open systems: Narratives and the second law of thermodynamics. In *Handbook of ecosystem theories and management*, eds. S. E. Jørgensen and F. Müller, 135–60. Boca Raton, FL: CRC Press.

Kellermann, A. C. Gätje, and E. Schrey. 1998. Ökosystemforschung im Schleswig-Holsteinischen Wattenmeer. In *Handbuch der ökosystemforschung*, eds. O. Fränzle, F. Müller, and W. Schröder (Hrsg.), Ch. V-4.1.1. Landsberg: Ecomed-Verlag.

Kerner, H.-F., L. Spandau, and J. Köppel. 1991. Methoden zur angewandten ökosystem-forschung entwickelt im MAB 6—Projekt Ökosystemforschung Berchtesgaden. MAB Mitteilungen 35.1 und 35.2, Bonn.

Kutsch, W., O. Dilly, W. Steinborn, and F. Müller. 1998. Quantifying ecosystem maturity—A case study. In *Eco targets, goal functions and orientors*, eds. F. Müller and M. Leupelt, 209–31. Berlin: Springer.

Kutsch, W. L., W. Steinborn, M. Herbst, R. Baumann, J. Barkmann, and L. Kappen. 2001. Environmental indication: A field test of an ecosystem approach to quantify biological self-organization. *Ecosystems* 4:49–66.

Lenhart, H., B. Burkhard, and W. Windhorst. 2006. Ökologische auswirkungen erhöhter schwebstoffgehalte als folge der baumaßnahmen von offshore windkraftanlagen (in German). *EcoSys Suppl Bd* 46:90–106.

Meyer, E. 1992. Die benthischen invertebraten in einem kleinen fließgewässer am beispiel eines schwarzwaldbaches: Biozönotische dtruktur, populationsdynamik, produktion und stellung im trophischen gefüge. Habil thesis, Konstanz.

Meyer, M. 2000. Entwicklung und formulierung von planungsszenarien für die landnutzung im bereich der Bornhöveder Seenkette. Diss., University of Kiel.

Meyer, U. 2001. Landschaftsökologische modellierung als auswertungsinstrument in der ökosystemaren umweltbeobachtung—Beispielsfall Biosphärenreservat Rhön. *Ecosys Suppl* 36:156.

Millennium Assessment Board. 2003. *Ecosystems and human well-being.* Washington, DC: Island Press.

Müller, F. 1997. State of the art in ecosystem theory. *Ecological Modeling* 100:135–61.

———. 2005. Indicating ecosystem and landscape organization. *Ecological Indicators* 5 (4): 280–94.

Müller, F., and B. Breckling. 1997. Der ökosystembegriff aus heutiger sicht. In *Handbuch der umweltwissenschaften*, eds. O. Fränzle, F. Müller, and W. Schröder, Ch. II-2.2. Landsberg: Ecomed-Verlag.

Müller, F., B. Breckling, M. Bredemeier, V. Grimm, H. Malchow, S. N. Nielsen, and E. W. Reiche. 1997a. Emergente ökosystemeigenschaften. In *Handbuch der ökosystemforschung*, eds. O. Fränzle, F. Müller, and W. Schröder (Hrsg.), Ch. III-2.5. Landsberg: Ecomed-Verlag.

———. 1997b. Ökosystemare selbstorganisation. In *Handbuch der ökosystemforschung*, eds. O. Fränzle, F. Müller, and W. Schröder (Hrsg.), Ch. III-2.4. Landsberg: Ecomed-Verlag.

Müller, F., and B. Burkhard. 2007. An ecosystem-based framework to link landscape structures, functions and services. In *Multifunctional land use—Meeting future demands for landscape goods and services*, eds. Ü. Mander, H. Wiggering, and K. Helming, 37–64. Berlin: Springer.

Müller, F., O. Fränzle, and C. Schimming. 2008. Ecological gradients as causes and effects of ecosystem organization. In *Ecosystem organization of a complex landscape*. Ecological Studies, vol. 202, eds. O. Fränzle, L. Kappen, H.-P. Blume, and K. Dierssen, 277–96. Berlin: Springer-Verlag.

Müller, F., R. Hoffmann-Kroll, and H. Wiggering. 2000. Indicating ecosystem integrity—From ecosystem theories to eco targets, models, indicators and variables, *Ecological Modeling* 130:13–23.

Müller, F., and M. Leupelt. 1998. *Eco targets, goal functions and orientors*. Berlin: Springer.

Müller, F., and B. L. Li. 2004. Complex systems approaches to study human–environmental interactions. In *Ecological issues in a changing world: Status, response and strategy*, eds. S. Hong, J. A. Lee, B. Ihm, A. Farina, Y. Son, E. Kim, and J. C. Choe, 31–46. Dordrecht, The Netherlands: Kluwer Academic Publishers.

Müller, F., and S. N. Nielsen. 2000. Ecosystems as subjects of self-organizing processes. In *Handbook of ecosystem theories and management*, eds. S. E. Jørgensen and F. Müller, 177–94. Boca Raton, FL: CRC Press.

Müller, F., J. Schrautzer, E.-W. Reiche, and A. Rinker. 2006. Ecosystem-based indicators in retrogressive successions of an agricultural landscape. *Ecological Indicators* 6:63–82.

Müller, F., and H. Wiggering. 2004. Umweltindikatoren als maßstäbe zur bewertung von umweltzuständen und entwicklungen. In *Umweltziele und indikatoren*, eds. H. Wiggering and F. Müller, 121–29. Berlin: Springer.

Müller, F., and W. Windhorst. 2000. Ecosystems as functional entities. In *Handbook of ecosystem theories and management*, eds. S. E. Jørgensen and F. Müller, 33–50. Boca Raton, FL: CRC Press.

Odum, E. P. 1969. The strategy of ecosystem development. *Science* 104:262–70.

Odum, H. T., M. T. Brown, and S. Ulgiati. 2000. Ecosystems as energetic systems. In *Handbook of ecosystem theories and management*, eds. S. E. Jørgensen and F. Müller, 283–302. Boca Raton, FL: CRC Press.

Patten, B. C. 1992. Energy, emergy and environs. *Ecological Modeling* 62:29–69.

Pöpperl, R. 1996. Functional feeding groups of a macroinvertebrate community in a Northern German lake outlet. (Lake Belau, Schleswig-Holstein). *Int Revue Ges Hydrobiol* 81:183–98.

Radermacher, W., R. Zieschank, R. Hoffmann-Müller, J. V. Noyhus, D. Schäfer, and S. Seibel. 1998. *Entwicklung eines indikatorensystems für den zustand der Umwelt in der Bundesrepublik Deutschland. Beiträge zu den umweltökonomischen Gesamtrechnungen*, Bd. 5. Wiesbaden: Statistisches Bundesamt.

Rapport, D. J. 1989. What constitutes ecosystem health? *Perspectives in Biology and Medicine* 33 (1): 120–32.

Rapport, D. J., and R. Moll. 2000. Applications of ecosystem theory and modeling to assess ecosystem health. In *Handbook of ecosystem theories and management*, eds. S. E. Jørgensen and F. Müller, 487–96. Boca Raton, FL: CRC Press.

Rapport, D. J., and A. Singh. 2006. An ecohealth-based framework for state of environment reporting. *Ecological Indicators* 6:409–28.

Reiche, E. W. 1996. WASMOD. Ein modellsystem zur gebietsbezogenen simulation von wasser- und stoffflüssen. *Ecosys* 4:143–63.

Reiche, E. W., F. Müller, I. Dibbern, and A. Kerrinnes. 2001. Spatial heterogeneity in forest soils and understory communities of the Bornhöved Lakes District. In *Ecosystem approaches to landscape management in Central Europe*, eds. J. Tenhunen, R. Lenz, and R. Hantschel, 147. Ecological Studies.

Schimming, C. G., and S. von Stamm. 1993. Arbeitsbericht des projektzentrums ökosystemforschung, anhang I: Untersuchungsmethoden. Interne Mitteilungen aus dem FE-Vorhaben Ökosystemforschung im Bereich der Bornhöveder Seenkette.

Schneider, E. D., and J. J. Kay. 1994. Life as a manifestation of the second law of thermodynamics. *Mathematical and Computer Modeling* 19 (6–8): 25–48.

Schönthaler, K., U. Mejer, W. Windhorst, M. Reichenbach, D. Pokorny, and D. Schuller. 2001. Modellhafte umsetzung und konkretisierung der konzeption für eine ökosystemare umweltbeobachtung am beispiel des länderübergreifenden biosphärenreservats rhön. Berlin: Umweltbundesamt.

Schönthaler, K., F. Müller, and J. Barkmann. 2003. Synopsis of systems approaches to environmental research—German contribution to ecosystem management. UBA-Texte 85/03

Schrautzer, J. 2002. Niedermoore Schleswig-Holsteins: Charakterisierung und beurteilung ihrer funktionen im landschaftshaushalt. Habil thesis, University of Kiel.

Schrautzer, J., A. Rinker, K. Jensen, F. Müller, P. Schwartze, and K. Dierssen. 2007. Succession and restoration of drained fens: Perspectives from Northwestern Europe. In *Linking restoration and ecological succession*, Springer Series on Environmental Management, eds. L. Walker, R. J. Hobs, and J. Walker, 90–120. Berlin: Springer.

Statistisches Bundesamt, Forschungsstelle Für Umweltpolitik der Fu Berlin, and Ökologiezentrum der Universität Kiel. 2002. Makroindikatoren des Umweltzustands. Beiträge zu den Umweltökonomischen Gesamtrechnungen, Bd. 10.

Steinborn, W. 2001. Quantifizierung von ökosystemeigenschaften als grundlage für die umweltbewertung. Diss., University of Kiel.

Svirezhev, Y. M., and W. Steinbron. 2001. Exergy of solar radiation: Thermodynamic approach. *Ecological Modeling* 145:101–10.

Ulanowicz, R. E. 2000. Ascendency: A measure of ecosystem performance. In *Handbook of ecosystem theories and management*, eds. S. E. Jørgensen and F. Müller, 303–16. Boca Raton, FL: CRC Press.

Ulgiati, S., M. T. Brown, M. Giampietro, R. A. Herendeen, and K. Mayumi, eds. 2003. Advances in energy studies—Reconsidering the importance of energy. Padua: SGIE Detoriali.

Vihervaara, P., T. Kumpula, A. Tanskanen, and B. Burkhard (in press). Ecosystem services—A tool for sustainable management of human-environmental systems. Case study Finnish Forest Lapland. *Ecological Complexity*.

Widey, G.-A. 1998. Forschungszentrum waldökosysteme der Universität Göttingen In *Handbuch der ökosystemforschung*, eds. O. Fränzle, F. Müller, and W. Schröder (Hrsg.), Ch. V-4.4. Landsberg: Ecomed-Verlag.

Wiggering, H. 2001. Zentrum für agrarlandschafts- und landnutzungsforschung. In *Jahresbericht 2000/2001*. Müncheberg.

Windhorst, W., F. Müller, and H. Wiggering. 2004. Umweltziele und ondikatoren für den ökosystemschutz. In *Umweltziele und indikatoren*, eds. H. Wiggering and F. Müller, 345–73. Berlin: Springer.

Woodley, S., J. Kay, and G. Francis. 1993. *Ecological integrity and the management of ecosystems*. Ottawa: St. Lucie Press.

World Commission on Environment and Development (WCED). 1987. *Our common future*. Oxford: Oxford University Press.

17

Integrated Indicators
for Evaluating Ecosystem Health:
An Application to Agricultural Systems

V. Niccolucci, R. M. Pulselli, S. Focardi, and S. Bastianoni

CONTENTS

17.1 Introduction

The concept of ecosystem health is strongly related with that of sustainability. It comprises both the biophysical and the human dimensions of the environment. As Burkhard et al. (2008) affirmed, "An ecosystem is often called healthy if it is stable and sustainable in the provision of goods and services used by human society. This implies that it has the ability to maintain its structure (organization) and function over time under external stress." In this chapter, health and sustainability were often used as synonymous.

Sustainability cannot be easily measured because it is not a physical phenomenon. A human dominated system could be sustainable or not sustainable due to the use of energy and materials and/or to the wastes it produces. According to F. M. Pulselli et al. (2008), the role of indicators is to understand where we stand in relation to sustainability criteria and to find out what we can do to reduce unsustainability, in case. Since the concept of sustainability is complex and not directly measurable, many indicators are needed to assess how far we have to go to reach this goal.

Indicators of ecosystem health and sustainability belong to different categories. Based on this diversity, Bastianoni et al. (2008) highlighted the importance of a joint use of many indicators in order to consider different aspects of a system's state. In this chapter, an integrated use of three different thermodynamics-based indicators was proposed in order to evaluate their capacity to achieve a comprehensive general evaluation of a system's health. An evaluation of an agro-ecosystem, a grape cultivation in a biological farm in central Italy, was presented as a case study.

The accounting methods chosen for calculating synthetic indicators were Exergy (and Eco-Exergy) Analysis (Szargut et al. 1998; Jørgensen 1982), Emergy Evaluation (Odum 1988), and Ecological Footprint (Wackernagel and Rees 1996), each of which has a characteristic viewpoint on system's state. Even if they consider different aspects, consonances and complementarities can be highlighted. In fact, the joint use of these methods allowed us to collect synthetic information about the general level of health and sustainability of the agro-ecosystem analyzed.

Exergy analysis investigates, from a holistic viewpoint, the potential capacity of a system to make work, based on its current state. We can say that exergy has a forward perspective because it evaluates the potential development of a system in the future and measures the distance of a system's state from thermodynamic equilibrium.

Emergy accounts for environmental resources that were drawn, directly and indirectly, from natural cycles and stocks for achieving the current organization of a system. It is based on a detailed inventory of the main inputs and primary sources that fed a system in a given time. Quantity and quality of inputs are considered based on the chain of processes and energy transformations that occurred in the past to make every product or service involved in a process available. We can say that emergy has a backward perspective on a system's organization (emergy means memory of energy). All inputs to a process or system, given in energy and mass units, are transformed into one form of energy, the solar energy, which is the primary energy that drives all processes in nature. Based on this procedure, the results enable the evaluation of the sustainability/unsustainability of a system through a synthetic unique balance.

Ecological Footprint is an area-based approach considered an alternative way to account for resources use and waste emission (Monfreda et al. 2004). Footprint philosophy shows similarity to emergy. The main object is to convert all inventoried inputs to a given system into a common denominator. In this case, the productive capacity of territorial and marine area required on a continuous basis to produce all consumed resources, as well as to absorb the related emissions was chosen (Wackernagel and Rees 1996). Land accounted included both the direct land use and the indirect one (which is essentially needed for energy transformations). Ecologically productive land can be considered as the collector of solar energy. The sum of all land requirements is a convenient way to appraise how big is the footprint, or the impact of the systems analyzed, with respect to the available productive capacity (biocapacity). The footprint tool presents a backward perspective.

A detailed explanation of these methods is presented in the next section. After discussing the ability of exergy-, emergy-, and Ecological Footprint–based indicators to provide information for evaluating natural ecosystem health with respect to human activities and impacts, we proposed an investigation of differences and assonances between different indicators in order to understand the degree of similarity/congruence (Bastianoni et al. 2008).

17.2 Methods

17.2.1 Exergy-Based Indicators

Exergy function comes from the discipline of thermodynamics applied to heat engines and, although this concept was already known in the nineteenth century, it was formalized with this name in the 1950s by Rant (1956). This function is now useful in solving engineering cost-optimization procedures. Exergy analysis (Szargut et al. 1998) and thermo-economic analysis (Evans 1980; Lozano and Valero 1993; Bejan et al. 1996), both based on exergy, are among the most efficient tools for design and yield optimization of energy conversion systems and energy policies. In the last years, exergy function, and the analysis based on it, was proposed by several authors as a tool able to describe systems more complex than a simple energy conversion. In particular, many examples of exergy-based analysis applications exist on territorial systems at different scales, such as nations (Wall 1990), provinces (Sciubba et al. 2008), and even on natural ecosystems (Bendoricchio and Jørgensen 1997; Zaleta-Aguilar et al. 1998).

Jørgensen's formulation of exergy (Jørgensen and Mejer 1977; Mejer and Jørgensen 1979; Jørgensen 1982; Jørgensen 2002) differs from the others. Exergy, or eco-exergy, was defined as the amount of work (entropy-free energy) an ecosystem can perform when it is brought to thermodynamic equilibrium with its environment (Jørgensen 2008). At the reference state of equilibrium, there is no longer any gradient, and all components are inorganic at the highest oxidation state as possible. As Jørgensen (2008) affirmed, "the reference state will correspond to the ecosystem without life forms and with all chemical energy utilized or as an inorganic soup."

At a practical level, exergy represents the maximum work that we can extract from a certain system. It is usually defined as a thermodynamic measure of the efficiency in the use of resources (expressed in energy or mass). Sciubba et al. (2008) conceived exergy as a measure of the energetic content of every material and immaterial flow directly or indirectly used in processes. It calculated the exergy embedded in a product by keeping track of all of the exergy inputs and outputs in the production chain or, in other words, by tracing back every output to the primary resource flows that originate it. Comparing different processes, the best, among others, is the one that, for the same final output, destroys the minimum amount of primary exergy.

Exergy can be expressed accordingly to Equation (17.1).

$$Ex = RT \sum_{i=0}^{n} c_i \ln \frac{c_i}{c_{i,0}} \tag{17.1}$$

where R is the gas constant, T is the temperature of the environment, c_i is the concentration of the ith component in a suitable unit (e.g., for phytoplankton mg l^{-1} or mg l^{-1} of a focal nutrient), and c_{i0} is the concentration of the ith component at thermodynamic equilibrium and n is the number of components.

A reference state should necessarily be defined when eco-exergy is calculated. Eco-exergy (EEx) is a measure of a system's deviation from chemical equilibrium and then a measure of its development. It includes the contributions from both biomass and information. Equation (17.1) can be rewritten as follows:

$$EEx = \sum_{i=1}^{n} \beta_i c_i \tag{17.2}$$

where β_i are weighting factors of the various components i of the ecosystem that account for the information that the organisms have embodied in their genes. c_i is the concentration as g m^{-2} of the ith species.

Eco-exergy was widely used as an ecosystem health indicator. Ecosystem health assessment has been applied to lakes, lagoons, coastal zones, as well as to different farming systems (Jørgensen 2008).

For this chapter, the exergy analysis for an agricultural product was performed according to Jørgensen (1982), Jørgensen (2000), Bastianoni et al. (2005), and Fonseca et al. (2000). Three main exergy indicators were calculated: the exergy consumption (Ex$_{in}$), the exergy of the output of a process (Ex$_{out}$), and the exergy stored in a system (Ex$_s$).

The exergy consumption (Ex_{in}) for grape production is obtained by calculating recursively two different contributions:

$$Ex_{in} = \sum_{i=1}^{n} [Ex_{chem}^i + Ex_{req}^i]$$

where Ex_{chem}^i accounts for the chemical exergy of ith input and Ex_{req}^i is the exergy requirement or the exergy consumed for making the ith input available. Ex_{in} represents the exergy consumption related to the creation/maintenance of a system. In other words, it represents a cost of production.

The exergy stored (Ex_s) is calculated by evaluating the genetic content in a given product or in a system's structure. According to Bastianoni et al. (2005), Ex_s is viewed as a result of the fluxes taking place in the system and is thus a result of the system function as a whole.

Ex_{out} is the exergy content of the output of a process and it is important for the analysis of the second law efficiency. From a sustainability viewpoint, Ex_{out} has to be limited at the level expressed by the variation of exergy stored in the same time; maintenance of ecosystems' function, represented by the exergy storage Ex_s, should be allowed by managing rate of exergy output from a system, Ex_{out}.

Efficiency indices can be calculated by combining the three exergy measures above. This would allow us to provide more synthetic information on the sustainability of the system under study, in the long run.

Ex_{out}/Ex_{in} is the most economically/energetically oriented ratio. It is defined by Szargut et al. (1988) as the ratio of exergy of useful products to feeding exergy, where the latter is "the exergy delivered to the system for steady state operations." It deals with the immediate return on investment, i.e., higher production with lower exergy expenditure.

The Ex_s/Ex_{in} is useful to define the level of organization (Ex_s) that is maintained by a unit of exergy inflow (Ex_{in}). It has the dimension of time. It is not strictly an efficiency index, but it offers important information on agro-ecosystems sustainability.

17.2.2 Emergy-Based Indicators

According to H. T. Odum (1988), the realm of emergy analysis is the quantitative understanding of the relationships between human-dominated systems and the biosphere. The concept of energy hierarchy is a key point for understanding emergy (Brown et al. 2004). Due to the second law of thermodynamics, every energy transformation uses many calories of available energy of one kind, at a lower quality or grade, to generate a few calories of another kind of energy at the higher grade (Odum 1996). As Odum (1973) observed, "the scale of energy goes from dilute sunlight up to plant matter, to coal, from coal to oil, to electricity and up to the high quality work of computer and human information processing."

Emergy (spelled with an "m") is the available energy (exergy) of one kind required to be used up previously, directly and indirectly, to generate the inputs for an energy transformation (Odum 1971). In particular, an emergy content represents the concentration of solar energy, a dilute form of energy, in a given product or service. It derives from the accounting of all the inputs to a process, each of which is an outcome from a series of previous transformations. By definition, emergy is the quantity of solar energy that has been used, whether directly or indirectly, in order to obtain a final product or service.

In practical terms, emergy evaluation is an environmental accounting method based on an energy and material flows inventory that considers the main inputs to given processes. Emergy allows the overcoming of the diversity of metrics used, whether they were mass or energy quantities, for quantifying inputs by normalizing to a common unit of measure, namely, the *solar emergy joule* or *solar emjoule* (sej). Normalization is possible thanks to transformation coefficients, namely, transformity or specific emergy, given in dimensions of sej/J and sej/g, respectively, that correspond to the emergy content per unit of product or service.

By definition, the solar emergy Em_k of the flow k coming from a given process is:

$$Em_k = \Sigma_i \, Tr_i \, E_i \, i = 1, \ldots, n \qquad (17.3)$$

where E_i is the actual energy content of the ith independent input flow to the process and Tr_i is the emergy per unit energy of the ith input flow.

Indeed, human ability to produce work depends on energy quality and quantity. A process feeds on quantities of energy and materials that are drawn directly from the environment (in the form of daily flows such as sun-wind-rain or geothermal heat force; short-term storage flows such as wood, soil, and water; or long-term storage flows of fossil fuels and minerals) or indirectly, in the form of goods and services purchased from the global economy. The more work done to produce something, or the more energy that is transformed, the higher the emergy content of what is produced. Empower is defined as the emergy per unit time. In the case of ecosystems or some human-dominated systems, it is expressed as emergy used per year.

Based on these criteria, the emergy evaluation was applied to different case studies as a way of classifying and comparing systems. For example, a summary of specific emergy values and empower calculated for different ecosystems was provided in the cited literature (Brown and Bardi 2001; Brown and Cohen, 2008). In particular Brown and Bardi (2001) provided calculations and empower values, expressed as emergy per square meter per year (sej m^{-2} yr^{-1}), in ecosystems such as forests, agricultural systems, wetlands, lakes, landscape scale ecosystems including humans, etc. In particular, they observed that terrestrial ecosystems have renewable empower densities in the range of about 40–50 × 10^9 sej m^{-2} yr^{-1}. Wetlands have empower densities about one order of magnitude higher, while lake and estuarine ecosystems have one order of magnitude higher than wetlands. Brandt-Williams (2001) presented emergy evaluations for 23 agricultural commodities raised in the state of Florida and for two fertilizers produced and used extensively. Some applications to fish farms were discussed by Ridolfi and Bastianoni (2008). Agricultural systems were also investigated in the cited literature: Pizzigallo et al. (2008) and Campbell (2008). Regional studies were presented by Campbell et al. (2005), Pulselli et al. (2008), Pulselli et al. (2007), Ulgiati et al. (1994), Campbell (1998), Ortega et al. (1999), Higgins (2003), Tilley and Swank (2003), Campbell et al. (2005).

17.2.3 Ecological Footprint-Based Indicators

The Ecological Footprint measures the global impact imposed on the earth by a population, an activity, or a product (Rees 1992). Formally, the Ecological Footprint (hereafter Footprint) of a certain population or a production activity is defined as the area (real and virtual) of productive land and water ecosystems required, on a continuous basis, to produce the resources consumed and to assimilate the wastes produced, wherever on the earth the relevant land/water may be located and with the prevailing technology and resources management schemes (Wackernagel and Rees 1996; Wackernagel and Kitzes 2008).

The Footprint is then compared with how much land and sea is available. This is captured in a second indicator called Biocapacity (hereafter BC), which measures the annual production of biologically provided resources for human use (Monfreda et al. 2004).

Six major biological productive land types are considered: cropland, grazing land, fishing grounds, forest area, built-up land, and energy land (or carbon footprint that is the amount of forest land required to capture those carbon dioxide emissions not sequestered by the oceans) (Wackernagel and Rees 1996; Kitzes et al. 2009).

Both BC and Footprint are expressed in terms of a common unit called global hectare (*g*ha), which refers to a hectare with world-average productivity for all productive land and water area in a given area, to make results globally comparable (Kitzes et al. 2007; Galli et al. 2007). It is a normalized unit useful to make a comparison among lands with different productivity (Monfreda et al. 2004).

Yield factor (YF) and Equivalence factor (EQF) are used to translate the hectare of a specific land type into global hectares (Monfreda et al. 2004). EQF adjusts for the relative productivity of the six categories of land and water area, while YF adjusts for local to global average productivity of the same land category.

The total Footprint is calculated as the sum of the Footprint of all the material consumed and waste generated.

The Footprint is widely used to give a measure of the (un)sustainability of consumption patterns at different scales: regional (see, for example, Folke et al. 1997; Bagliani et al. 2008), national (see, for example, Erb 2004; Medved 2006; Moran et al. 2008), and global (Van Vuuren and Bouwman 2005; WWF 2008). Footprint has also been analyzed as temporal series together with economic indicators such as gross domestic product (GDP; Jorgenson and Burns 2007) and Index of Sustainable Economic Welfare (ISEW; Niccolucci et al. 2007), or incorporated in thermodynamic-based methods (Zhao et al. 2005; Chen and Chen 2006; Nguyen and Yamamoto 2007).

Up-to-date industrial and agricultural Footprint applications are still rare. Studies on cultivation of tomatoes (Wada 1993), conventional versus organic wine farming (Niccolucci et al. 2008), and shrimp and tilapia aquaculture (Kautsky et al. 1997) have been carried out to highlight appropriation of natural capital, efficiency of natural resource use, and environmental pressure. Evaluations of the environmental impact of farms (van der Werf et al. 2007) and dairy production (Thomassen and de Boer 2005) as well as assessment of economic and ecological carrying capacity of crops (Cuadra and Björklund 2007) proposed the Footprint jointly with other methods, such as Life Cycle Assessment, Emergy Analysis, and Economic Cost and Return Estimation.

The Footprint calculation performed in this study is based on a "life cycle approach." All relevant inputs, from cradle to gate, were accounted (on the basis of their lifetime) to give an estimation of the environmental impacts. The first step inputs were converted into relative bioproductive areas by means of specific conversion factors available in Chambers et al. (2000). When conversion factors were not available, energy intensity coefficients were used to convert data into energy units from Gabi4 database. A conversion into the equivalent emission of CO_2 and then into the area of forest needed for sequestration was then performed. A world-average carbon absorption factor of 0.271 gha $t_{CO_2}^{-1}$ was used (Global Footprint Network 2008). Inputs that were not completely consumed in the annual production, such as materials and machineries, were considered on the basis of their lifetime (from 10 to 20 years for machineries and from 5 to 10 years for wooden poles). The contribution of human labor was included in the Footprint account. The Footprint of an average Italian person (WWF 2008) was allocated on the basis of the number of work-hours per year.

The Footprint of the vineyard (EF_V) was assessed using the "calculated area" method (Kitzes et al. 2007). The area required to grow the grapes was calculated as reported below:

$$EF_V = \frac{T}{Y_l} * YF * EQF = A_l * YF * EQF$$

where T was the annual quantity of grapes produced (in t) and vinified by the producer, and Y_l was the local grape yield (t ha^{-1}). The EQF for cropland was obtained from the WWF Living Planet Report (2008), while YF was calculated by comparing farm grape yield (Y_l) with the world-average grape yield for the same year (Y_w), extracted from the Food and Agriculture Organization (FAO; FAOSTAT 2008).

Results were expressed as the total area with world-average productivity required per unit of grape (gha yr kg^{-1}) as well as total area with world-average productivity required per unit area of vineyard.

17.3 Results and Discussion

An inventory of the main flows involved in the process of grape cultivation is presented in Table 17.1 given in energy and mass quantities. This includes the natural resources that directly converge to the system, and other inputs of purchased goods and services. Procedures were aggregated into two main phases in the process and were assessed, step by step, considering all the

TABLE 17.1

The Inventory of 1 ha of Grape Production in Tuscany

Input	Amount	Unit
Vineyard	1	ha
Natural resources		
Solar energy	4.16×10^{14}	J
Rain	8.18×10^{10}	g
Geothermal heat	3.51×10^{11}	J
Soil erosion	5.02×10^{10}	J
Phase 1: vine planting		
Iron	0.94×10^{1}	kg
Steel	1.59×10^{2}	kg
Wood	2.04×10^{3}	kg
Manure	1.67×10^{2}	kg
Diesel	3.97×10^{9}	J
Phase 2: grape product		
Iron	2.59×10^{3}	kg
Fiberglass	0.11×10^{1}	kg
Diesel	7.06×10^{10}	J
Pesticides	1.09×10^{3}	kg
Manure	5.00×10^{3}	kg
Human work	1.54×10^{9}	J
Grapes harvested	7.00×10^{7}	g
	1.84×10^{11}	J

activities made in the field: (1) The phase of vine planting includes field deep ploughing, field ploughing, field manuring, vine planting, and growth and maintenance of the vineyard for two years after which vine will be ready for the production of grapes. In fact, vine planting is made once every 30 years, the estimated lifetime of the vineyard. Thus data were allocated considering the entire lifetime in order to estimate a correspondent quantity per year. (2) The phase of grape production was considered relative to a year of work that corresponds to a whole cycle, starting and finishing in September. This includes maintenance of the vineyard (including treatments of both the field and the vine), treatment of vine with pesticides when necessary (mainly based on sulphur and copper as allowed in biological production), and grape harvest and transport to the wine cellar.

In general, the use of machineries was estimated considering their weight, material (mainly steel and iron), and lifetime. The correspondent inflow is thus given in quantity of iron and steel relative to the time of use. Since we considered a biological type of production, there are no special products for treating plants. In this case study, we assumed that farmers use manure instead of high-grade chemical fertilizers, and sulphur and copper instead of more complex chemical remedies or pesticides.

All results deriving from the three methods Exergy, Emergy, and Ecological Analysis are presented in Table 17.2.

Results were compared with other Italian grape productions (for Exergy and Emergy analysis) or similar agricultural product (for Emergy and Ecological Footprint analysis), as given in literature. In particular, results for other two types of grape production were provided in Table 17.2 referring to agricultural systems in different regions of Italy—Piedmont (grape B), Tuscany (grape C)—that produce grapes for vinification. Both grape B and grape C (presented in Bastianoni et al. 2005) were produced with conventional methods while the grape analyzed in this study was produced according to biological procedures.

17.3.1 Exergy Analysis

The Exergy input, Ex_{in}, of grape production was found to be 4.93×10^{11} J yr^{-1} ha^{-1}. Solar radiation was the most significant driving force, corresponding to 84% of the total. When solar exergy is not included the most relevant inputs are iron and steel (incorporated in the machinery).

The exergy stored, Ex_s, was calculated by evaluating the genetic content in the grape harvested (1.07×10^{13} J ha^{-1}) and in the vine's structure including branches, leaves, roots, and other elements (1.05×10^{14} J ha^{-1}) and the general exergy content in the organic matter contained in the first meter of soil (2.34×10^{12} J ha^{-1}). The β weighting factors for assessing the exergy stored in the agricultural system (Ex_s) were extracted by Fonseca et al. (2000). Ex_s was thus found to be 1.18×10^{14} J ha^{-1}. The exergy output, Ex_{out}, is the exergy content in the grape harvested and corresponds to 1.26×10^{10} J yr^{-1} ha^{-1}.

TABLE 17.2

Summary of the Results from Exergy, Emergy, and Ecological Footprint Analysis of 1 ha of Grape Production

	Unit	Grape A (This study)	Grape B[a] (Piedmont)	Grape C[a] (Tuscany)
Exergy				
Ex_{in}	J ha^{-1} yr^{-1}	4.93×10^{11}	5.21×10^{11}	5.02×10^{11}
Ex_{in} (*without solar radiation*)	J ha^{-1} yr^{-1}	7.70×10^{10}	1.32×10^{11}	4.34×10^{10}
Ex_{out}	J ha^{-1} yr^{-1}	1.26×10^{10}	1.17×10^{10}	1.32×10^{10}
Ex_s	J ha^{-1}	1.08×10^{14}	1.22×10^{14}	6.17×10^{13}
Exergy-based ratios				
Ex_{out}/Ex_{in}		0.026	0.022	0.026
Ex_{out}/Ex_{in} (*without solar radiation*)		0.163	0.088	0.304
Ex_s/Ex_{in}	yr	220.13	234.96	122.96
Ex_s/Ex_{in} (*without solar radiation*)	yr	1048.35	926.02	1422.10
Emergy				
Em	sej ha^{-1} yr^{-1}	7.46×10^{15}	5.73×10^{15}	7.05×10^{15}
Transformity	sej J^{-1}	4.05×10^5	3.34×10^5	3.64×10^5
Emergy-based ratios				
ELR		1.41	2.64	2.14
ED	sej m^{-2} yr^{-1}	7.46×10^{11}	5.73×10^{11}	7.05×10^{11}
Exergy/Emergy-based ratios				
Ex_{out}/Em	J sej^{-1}	1.69×10^{-6}	2.04×10^{-6}	1.87×10^{-6}
Ex_s/Em	J yr sej^{-1}	0.014	0.021	0.009
Footprint				
EF	gha yr ha^{-1}	3.05	—	—
	gm^2 yr kg^{-1} ha^{-1}	4.36	—	—
BC	gha yr ha^{-1}	1.48	—	—
EF/BC		2.01	—	—

[a] Bastianoni et al. (2005).

Results can be used for comparing processes and evaluating the environmental efficiency and the state of health of different agro-ecosystems. Here we highlighted some differences between the grape production analyzed (grape A) and the other productions previously studied (Bastianoni et al. 2005), namely, grape B and grape C. One was due to the contribution of solar radiation, which was less relevant for grape B (75%), with respect to grapes A and C (92% and 84%, respectively). This probably depends on different location, climatic condition, and exposure to the sun of each agricultural system. Differences were also due to other parameters. In particular, in the production of grape B, a much higher quantity of pesticides was used. Regarding to the Ex_{out}, grapes A and C presented higher values essentially due to a lower quantity and size of grape. The exergy storage (Ex_s), or the exergy stored in the soil, plants, and grapes, was higher for grapes A and B. This was particularly due to a higher number of heavier plants per hectare with respect to grape C that presents a more equilibrated exergy repartition among the three components (soil, plants, and grapes).

Exergy-based indices measure the level of organization of a system (see Ex_s/Ex_{in}) and detect the efficiency of internal processes in achieving and maintaining a system's organization and providing output (Ex_{out}/Ex_{in}). The first exergy-based ratio is an index of efficiency in using resources. It was obtained by dividing the exergy output to the exergy input needed to obtain the final output (Ex_{out}/Ex_{in}). Results showed that less than 3% of the total exergy consumed (Ex_{in}) was converted into an output (Ex_{out}) for all the systems analyzed. Nevertheless, when solar radiation is not accounted there is a more heterogeneous situation. Grape B showed the highest efficiency (30%), twice of grape C (16%) and more than three times of grape A (8%).

A second exergy-based indicator (Ex_s/Ex_{in}) showed that grape A presented a higher level of organization per unit of exergy output with respect to the other two grape productions. Nevertheless, when solar exergy is not included in Ex_{in}, outcomes show an inverse condition. Both of the perspectives are important: the first "measures" the global amount of structure in the agro-ecosystem, while the second focuses on the investments required to maintain that organizational level.

17.3.2 Emergy Evaluation

Results of the emergy evaluation show that the total emergy used for grape collection corresponds to a total flow of 7.46×10^{15} sej yr^{-1} ha^{-1}, 38% of which is due to the direct flow of natural resources (rain, geothermal heat, soil organic content); 27% of the total emergy is related to the procedures in the phase of grape production and only 9% depends on the initial investment for vine planting. Human labor corresponds to 26%.

The resources used were classified according to their renewability-nonrenewability. In some cases these classes are given in percentages because renewable resources such as, for example, manure or wood, were purchased as finite products from an industrial process including at least collection, treatment, and transport.

In Figure 17.1, resources were gathered into aggregated groups in order to show the relevance, in terms of emergy, of natural resources (rain, geothermal heat, soil organic content), human labor, technology (machinery and fuel), materials (wood), and products for cultivating the vineyard (manure and pesticides). Also, the diagram shows an aggregation into renewables and nonrenewables according to the classification made in the figure. Since renewable resources were assumed to be locally available and nonrenewables

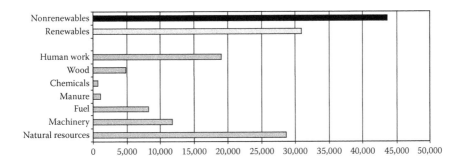

FIGURE 17.1
Diagram of emergy flows in an agricultural system (a cultivation of grape in Tuscany). Emergy units are 10^{12} sej yr^{-1} ha^{-1}.

purchased from other systems, we assessed that local renewable resources (L) are 3.10×10^{15} sej yr^{-1} ha^{-1}, which corresponds to 42% of the total emergy flow, while the purchased nonrenewable resources (F) are 4.36×10^{15} sej yr^{-1} ha^{-1}, the other 58%.

Based on these classes, we assessed some indices that can provide synthetic information for evaluating the whole system. The Environmental Loading Ratio (ELR) is the ratio between renewable and nonrenewable emergy flows and was about 1.41. This low value (with respect to traditional productions in agricultural systems) of ELR is due to a biological grape production that presents a certain equilibrium between natural availability of renewable resources and the exploitation of nonrenewables. An industrial production based, for example, on a high use of chemical fertilizers (instead of manure) and pesticides and an increasing mechanization of most of the processes, would enhance the use of purchased nonrenewable products (particularly fuels and chemicals) and provide a higher value of this index.

With respect to the other Italian grapes (see Table 17.2), all emergy values exhibit low variability. Grape B (Piedmont) shows the lower emergy flow and transformity to demonstrate a lower emergy demand but a higher ELR (2.64), which means that the emergy flows used are prevalently nonrenewable. With respect to grape B, the production of grape A needs a higher amount of emergy but a higher percentage of renewability.

The Empower Density value was 7.46×10^{11} sej m^{-2} yr^{-1}. This represents a concentration or density of flows that feed a spatial unit of an agricultural system in a unit of time corresponding to a year. In Table 17.3 there are emergy values for other agricultural products. Empower density of grapes (biological production) was at the lowest values, with oranges and corn (type 2). As shown in the table, tomatoes are energy-intensive products mainly due to inputs and services such as fuels, plastic, and building management.

Also, transformity values exhibit a high variability. The transformity of grapes was assessed considering an average grape production of 70,000 kg yr^{-1}, and was found to be 4.05×10^5 sej J^{-1}. Cucumbers, oranges, potatoes, corn

TABLE 17.3

Empower Density and Specific Emergy for Several Agricultural Products

Agricultural Product	Empower Density ($\times 10^{15}$ sej ha^{-1})	Transformity ($\times 10^5$ sej J^{-1})
Oranges[a]	9.44	1.09
Cabbages[a]	12.1	2.71
Corn (type 1)[a]	13.1	1.26
Green beans[a]	13.5	12.0
Cucumbers[a]	17.8	0.68
Lettuce[a]	15.8	8.45
Potatoes[a]	15.2	1.78
Tomatoes[a]	39.0	8.57
Corn (type 2)[b]	6.12	2.18
Grapes (biological)[c]	7.46	4.05[c]

[a] Brandt-Williams (2001).
[b] Campbell (2008).
[c] This study.

(1 and 2), and cabbages show lower values than grapes. Lettuce, tomatoes, and green beans have higher values. Grapes (biological production) have thus an intermediate value among these products.

17.3.3 Ecological Footprint

The Footprint approach evaluates the global nature's effort to sustain an agricultural production of grapes in terms of area (i.e., *gha*) needed to make available all resources consumed and to absorb the carbon dioxide emissions associated with the resources used. Results showed that 3.05 *gha* are required to produce grapes relative to the organic procedures followed by the farm. This corresponds to 4.36 *gm²* (1 ha is equal to 10,000 m²) per kilogram of grapes.

The ratio of the total Footprint to the bio-productivity of the vineyard measures how much the overall nature demand exceeds the local supply of resources. This is 2.06 and indicates an area almost twice the vineyard. The extra area ensures the need of machinery (iron and steel), fuel, and chemicals. Human labor was also relevant (6.25%) as typical for this kind of production. The greater the extra area requirement, the higher the dependence from imported (and generally nonrenewable) resources. Obviously, the total Footprint to local supply of biocapacity should be kept as low as possible to reduce the use of virtual lands that generally are hidden in energetic resources.

Figure 17.2 reports the total Footprint by land category. As seen before, the total Footprint is mostly due to cropland (52%) and the remainder is attributed to carbon footprint or energy land (38%) and forest (wood) component (9%).

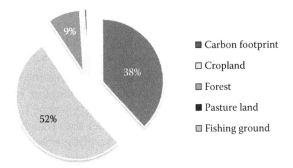

FIGURE 17.2
Ecological Footprint for grape production by land categories.

TABLE 17.4

Ecological Footprint for Some Agricultural Products

Crop	EF (gha ha⁻¹)
Beans[a]	4.51
Tomatoes[a]	9.97
Cabbage[a]	11.31
Maize[a]	6.51
Pineapple[a]	6.24
Coffee[a]	6.38
Grapes (biological production)[b]	3.05
Maize (biological production)[c]	4.47

[a] Our elaboration on Cuadra and Björklund data (2007).
[b] This study.
[c] Our estimation.

Unfortunately, the literature does not offer a wide overview on Footprint application on agricultural production to make useful comparison. Cuadra and Björklund (2007) provided a Footprint application, among the methods used, to six agricultural crop production systems in Nicaragua: common bean (*Phaseolus vulgaris* L.), tomato (*Lycopersicum esculentum* L. Mill), cabbage (*Brassica oleraceae* L. var. *capitata*), maize (*Zea mays* L.), pineapple (*Ananas comosus* L. Merr.), and coffee (*Coffea arabica* L.). Results are shown in Table 17.4. Tomato and cabbage resulted to be the most Footprint-intensive product demanding more resources, while beans and grapes (biological cultivation) were found to be more "sustainable."

17.3.4 Comparing Indicators

An integrated or joint use of the three methods above can enrich our comprehension of the health/sustainability of the systems analyzed. For example, since emergy is a measure of the environmental costs sustained by a system brought to its current state, the combination with exergy would add meaning to exergy indices (Bastianoni et al. 2005).

The ratio Ex_{out}/Em measures how much output can be obtained using a unit of solar emergy. Since the emergy flow represents the work of the biosphere, in terms of solar energy, to produce a product or to sustain a system, this ratio is a measure of efficiency in emergy terms. In processes with the same output, the higher the ratio, the higher the efficiency. In this sense this ratio is the reverse of a transformity, providing how much exergy output per unit of emergy used. Another index is defined as the ratio between the exergy stored in a product and the solar emergy needed to produce that product, Ex_s/Em. The ratio measures the level of organization supported by a unit of solar emergy. Both these ratios are higher for grape B. This means that efficiency is higher for this grape's production because a lower work of nature is needed to produce the same unit of product.

Emergy Analysis and Ecological Footprint could also be used jointly to assess the sustainability of an agricultural system production even if they take into account different aspects of a system. Results from these two methods provide information that is often complementary and consistent. For example, values of empower density in Table 17.3 are coherent with Ecological Footprint values in Table 17.4, even considering that values refer to specific production processes analyzed in different areas.

Similarities are mostly due to the fact that both the analyses are based on an energy and material flows inventory. This inventory usually considers similar but not the same parameters for both analyses. Nevertheless, the interpretation of results is different and some aspects can be highlighted.

Ecological Footprint evaluates impacts due to resource use and waste production relative to ecosystem capacity. This implies constraints in a system's development. Thus Footprint values refer to a threshold, given by the level of bioproductivity of the agricultural system that is not detected by emergy.

Emergy evaluation considers different classes of resources, such as renewable or nonrenewable and local or imported. These are not distinguished within an Ecological Footprint assessment. Moreover, with respect to the emergy evaluation, Ecological Footprint does not take into account the depletion of nonrenewable resources (i.e., minerals, metals, fossil fuels), the use of freshwater resources, or soil erosion and other processes that degrade bioproductive land.

Since Ecological Footprint, at least in principle, accounts for the area of ecosystems that provide environmental resources and absorbs emissions with respect to a system's performance, the information given refers to both Daly's sustainability principles (Daly 1990), the one about limits of exploitation of natural resources due to their rate of production, and the other one about the rate of absorption by nature of human wastes and emissions. With respect to these principles, emergy focuses on the problem of resource use and does not consider that of emissions of wastes, thus referring to only the first of Daly's two principles.

17.4 Conclusion

Indicators based on exergy, emergy, and Ecological Footprint concepts can inform on the level of health and sustainability of systems, such as ecosystems and agricultural systems. In particular, grape production was presented as a case study. This allowed us to discuss the three methods and to briefly introduce and compare their theoretical basis and their capacity to provide information on different aspects of an ecosystem's health. In particular, correlations and complementarities of these methods were found and showed through the elaboration and discussion of different indices.

In synthesis, exergy analysis gives a measure of the state (of health) of a system, its organization, and its ability to provide an output. Exergy-based indices give information on the environmental efficiency of a system considering its level of complexity.

Emergy evaluation considers all the energy inflows that converge into a system and allow it to achieve and maintain in time a certain state of health. Through a combination with exergy, indices that combine exergy and emergy values were provided in order to estimate the state of a system with respect to environmental costs for achieving and maintaining it. In particular, the level of organization (the exergy stored) of an agroecosystem, a vineyard, and its final output (the exergy output), the grapes harvested, was evaluated relative to the natural resources used to feed it (the emergy used).

Ecological Footprint measures the quantity of earth ecosystems needed to support a system (meaning to feed it and to absorb its wastes), in terms of area. This does not account for the whole energy inflows to a system, as exergy and emergy do, but for the capacity of an ecosystem to exploit those inflows and withdraw net energy from them. For example, while emergy accounts for the whole solar energy irradiating a given area, Ecological Footprint considers the capacity of the ecosystem in that area to capture solar energy through photosynthesis and make it available within natural cycles.

We highlighted how the three viewpoints on a system's organization (exergy), natural resource demand (emergy), and bioproductive land requirement (Ecological Footprint) provide different information on ecosystem health. Differences between these three methods determine a high complementarity of information.

References

Bagliani, M., Galli, A., Niccolucci, V., and Marchettini, N. 2008. Ecological footprint analysis applied to a sub-national area: The case of the Province of Siena (Italy). *Journal of Environmental Management* 86 (2): 354–64.

Bastianoni, S., Nielsen, S. N., Marchettini, N., and Jørgensen, S. E. 2005. Use of thermodynamic functions for expressing some relevant aspects of sustainability. *International Journal of Energy Research* 29:53–64.

Bastianoni, S., Pulselli, F. M., Focardi, S., Tiezzi, E. B. P., and Gramatica, P. 2008. Correlations and complementarities in data and methods through Principal Components Analysis (PCA) applied to the results of the SPIn-Eco Project. *Journal of Environmental Management* 86 (2): 419–26.

Bejan, A., Tsatsaronis, G., and Moran, M. J. 1996. *Thermal design and optimization*. New York: Wiley.

Bendoricchio, G., and Jørgensen, S. E. 1997. Exergy as goal function of ecosystems dynamic. *Ecological Modelling* 102 (1): 5.

Brandt-Williams, S. L. 2001. *Handbook of emergy evaluation*. A compendium of data for emergy computation issued in a series of folios. Folio #4—Emergy of Florida agriculture. Gainesville, FL: Center for Environmental Policy, University of Florida.

Brown, M. H., and Cohen, M. T. 2008. Emergy and network analysis. In Jørgensen S. E., Fath B. D. (eds.). *Encyclopedia of ecology—Ecological indicators*, 1229–38. Amsterdam: Elsevier.

Brown, M. T., and Bardi, E. 2001. *Handbook of emergy evaluation*. A compendium of data for emergy computation issued in a series of folios. Folio #3—Emergy of ecosystems. Gainesville, FL: Center for Environmental Policy, University of Florida.

Brown, M. T., Odum, H. T., and Jørgensen, S. E. 2004. Energy hierarchy and transformity in the universe. *Ecological Modelling* 178:17–28.

Burkhard, B., Muller, F., and Lill, A. 2008. Ecosystems health indicators. In Jørgensen S. E, Fath B. D. (eds.). *Encyclopedia of ecology—Ecological indicators*, 1132–38. Amsterdam: Elsevier.

Campbell, D. E. 1998. Emergy analysis of human carrying capacity and regional sustainability: An example using the state of Maine. *Environmental Monitoring Assessment* 51:531–69.

———. 2008. Emergy and its importance. Environmental Research Brief. US EPA, ORD, NHEERL, Atlantic Ecology Division, Narragansett, RI. EPA/600/S-08/003, Contribution No. AED-08-025.

Campbell, D. E., Brandt-Williams, S. L., and Meisch, M. E. A. 2005. Environmental accounting using emergy: Evaluation of the State of West Virginia. Narragansett, RI: USEPA.

Chambers, N., Simmons, C., and Wackernagel, M. 2000. *Sharing nature's interest: Ecological Footprint as an indicator of sustainability*. London: Earthscan Publications Ltd.

Chen, B., and Chen, G. Q. 2006. Modified ecological footprint accounting and analysis based on embodied exergy—A case study of the Chinese society 1981–2001. *Ecological Economics* 61:355–76.

Cuadra, M., and Björklund, J. 2007. Assessment of economic and ecological carrying capacity of agricultural crops in Nicaragua. *Ecological Indicators* 7 (1): 133–49.

Daly, H. E. 1990. Toward some operational principles of sustainable development. *Ecological Economics* 2:1–6.

Erb, K. H. 2004. Actual land demand of Austria 1926–2000: A variation on Ecological Footprint assessments. *Land Use Policy* 21:247–59.

Evans, R. B. 1980. Thermoeconomic isolation and essergy analysis. *Energy* 5 (8–9): 805–21.

FAOSTAT. 2008. www.faostat.fao.org

Folke, C., Jansson, Å., Larsson, J., and Costanza, R. 1997. Ecosystems appropriation by cities. *Ambio* 26:167–72.

Fonseca, J. C., Marques, J. C., Paiva A. A., Freitas A. M., Madeira V. M. C., and Jørgensen, S. E. 2000. Nuclear DNA in the determination of weighing factors to estimate exergy from organisms biomass. *Ecological Modelling* 126 (2–3): 179–89.

Galli, A., Kitzes, J., Wermer, P., Wackernagel, M., Niccolucci, V., and Tiezzi, E. An exploration of the mathematics behind the ecological footprint. *International Journal of Ecodynamics* 2 (4): 250–57.

Global Footprint Network. 2008. National Footprint and Biocapacity Accounts. Global Footprint Network, Oakland, CA (http://www.footprintnetwork.org).

Higgins, J. B. 2003. Emergy analysis of the Oak Openings region. *Ecological Engineering* 21:75–109.

Jørgensen, S. E. 1982. Editorial. *Ecological Modelling* 14 (3–4): 153.

———. 2000. Exergy of an isolated living system may increase. Advances in energy studies, Porto Venere, Italy, SGE editoriali, Padua.

———. 2002. Explanation of ecological rules and observation by application of ecosystem theory and ecological models. *Ecological Modelling* 158 (3): 241.

———. 2008. Exergy. In Jørgensen S. E., Fath B. D. (eds.). *Encyclopedia of ecology— Ecological indicators*, 1498–1509. Amsterdam: Elsevier.

Jørgensen, S. E., and Mejer H. F. 1977. Ecological buffer capacity. *Ecological Modelling* 3:39–61.

Jorgenson, A. K., and Burns, T. J. 2007. The political-economic causes of change in the ecological footprints of nations, 1991–2001: A quantitative investigation. *Social Science Research* 36 (2): 834–53.

Kautsky, N., Berg, H., Folke, C., Larsson, J., and Troell, M. 1997. Ecological footprint for assessment of resource use and development limitations in shrimp and tilapia aquaculture. *Aquacult Res* 28:753–66.

Kitzes, J., Galli, A., Bagliani, M., Barrett, J., Dige, G., Ede, S., Erb, K., Giljum, S., Haberl, H., Hails, C., Jolia-Ferrier, L., Jungwirth, S., Lenzen, M., Lewis, K., Loh, J., Marchettini, N., Messinger, H., Milne, K., Moles, R., Monfreda, C., Moran, D., Nakano, K., Pyhälä, A., Rees, W., Simmons, C., Wackernagel, M., Wada, Y., Walsh, C., Wiedmann, T. 2009. A research agenda for improving national Ecological Footprint accounts. *Ecological Economics* 68 (7): 1991–2007.

Kitzes, J. F., Peller, A., Goldfinger, S., and Wackernagel, M. 2007. Current methods for calculating national ecological footprint accounts. *Science for Environment & Sustainable Society* 4 (1): 1–9.

Lozano, M. A., and Valero A. 1993. Theory of the exergetic cost. *Energy* 18 (9): 939.

Medved, S. 2006. Present and future ecological footprint of Slovenia—The influence of energy demand scenarios. *Ecological Modelling* 192:25–36.

Mejer, H. F., and Jørgensen, S. E. 1979. Energy and ecological buffer capacity. State of the art of ecological modelling. Environmental Sciences and Applications, Proc. 7th Conf. Ecological Modelling, Copenhagen.

Monfreda, C., Wackernagel, M., and Deumling, D. 2004. Establishing national natural capital accounts based on detailed ecological footprint and biological capacity assessments. *Land Use Policy* 21:231–46.

Moran, D. D., Wackernagel, M., Kitzes, J., Goldfinger, S.H., and Boutaud, A. 2008. Measuring sustainable development—Nation by nation. *Ecological Economics* 64:470–74.

Nguyen, H. X., and Yamamoto, R. 2007. Modification of ecological footprint evaluation method to include non-renewable resource consumption using thermodynamic approach. *Resources, Conservation & Recycling* 51 (4): 870–84.

Niccolucci, V., Galli, A., Kitzes, J., Pulselli, R. M., Borsa, S., and Marchettini, N. 2008. Ecological Fooprint analysis applied to the production of two Italian wines. *Agriculture, Ecosystems and Environment* 128:162–66.

Niccolucci, V., Pulselli, F. M., and Tiezzi, E. 2007. Strengthening the threshold hypothesis: Economic and biophysical limits to growth. *Ecological Economics* 60:667–72.

Odum, H. T. 1973. Energy ecology and economics. Royal Swedish Academy of Science. *Ambio* 2 (6): 220–27.

———. 1971. *Environment, power and society*. New York: Wiley.

———. 1996. *Environmental accounting: Emergy and environmental decision making*. New York: Wiley.

———. 1988. Self organization, transformity and information. *Science* 242:1132–39.

Ortega, E., Safonov, P., and Comar, V., eds. 1999. *Introduction to ecological engineering with Brazilian case studies*. Campinas, Brazil: UNICAMP.

Pizzigallo, A. C. I., Granai, C., and Borsa, S. 2008. The joint use of LCA and emergy evaluation for the analysis of two Italian wine farms. *Journal of Environmental Management* 86:396–406.

Pulselli, F. M., Bastianoni, S., Marchettini, N., and Tiezzi, E. 2008. *The road to sustainability. GDP and future generations*. Southampton, UK: WIT press.

Pulselli, R. M., Pulselli, F. M., and Rustici, M. 2008. The emergy accounting of the Province of Siena: Towards a thermodynamic geography for regional studies. *Journal of Environmental Management* 86:342–53.

Pulselli, R. M., Rustici, M., and Marchettini, N. 2007. An integrated holistic framework for regional studies: Emergy-based spatial analysis of the province of Cagliari. *Environmental Monitoring and Assessment* 133:1–13.

Rant, Z. 1956. Exergy the new word for technical available work. *Forschungen im Ingenieurwesen* 22:36.

Rees, W. E. 1992. Ecological Footprints and appropriated carrying capacity: What urban economics leaves out. *Environment and Urbanization* 4:121–30.

Ridolfi, R., and Bastianoni, S. 2008. Emergy. In Jørgensen S. E., Fath B. D. (eds.). *Encyclopedia of ecology—Ecological indicators* 1218–28. Amsterdam: Elsevier.

Sciubba, E., Bastianoni, S., and Tiezzi, E. 2008. Exergy and extended exergy accounting of very large complex systems with an application to the province of Siena, Italy. *Journal of Environmental Management* 86 (2): 372–82.

Szargut, J. D., Morris, D. R., and Steward, F. R. 1998. Exergy analysis of thermal, chemical, and metallurgical processes. New York: Hemisphere.

Thomassen, M. A., and de Boer, I. J. M. 2005. Evaluation of indicators to assess the environmental impact of dairy production systems. *Agriculture, Ecosystems and Environment* 111:185–99.

Tilley, D. R., and Swank, W. T. 2003. Emergy-based environmental system assessment of a multi-purpose temperate mixed-forest watershed of the southern Appalachian Mountains, USA. *Journal of Environmental Management* 69:213–27.

Ulgiati, S., Odum, H. T., and Bastianoni, S. 1994. Emergy use, environmental loading and sustainability. An emergy analysis of Italy. *Ecological Modelling* 73:215–68.

van der Werf, H. M. G., Tzilivakis, J., Lewis, K., and Basset-Mens, C. 2007. Environmental impacts of farm scenarios according to five assessment methods. *Agriculture, Ecosystems and Environment* 118 (1–4): 327–38.

Van Vuuren, D. P., and Bouwman, L. F. 2005. Exploring past and future changes in the ecological footprint for world regions. *Ecological Economics* 52:43–62.

Wackernagel, M., and Kitzes, J. 2008. Ecological Footprint. In Jørgensen, S. E., and Fath, B. D. (eds.). *Encyclopedia of ecology*, 1031–37. Amsterdam: Elsevier B.V.

Wackernagel, M., and Rees, W. E. 1996. *Our Ecological Footprint: Reducing human impact on the earth*. Gabriola Island, British Columbia, Canada: New Society Publishers.

Wada, Y. 1993. The appropriated carrying capacity of tomato production: The Ecological Footprint of hydroponic greenhouse versus mechanized open field operations. M.A. thesis, School of Community and Regional Planning, University of British Columbia, Vancouver, Canada.

Wall, G. 1990. Exergy conversion in the Japanese society. *Energy* 15:435.

WWF (World-Wide Fund for Nature International, Global Footprint Network, ZSL Zoological Society of London). 2008. Living Planet Report 2008. Gland, Switzerland: WWF. www.panda.org/livingplanet

Zaleta-Aguilar, A., L. Ranz, et al. 1998. Towards a unified measure of renewable resources availability: The exergy method applied to the water of a river. *Energy Convers Manage* 39 (16–18): 1911.

Zhao, S., Li, Z., and Li, W. 2005. A modified method of ecological footprint calculation and its application *Ecological Modelling* 185:65–75.

18

Ecological Indicators to Assess the Health of River Ecosystems

Carles Ibáñez, Nuno Caiola, Peter Sharpe, and Rosa Trobajo

CONTENTS

18.1 Introduction

The systematic use of biological responses to evaluate changes in the environment with the intent to use this information in a water quality control program is defined as biological assessment (Matthews et al. 1982). The biological response is measured by using biological indicators, and river ecosystems were one of the first where they were used as an alternative or a complement of assessment systems based on physicochemical indicators. In the first decades of the twentieth century biological assessment of rivers mostly used simple techniques related to organic waste pollution (Hellawell 1978). This approach was used in the early 1900s by German aquatic ecologists in the development of the Saprobic indices to assess the effect of organic pollution on streams (Kolkwitz and Marsson 1902), and it was also applied in the United States (Forbes and Richardson 1913; Ellis 1937). In the last three decades of the twentieth century a number of approaches were developed to evaluate the ecological effects of stress on stream ecosystems using organisms (Descy 1979; Karr 1981; Armitage et al. 1983; Johnson et al. 1993).

However, the traditional quality assessment approaches failed after the impairment of rivers due to organic pollution got mixed with other environmental disturbances, and no longer provided a sufficient tool for integrated water management due to their restricted approach (Verdonschot and Moog 2006). Thus, to assess a river ecosystem a great variety of parameters reflecting its structure and functioning and different types of disturbance should be used (Karr et al. 1986; Allan 1995). Recently, it has become increasingly common to use multiple organism groups in bioassessment (Johnson et al. 2006), and this is one of the innovative aspects of the Water Framework Directive (WFD) legislation (European Commission 2000).

The main goals of this chapter are (a) to summarize the state of the art concerning the use of biological indicators in river ecosystems, focusing on the most widely used groups, which are the benthic diatoms, the benthic macroinvertebrates, and the fish communities; and (b) to compare the features of the most relevant official systems of river ecosystem assessment, which are those used in the United States and the European Union.

18.2 Biological Indicators

The use of biological indicators for monitoring rivers has a long history, especially in Europe and the United States (see Furse et al. 2006 and Barbour et al. 1999). The use of several (complementary) indicators is based on the premise that using multiple organism groups/assemblages can help to distinguish the effects of human-induced stress more efficiently (with less uncertainty) and more effectively (by detecting the effects of multiple stressors).

Within this approach, the most widely used groups for the assessment of river ecosystem health have been the phytobenthos (benthic diatoms), benthic macroinvertebrates, and fishes (Barbour et al. 1999; Johnson et al. 2006), and they are the main indicators used in the official bioassessment systems in the United States and the European Union (Furse et al. 2006; Hughes and Peck 2008). For example, benthic diatoms have been used for assessing the effects of acidification and eutrophication (Potapova and Charles 2007), fish are suitable indicators of flow and habitat alterations (Bain et al. 1988), and benthic invertebrates are commonly used for monitoring the effects of organic pollution, acidification, and hydromorphological impacts (Verdonschot and Moog 2006). In the next sections a review and synthesis of the use of these groups as biological indicators for the assessment of river ecosystem health are carried out.

18.2.1 Phytobenthos

Diatoms are microscopic siliceous unicellular algae. Diatoms are good environmental indicators, since they are present in almost all aquatic habitats (e.g., in all river types) and they respond directly and rapidly to many environmental changes. This response can vary according to species physiology, and this species-specific sensitivity to parameters leads to a large panel of assemblage composition according to the river ecological conditions (Tison et al. 2008). Specifically, they have been shown to be effective indicators of physicochemical water characteristics such as pH, salinity, and nutrients (Potapova and Charles 2007), and they have been widely used as indicators of eutrophication and water river pollution as well as of the integrity of biological habitats (Sabater and Admiraal 2005). The bioindicative properties of diatoms are now used in routine river quality assessment programs in Europe, North America, and elsewhere as well as in paleoecology.

A great number of methods based on the use of diatoms for assessing river and stream health have been developed. In fact, the use of microalgae as biological indicators for assessing aquatic ecosystem health could be dated from the early twentieth century with the saprobic index of Kolkwitz and Marsson (1902), who developed a model with five pollution states based on the presence of certain key species (i.e., indicator species, among them algal taxa) as indicative of polluted conditions. However, with the time it became clear that those species taken to be indicative of pollution also occur in non-polluted waters (i.e., they are not restricted to polluted waters) (Round 1981).

Since the 1960s the diatom community characteristics have been used to assess the ecological health of rivers and streams to diagnose causes of degradation, considering not only the presence of the diatom taxa but also their proportion in the community.

Rank-diatom abundance curves (e.g., Lobo and Kobayasi 1990), which apparently can inform on the ecosystem quality to assess river condition, was disappointing or completely ineffective (Ector and Rimet 2005; Lavoie et al. 2009) due to the multiple and differing effects of pollutants on diatom species richness and evenness.

Lange-Bertalot (1979) developed a practical method to assess water quality using both the specific identity and the relative abundance of constituent diatom species in the assemblage for European rivers. The Lange-Bertalot classification divided 62 species into three groups depending on their resistance, sensitivity, or indifference to pollution. The river water quality diagnosis is based on the proportional representation of these three groups. SHE (Steinberg and Schiefele 1988) and the Japanese DAIpo (Diatom Assemblage Index to Organic Pollution) (Watanabe et al. 1988) are two other indices based on the same approach as Lange-Bertalot's method.

Many of the diatom indices that are now used in routine biomonitoring programs of rivers and streams in numerous European countries are derived from the weighted average formula of Zelinka and Marvan (1961), which considers the sum of the different species abundance influenced by their sensitivity to the described disturbance and by their indicator value:

$$ID = \Sigma\, A_j\, I_j\, S_j\, /\, \Sigma\, A_j\, I_j$$

where
  A_j = relative abundance of the species j
  I_j = indicator value of the species j (tolerance)
  S_j = sensitivity value of the species j (optimum)

The optimum and tolerance values of each species need to be a priori determined. In the earlier works these values were determined from summaries of large amounts of information scattered in small-scale observational or experimental studies, while in more recent works these values are determined from large-scale consistent diatom datasets and appropriate numerical techniques (Potapova and Charles 2007).

Some of the most used diatom-based indices in rivers' routine sampling in many European countries are the Specific Pollution Index (IPS; Coste 1982), the Biological Diatom Index (IBD; Lenoir and Coste 1996), the Trophic Diatom Index (TDI; Kelly and Whitton 1995), the Sládeček Index (SLA; Sládeček 1986), the Generic Diatom Index (GDI; Rumeau and Coste 1988), the European Economic Community (EEC) (Descy and Coste 1991), and the Eutrophication Pollution Index Diatoms (EPI-D; Dell'Uomo 1996).

The differences among these indices are related to the number of considered taxa, the taxonomic resolution (genera, species, varieties), the sensitivity (optimum), indicator (tolerance) values that were attributed to each taxa, and the water quality information provided (e.g., trophic state, organic pollution, etc.) (Lavoie et al. 2009). The species' autoecologies considered by these indices can be provided with the software OMNIDIA (Lecointe et al. 1993) and the indices values are calculated automatically. Some indices like IBD in France, IPS in Luxemburg, or the TDI in England are routinely used to assess biological quality of rivers on national networks. For a detailed list of the diatom indices that are already applied in several European countries, see Ector and Rimet (2005).

Diatom-based Indices of Biotic Integrity (IBI) are mostly restricted to the United States. These indices are referred to as multimetric indices because they are composed (average or sum) of more than one metric, where structural metrics such as diversity indices, relative abundance of species, and pollution-sensitive species or functional groups are included.

Some of the diatom-based IBI developed and applied for monitoring stream integrity in some states of the United States are the Kentucky Diatom Pollution Tolerance Index (KYDPTI; Wang et al. 2005), the Montana Diatom

Pollution Index (MTDPI; Kydow 1993), the Interior Plateau Ecoregion (IPE; Bahls 1993), and the Periphyton Index of Biotic Integrity (PIBI; Hill et al. 2000). Very recently a diatom IBI for Québec streams was developed (Lavoie et al. 2009) following the methods in Wand et al. (Bahls 1993). The also recent Ecological Distance Index (EDI; Tison et al. 2008) is the first attempt to include into the already existing European indices this multimetric approach aiming to assess river ecological status, taking into account several community attributes significantly correlated with different measures of human disturbances, while fulfilling the reference criteria concept defined by the European Water Framework Directive (European Commission 2000; Tison et al. 2008).

18.2.2 Benthic Macroinvertebrates

Macroinvertebrates are frequently used in biomonitoring of streams and rivers worldwide due to their characteristics of high abundance in most rivers and streams in a wide range of habitats. Furthermore, they belong to different trophic levels, thus making them good potential indicators of disturbances affecting communities' structure (Metcalfe-Smith 1996; Barbour et al. 1999). They also present different degrees of tolerance and response to environmental stressors, like eutrophication and heavy metal pollution (Kiffney and Clements 2003). On the other hand, benthic macroinvertebrates are good indicators of local conditions, since they have a limited movement and live long enough to integrate the effect of temporal variability and perturbations; some species and life stages have a short life span, and they respond rapidly to stressors. In addition, sampling methods are relatively easy and they require little personal and economic effort. Macroinvertebrates are good indicators at both species and community levels, and the assessment methods are mostly based on species composition and abundance, but also on functional and trophic variables.

River monitoring programs in most countries include the benthic macroinvertebrate community, though the design and performance of individual methods vary significantly, due to different traditions in stream assessment (Birk and Hering 2006). While in many Central and Eastern European countries modifications of the Saprobic System have been applied for decades as standard methods (Birk and Schmedtje 2005), other countries rely on the Biological Monitoring Working Party score (BMWP 1978), which has been adjusted for use in various countries (e.g., IBMWP in Spain; see Alba-Tercedor and Pujante 2000). In many cases each country has developed individual assessment methods, like RIVPACS in the United Kingdom (Wright et al. 2000), IBGN in France (AFNOR 1982), and BBI in Belgium (De Pauw and Vanhooren 1983). In most of the European Union countries efforts are being made to adapt the national programs to the new requirements of the WFD; however, different approaches are being used, since in some countries a single stressor is overwhelming, while in other regions different stressors are of equal importance (Birk and Hering 2006).

The conventional approach has been the use of individual species composition and/or abundance measures mostly related to a strong stressor (e.g., organic pollution), such as the Saprobic indices. Such a single biological parameter was interpreted with a summary statement about the water quality by using an index or metric score. This approach is limited in that the key parameter emphasized may not reflect the overall ecological status (Verdonschot and Moog 2006). Biotic indices and scores combine a diversity measure and a pollution tolerance measure (mostly organic), so they actually use two types of metrics (De Pauw et al. 1992).

The next step in metric development was to combine a number of different metrics, each of which provides information on an ecosystem feature and when integrated, performs as an overall indicator of ecological conditions of a water body. The scores of the individual metrics are aggregated to calculate the multimetric index. The multimetrics establish relative values for each single metric based on comparison of values for the best available habitat to those areas that are strongly disturbed (see Verdonschot 2000). Such multimetric assessments provide detection capability over a broader range and nature of stressors and give a more complete picture of ecological conditions than single biological indicators (Verdonschot and Moog 2006). The Index of Biotic Integrity (IBI) was likely the first multimetric index, and was based on fish communities (Karr 1981). Later on, other multimetric indices were developed to include benthic macoinvertebrate communities (Kerans and Karr 1994; Barbour et al. 1996).

Eight major groups of metrics can be distinguished (see Verdonschot and Moog 2006):

- Richness measures (e.g., number of taxa, number of Chironomidae taxa); these metrics are considered to be sensitive to organic pollution.

- Composition measures (e.g., number of individuals, number of intolerant taxa, percentage of dominant taxon, percentage of Oligochaeta); these metrics are considered to increase in dominance due to pollution or disturbance.

- Diversity measures (e.g., Shannon–Wiener index, sequential comparison index); these metrics are considered to decrease with increasing disturbance.

- Similarity/loss measures (e.g., number of taxa in common, Bray–Curtis index); these metrics use comparisons between reference and disturbed sites.

- Tolerance/intolerance measures (e.g., Saprobic index, BMWP score); these metrics are based on tolerance values of taxa sensitive to stressors.

- Functional and trophic measures (e.g., percentage of functional feeding groups, percentage of habitat preferences); these metrics use the alteration in food types, habitats, and environmental conditions under different types of disturbance.

- Life strategy metrics, which use the biological life strategy features (e.g., length of the life cycle, number of eggs or diapauses).
- Condition metrics, which use features of the condition of specimens (e.g., percentage of diseased or deformed individuals).

Within the AQEM and STAR projects multimetric indices have been developed and intercalibrated for various river types throughout Europe. The experiences of these projects clearly show that to enhance comparability between assessment systems the procedure of developing and applying a multimetric index needs to be standardized. Several software packages (e.g., ECOPROF, Moog et al. 2001) aid the quick derivation of metrics from taxa lists, among which the AQEM River Assessment Program (Hering et al. 2004) provides a tool for calculating more than 200 macroinvertebrate metrics. In order to reduce the long list of metrics that are processed by software packages, filter procedures have to be applied.

18.2.3 Fish Fauna

18.2.3.1 Fish as Indicators

Fish are known to be good indicators of ecological status of aquatic ecosystems as they live permanently in water, occupy a wide range of ecological niches, and operate over a variety of spatial scales. Moreover, the high longevity of some fish species enables the detection of disturbances over a long time frame. Fish species are relatively easy to identify and their taxonomy, ecological requirements, and life history traits are generally better known than those for other species groups, making their assessment easier, cheaper, and more accurate. Another advantage of the use of fish to assess the ecological status of rivers is the fact that some species have developed complex migration patterns, making them sensitive to the presence of dams and weirs. Fish usually occupy high trophic levels integrating disturbances affecting lower trophic levels. Fish also have more charisma than other species groups and provide valuable economic resources, being thus important to public awareness.

18.2.3.2 Index of Biotic Integrity: Species Guild Approach

Biotic integrity, as defined by Karr and Dudley (1981) is the capacity of an ecosystem to hold a biological community regarding its species richness, structure, and function, comparable to a non-altered condition. Based on this principle, Karr (1981) developed the IBI. The IBI is a fish-based method to assess the ecological status of rivers. In this approach it is assumed that fish communities respond to human alterations of aquatic ecosystems in a predictable and quantifiable manner. An IBI is, thus, a tool to quantify human pressures by analyzing alterations in the structure of fish communities.

The original IBI (Karr 1981) uses several aspects of fish communities such as species richness and composition, abundance, tolerance to environmental conditions, trophic levels composition, habitat requirements, reproduction traits, length and age structure, migratory behavior, and fishes' health. Each component is measured by quantifiable variables named metrics (e.g., number of species, total biomass, abundance of intolerant species, percentage of omnivores, percentage of lithophilic species, percentage of specialized spawners, number of length classes in the population of a species, number of long-distance migratory species, proportion of individuals with injuries).

These multimetric indices have been used and adapted worldwide since their first version (Karr 1981). Some examples are the development of fish-based indices adapted to different river systems of the United States (Fausch et al. 1984; Leonard and Orth 1986; Karr et al. 1987; Bramblett and Fausch 1991; Osborne and Wiley 1992; Shields et al. 1995; Paller et al. 1996). In Europe, although the development of these multimetric indices started later, it became an extended practice in many countries. This is the case of France (Oberdorff et al. 2002), Belgium (Kestemont et al. 2000), Austria (Schmutz et al. 2000), Spain (Sostoa et al. 2004), and an attempt of a European index (Pont et al. 2006).

The strength and accuracy of fish-based multimetric indices rely on the species guild approach used for their development. The use of functional ecological guilds reflects the functional relationships between fish community structure and the functional complexity of aquatic habitats. The guild concept denotes that the structure of the fish community is determined by the functional structure of the aquatic habitat, in terms of habitat available and prevalent hydrological processes (Noble et al. 2007). Therefore, any disturbance in the functionality or structure of the riverine habitat will be reflected by responses in the functional structure of the fish community. Moreover, the guild approach provides an operational unit between the individual species and the community as a whole (Root 1967), giving the potential to overcome zoogeographic problems of species distributions when considering fish communities over large geographic scales.

18.2.3.3 Metrics Selection and Scoring

To select the metrics that will form the multimetric index, a large set of candidate metrics is established, based on different criteria (e.g., literature, legislation objectives and directives, knowledge of fish species and/or aquatic ecosystems to be evaluated, etc.). The responses of each metric to the human degradation gradient (e.g., nutrient loading, water contaminants, hydromorphological condition, etc.) must be evaluated. The metrics' responses to human disturbance can be positive (higher metric value at higher disturbances), negative (lower metric value at higher disturbances), or

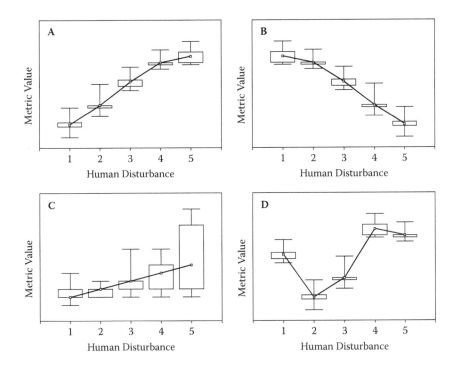

FIGURE 18.1
Theoretical trends and distributions of a fish-based metric value along human disturbance. Examples (a) and (b) represent negative and positive significant responses (e.g., ANOVA $p < 0.05$) of the metrics to human disturbance. Example (c) represents a positive trend in the response of the metric without statistical significance. The metric represented in example (d) has an unpredictable response to human disturbance.

unpredictable (Figure 18.1). The selection of a metric will depend on the significance and predictability of its response to human degradation. It is usual to perform autocorrelation analysis within the selected metrics in order to eliminate redundancies.

Although metrics are usually continuous variables, it is very useful to transform them into categories to fulfill water management criteria and discriminate between three or five ecological status classifications. Although there is not a unique system to determine the separation thresholds between the ecological status classes, measuring the deviation from the excellent expected scenario at reference sites is a standard procedure (Roset et al. 2007). The measurement of the deviation varies from plain expert judgment criteria (Karr 1981) to more or less complex statistical techniques such as box-plot percentile overlapping (Sostoa et al. 2004; Ferreira et al. 2007), adjustment of a mean trend line and percentile thresholds along an environmental gradient (Karr et al. 1986; Breine et al. 2004), or the deviation between the observed and the predicted value of a metric by means of analyses of residuals (Oberdorff et al. 2002; Pont et al. 2006).

18.3 European versus U.S. Approaches

The U.S. government and the European Union Council developed legal frameworks to assist in the management of water bodies in a sustainable manner: the U.S. Clean Water Act (CWA) and the E.U. Water Framework Directive (WFD).

The U.S. CWA was enacted by the U.S. Congress in 1972 with the ultimate goal of restoring and maintaining the biological, physical, and chemical quality of U.S. water resources (Shapiro et al. 2008). Section 305(b) of the CWA requires the U.S. Environmental Protection Agency (USEPA) to report on the status and extent of aquatic resources in the United States (Olsen and Peck 2008). Individual states and American Indian tribes are responsible for implementing the CWA directives.

The E.U. WFD, approved by the European Parliament in 2001, has the aim of protecting inland surface waters (rivers and lakes), transitional waters (estuarine systems), coastal waters, and groundwater. It requires that all inland and coastal waters within defined river basin districts must reach at least "good" status by 2015 and defines how this should be achieved through the establishment of environmental objectives and ecological targets for surface waters (WFD Article 4). For this purpose, the WFD requires member states to assess the ecological quality status of their water bodies (WFD Article 8).

Both the U.S and E.U. legislations have some common issues regarding the ecological indicators that must be applied in order to assess rivers' health. In both cases biological quality elements supported by hydromorphological and physicochemical quality elements are taken into account. The biological indicators considered, of obligatory use in the case of the E.U. WFD, are the aquatic flora (e.g., periphyton and/or macrophytes), the benthic invertebrate fauna, and the fish fauna. Although the development of the biological-based assessment methods was performed in a different manner regarding competent authorities (the Environmental Protection Agency for the whole United States and each member state in the case of the European Union), the approach to this problem was quite the same. This was a spatially based approach. The principle behind this approach is that rivers within the same river basin district (e.g., river basin, group of small watersheds) are understood as a sequence of distinct segments with homogeneous abiotic and biotic characteristics. Thus, the entire river network is classified into distinct types. For each river type, the basic functional unit, undisturbed conditions are formulated and the deviation from these conditions provides the measure of the ecological status (Figure 18.2). This way, it is possible to develop regionalization schemes that reduce the effect of natural environmental variation on indicator values and still provide large

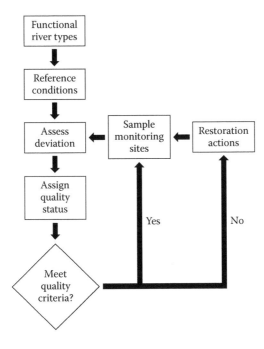

FIGURE 18.2
Simplified scheme of the spatially based criteria used in both the U.S. and E.U. approaches to develop and apply ecological indicators to assess the ecological status of aquatic systems.

enough sample sizes to allow statistically valid assessments within each targeted ecoregion. In the case of the E.U. WFD, this regionalization can be performed using two different sets of environmental descriptors. The first set (system A) includes three descriptors (altitude, catchment size, and geology) whereas the other set (system B) includes two subsets: (i) obligatory factors (the same as system A plus the latitude/longitude information); (ii) optional factors (distance from river source, energy of flow, mean water width, depth and slope, form and shape of main riverbed, river discharge, valley shape, transport of solids, acid neutralizing capacity, mean substratum composition, chloride, air temperature range, mean air temperature, and precipitation).

There is, nevertheless, an important distinction between the U.S. CWA and the E.U. WFD. The U.S. CWA application is based on the water quality standards (WQS) program. The WQS program gives the individual states and American Indian tribes the right to set water quality standards for waters within their territory and provides the legal basis for determining if an aquatic resource is impaired (USEPA 2009). These water quality standards set the uses for these water bodies. Once a use has been accepted into law, then specific water quality criteria must be met to achieve the designated use—for example, different standards exist for waterways designated as drinking-water streams versus warm-water fisheries, etc. The E.U. WFD main principle is that water is not a commercial product like any other but, rather, a heritage that must be protected, defended, and treated as such. Based on this principle, all surface water bodies must achieve a "good" ecological status and, therefore, the uses of the water resources must be compatible with the achievement of the ecological requirements. However, if a body of water has strong anthropic pressures and therefore cannot meet the requirements to achieve the "good ecological status," the E.U. WFD allows member states to declare it as "heavily modified." In such cases, the European directive is quite similar to the U.S. CWA, i.e., the uses of the water body set the quality standards to be achieved. The "good ecological potential" will then be the management objective instead of the good ecological status. There is another important principle in the E.U. WFD, that is the obligation to prevent any water body from getting further deteriorated (WFD Article 1), even if it is declared heavily modified.

Although the E.U. WFD is in the process of implementation (an intercalibration process between E.U. state members' ecological indices is going on) and the first results will only be discussed in 2015, the U.S. CWA has a much longer life and therefore a critical review is being performed. To address criticisms related to a lack of consistent monitoring and reporting of ecological conditions across various tribes and states, the USEPA revised its strategy in 2000 and launched two comprehensive surveys— EMAP-West and the Wadable Streams Assessment (WSA; Hughes and Peck 2008). EMAP-West is a research effort designed to demonstrate the feasibility of implementing a survey of streams and rivers to produce a detailed and accurate assessment of waterway health in the western United States (Hughes and Peck 2008; Stoddard et al. 2005). The EMAP-West program includes 42% of the land area of the conterminous United States, ranging from the rain forests of the Olympic Peninsula to the arid climates of the Sonoran and Mohave Deserts, and covers the states of California, Oregon, Washington, Idaho, Montana, North and South Dakota, Wyoming, Colorado, Utah, Nevada, and Arizona (Stoddard et al. 2005). In this exercise a more accurate and standard procedure to define quality thresholds was performed (Table 18.1).

TABLE 18.1

Example Thresholds for Ecological Indicators in Both Healthy and Impaired Reaches Are Listed, Based on the Most Recent EMAP-West Report

Indicator	Most Disturbed Threshold	%	Least Disturbed Threshold	%
Plains region				
Aquatic vertebrate IBI	<35	25th	≥45	50th
Macroinvertebrate IBI	<41	25th	≥51	50th
Riparian disturbance	>1.3	75th	≤1.0	50th
Habitat complexity	<0.125	25th	≥0.359	50th
Streambed stability	< −2.5 or >0.3	10th	≥ −1.7 and ≤ −0.5	25th
Riparian vegetation	<0.15	10th	≥0.35	35th
Xeric region				
Aquatic vertebrate IBI	<29	5th	≥40	25th
Macroinvertebrate IBI	<47	5th	≥56	25th
Riparian disturbance	>0.9	90th	≤0.7	75th
Habitat complexity	<0.132	10th	≥0.270	35th
Streambed stability	< −1.7 or >0.3	10th	≥ −0.9 and ≤ −0.1	25th
Riparian vegetation	<0.32	5th	≥0.60	25th
Mountains region				
Aquatic vertebrate IBI	<37	5th	≥62	25th
Macroinvertebrate IBI	<57	5th	≥71	25th
Riparian disturbance	>0.95	95th	≤0.35	75th
Habitat complexity	<0.18 (NRock)	5th	≥0.34 (NRock)	25th
	<0.14 (PNW)	5th	≥0.33 (PNW)	25th
	<0.31 (SRock)	5th	≥0.56 (SRock)	25th
	<0.10 (SWest)	5th	≥0.37 (SWest)	25th
Streambed stability	< −1.8 or >0.1 (NRock)	5th	≥ −1.1 and ≤ −0.4 (NRock)	25th
	< −1.3 or >0.6 (PNW)	5th	≥ −0.7 and ≤0.1 (PNW)	25th
	< −1.6 or >0.3 (SRock)	5th	≥ −0.9 and ≤ −0.2 (SRock)	25th
	< −1.3 or >0.6 (SWest)	5th	≥ −0.6 and ≤0.1 (SWest)	25th
Riparian vegetation	<0.23	5th	≥0.67	25th

Source: Stoddard, J. L., Peck, D. V., Paulsen, S. G., Van Sickle, J., Hawkins, C. P., Herlihy, A. T., Hughes, R. M., Kaufmann, P. R., Larsen, D. P., Lomnicky, G., Olsen, A. R., Peterson, S. A., Ringold, P. L., and Whittier, T. R. 2005. *An ecological assessment of western streams and rivers.* EPA 620/R-05/005, 1–48.

References

AFNOR. 1982. Essais des eaux. Détermination de l'indice biologique global normalisé (IBGN). Association Francaise de Normalisation NF T 90–350 France.

Alba-Tercedor, J., and Pujante, A. M. 2000. Running water biomonitoring in Spain: Opportunities for a predictive approach. In *Assessing the biological quality of fresh waters. RIVPACS and similar techniques.* Ambleside: Freshwater Biological Association, 175–85.

Allan, J. D. 1995. *Stream ecology. Structure and function of running waters.* London: Chapman and Hall.

Armitage, P. D, Moss, D., Wright, J. F., and Furse, M. T. 1983. Performance of a new biological water quality score system based on macroinvertebrates over a wide range of unpolluted running water sites. *Water Research* 17:333–47.

Bahls, L. L. 1993. *Periphyton bioassessment methods for Montana streams.* Helena, Montana: Montana Department of Health and Environmental Sciences.

Bain, M. B., Finn, J. T., and Booke, H. E. 1988. Streamflow regulation and fish community structure. *Ecology* 69:382–92.

Barbour, M. T., Gerritsen, J., Griffith, G. E., Frydenborg, R., McCarron, E., White, J. S., and Bastian, M. L. 1996. A framework for biological criteria for Florida streams using benthic macroinvertebrates. *Journal of the North American Benthological Society* 15:185–211.

Barbour, M. T., Gerritsen, J., Snyder, B. D., and Stribling, J. B. 1999. Rapid bioassessment for use in streams and wadeable rivers: Periphyton, benthic macroinvertebrates and fish. Washington, DC: U.S. Environmental Protection Agency, Office of Water.

Birk, S., and Hering, D. 2006. Direct comparison of assessment methods using benthic macroinvertebrates: A contribution to the EU Water Framework Directive intercalibration exercise. *Hydrobiologia* 566:401–15.

Birk, S., and Schmedtje, U. 2005. Towards harmonization of water quality classification in the Danube River Basin: Overview of biological assessment methods for running waters. *Archiv für Hydrobiologie* 16:171–96.

BMWP (Biological Monitoring Working Party). 1978. Final report of the Biological Monitoring Working Party: Assessment and presentation of the biological quality of rivers in Great Britain. London: Department of the Environmental Water Data Unit.

Bramblett, R. G., and Fausch, K. D. 1991. Variable fish communities and the Index of Biotic Integrity in a Western great plains river. *Transactions of the American Fisheries Society* 120:752–69.

Breine, J., Simoens, I., Goethals, P., Quataert, P., Ercken, D., Van Liefferinghe, C., and Belpaire, C. 2004. A fish-based index of biotic integrity for upstream brooks in Flanders (Belgium). *Hydrobiologia* 522:133–48.

Coste, M. 1982. Etude des methods biologiques d'appréciation quantitative de la qualité des eaux. Rapport Q. E. Lyon—Agence de l'Eau Rhône-Méditerranée-Corse.

De Pauw, N., Ghetti, P. F., and Manzini, D. P. 1992. Biological assessment methods for running waters. In *River water quality: Ecological assessment and control*, eds. Newman et al., 217–48. Brussels: C.C.E.

De Pauw, N., and Vanhooren G. 1983. Method for biological quality assessment of watercourses in Belgium. *Hydrobiologia* 100:153–68.

Dell'Uomo, A. 1996. Assessment of water quality of an Apennine river as a pilot study. In *Use of algae for monitoring rivers II*, eds. Whitton, B. A., and Rott, E., 65–73. Institut für Botanik, Universität Innsbruck.

Descy, J.-P. 1979. A new approach to water quality estimation using diatoms. *Nova Hedwigia* 64:305–23.

Descy, J.-P., and Coste, M. 1991. A test of methods for assessing water quality based on diatoms. *Verhandlungen International Verein Limnology* 24:2112–16.

Ector, L., and Rimet, F. 2005. Using bioindicators to assess rivers in Europe: An overview. In *Modelling community structure in freshwater ecosystems*, eds. Lek, S., Scardi, M., Verdonschot, P. F., Descy, J. P., and Park, Y. S., 7–19. Berlin: Springer Verlag.

Ellis, M. M. 1937. Detection and measurement of stream pollution. *Bulletin of Bureau of Fisheries* 48:365–437.

European Commission. 2000. Directive 2000/60/EC of the European Parliament and of the Council—Establishing a framework for community action in the field of water policy. Brussels, Belgium.

Fausch, K. D., Karr, J. M., and Yant, P. R. 1984. Regional application of an index of biotic integrity based on stream fish communities. *Transactions of the American Fisheries Society* 113:39–55.

Ferreira, T., Caiola, N., Casals, F., Oliveira, J. M., and Sostoa, A. 2007. Assessing perturbation of river fish communities in the Iberian Ecoregion. *Fisheries Management and Ecology* 14:519–30.

Forbes, S. A., and Richardson, S. E. 1913. Studies on the biology of the upper Illinois River. *Bulletin of the Illinois State Laboratory of Natural History* 9:481–574.

Furse, M. T., Hering, D., Brabec, K., Buffagni, A., Sandin, L., and Verdonschot, P. F. M. 2006. Developments in hydrobiology. The ecological status of European rivers: Evaluation and intercalibration of assessment methods. Dordrecht: Springer.

Hellawell, J. M. 1978. *Biological surveillance of rivers. A biological monitoring handbook.* Stevanage: NERC.

Hering, D., Moog, O., Sandin, L., and Verdonschot, P. F. M. 2004. Overview and application of the AQEM assessment system. *Hydrobiologia* 516:1–20.

Hill, B. H., Herlihy, A. T., Kaufmann, P. R., Stevenson, R. J., McCormick, F. H., and Burch Johnson, C. 2000. Use of periphyton assemblage data as an index of biotic integrity. *Journal of North American Benthological Society* 19 (1): 50–57.

Hughes, R. M., and Peck, D. V. 2008. Acquiring data for large aquatic resource surveys: The art of compromise among science, logistics, and reality. *Journal of the North American Benthological Society* 27:837–59.

Johnson, R. K., Hering, D., Furse, M. T., and Clarke, R.T. 2006. Detection of ecological change using multiple organism groups: Metrics and uncertainty. *Hydrobiologia* 566:115–37.

Johnson, R. K., Wiederholm, T., and M. Rosenberg, D. 1993. Freshwater biomonitoring using individual organisms, populations and species assemblages of benthic macroinvertebrates. In *Freshwater biomonitoring and benthic macroinvertebrates,* eds. Rosenberg, D. M., and Resh, V. H., 40–158. New York: Chapman and Hall.

Karr, J. R. 1981. Assessment of biotic integrity using fish communities. *Fisheries* 6:21–27.

Handbook of Ecological Indicators

Karr, J. R., and Dudley, D. R. 1981. Ecological perspective on water quality goals. *Environmental Management* 5:55–68.

Karr, J. R., Fausch, K. D., Angermeier, P. R., Yant, P. R., and Schlosser, I. J. 1986. *Assessing biological integrity in running waters: A method and its rationale.* Illinois Natural History Survey Special Publication 5.

Karr, J. R., Yant, P. R., Fausch, K. D., and Schlosser, I. J. 1987. Spatial and temporal variability of the index of biotic integrity in three midwestern streams. *Transactions of the American Fisheries Society* 116:1–11.

Kelly, M., and Whitton, B. A. 1995. The trophic Diatom Index: A new index for monitoring eutrophication in rivers. *Journal of Applied Phycology* 7:433–44.

Kerans, B. L., and Karr, J. R. 1994. A benthic Index of Biotic Integrity (B-IBI) for rivers of the Tennessee Valley. *Ecological Applications* 4:768–85.

Kestemont, P., Didier, J., Depiereux, E., and Micha, J. C. 2000. Selecting ichtyological metrics to assess river basin ecological quality. *Archiv für Hydrobiologie* 121:321–48.

Kiffney, P. M., and Clements, W. H. 2003. Ecological effects of metals on benthic invertebrates. In *Indicator patterns using aquatic communities*, ed. Simon, T. P., 135–54. Boca Raton, FL: CRC Press.

Kolkwitz, R., and Marsson, M. 1902. Grundsätze für die biologische Beurteilung des Wassers nach seiner Flora und Fauna. *Mitt Aus d Kgl Prüfungsanstalt für Wasserversorgung und Abwässerbeseitigung* 1:33–72.

Kydow (Kentucky Division of Water). 1993. *Methods for assessing biological integrity of surface waters.* Frankfort, KY: Kentucky Department of Environmental Protection.

Lange-Bertalot, H. 1979. Pollution tolerance of diatoms as a criterion for water quality estimation. *Nova Hedwigia*, Beiheft 64:285–304.

Lavoie, I., Hamilton, P. B., Wang, Y.-K., Dillon, P. J., and Campeau, S. 2009. A comparison of stream bioassessment in Québec (Canada) using six European and North American diatom-based indices. *Nova Hewigia*, Beiheft 135:37–56.

Lecointe, C., Coste, M., and Prygiel, J. 1993. "OMNIDIA": A software for taxonomy, calculation of diatom indices and inventories management. *Hydrobiologia* 269/270:509–13.

Lenoir, C., and Coste, M. 1996. Development of a practical diatom index of overall water quality applicable to the French National Water Board network. In *Use of algae for monitoring rivers II*, eds. Whitton, B. A., and Rott, E., 29–45. Institut für Botanik, Universität Innsbruck.

Leonard, P. M., and Orth, D. J. 1986. Application and testing of an index biotic integrity in small, coolwater streams. *Transactions of the American Fisheries Society* 115:401–14.

Lobo, E., and Kobayasi, A. H. 1990. Shannon's diversity index applied to some freshwater diatom assemblages in the Sakawa river system (Kanagawa Pref., Japan) and its use as an indicator of water quality. *Japanese Journal of Phycology (Sörui)* 38:229–43.

Matthews, R. A., Buikema, A. L., Cairns, J., Jr., and Rodgers, J. H., Jr. 1982. Biological monitoring: Part IIa: Receiving system functional methods, relationships and indices. *Water Research* 16:129–39.

Metcalfe-Smith, J. L. 1996. Biological water-quality assessment of rivers: Use of macroinvertebrate communities. In *River restoration*, eds. Petts, G., and Calow, P., 17–59. Oxford: Blackwell Science.

Moog, O., Schmidt-Kloiber, A., Vogl, R., and Koller-Kreimel, V. 2001. *ECOPROF-Software. Wasserwirtschaftskataster, bun desministeriums für land- and forstwirtschaft.* Vienna: Umwelt & Wasserwirtschaft.

Noble, R., Cowx, I., Goffaux, D., and Kestemont, P. 2007. Assessing the health of European rivers using functional ecological guilds of fish communities: Standardising species classification and approaches to metric selection. *Fisheries Management and Ecology* 14:381–92.

Oberdorff, T., Pont, D., Hugueny, B., and Porcher, J. P. 2002. Development and validation of a fish-based index for the assessment of "river health" in France. *Freshwater Biology* 47:1720–34.

Olsen, A. R., and Peck, D. V. 2008. Survey design and extent estimates for the Wadeable Streams Assessment. *Journal of the North American Benthological Society* 27:822–36.

Osborne, L. L., and Wiley, M. J. 1992. Influence of tributary spatial position on the structure of warmwater fish communities. *Canadian Journal of Fisheries and Aquatic Sciences* 49:671–81.

Paller, M. H., Reichert, M. J. M., and Dean, J. M. 1996. Use of fish communities to assess environmental impacts in South Carolina coastal plain streams. *Transactions of the American Fisheries Society* 125:633–44.

Pont, D., Hugueny, B., Beier, U., Goffaux, D., Melcher, A., Noble, R., Rogers, C., Roset, N., and Schmutz, S. 2006. Assessing river biotic condition at a continental scale: A European approach using functional metrics and fish assemblages. *Journal of Applied Ecology* 43:70–80.

Potapova, M., and Charles, F. D. 2007. Diatom metrics for monitoring eutrophication in rivers of the United States. *Ecological Indicators* 7:48–70.

Root, R. B. 1967. The niche exploitation pattern of the blue-gray gnatcatcher. *Ecological Monographs* 37: 317–350.

Roset, N., Grenouillet, G., Goffaux, D., Pont D., and Kestemont, P. 2007. A review of existing fish assemblage indicators and methodologies. *Fisheries Management and Ecology* 14:393–405.

Round, R. 1981. *Ecology of algae.* Cambridge: Cambridge University Press.

Rumeau, A., and Coste, M. 1988. Initiation à la systématique des diatomées d'eau douce pour l'utilisation pratique d'un indice diatomique générique. *Bulletin Francais de la Peche et de la Piscicultur* 309:1–69.

Sabater, S., and Admiraal, W. 2005. Biofilms as biological indicators in managed aquatic ecosystems. In *Periphyton: Ecology, exploitation and management*, eds. Azim, M. E., Verdegem, M. C. J., van Dam, A. A., and Beveridge, M. C. M., 159–77. Wallingford: CAB International.

Schmutz, S., Kaufmann, M., Vogel, B., Jungwirth, M., and Muhar, S. 2000. A multi-level concept for fish-based, river-type-specific assessment of ecological integrity. *Hydrobiologia* 422/423:279–89.

Shapiro, M. H., Holdsworth, S. M., and Paulsen, S. G. 2008. The need to assess the condition of aquatic resources in the US. *Journal of the North American Benthological Society.* 27:808–11.

Shields, F. D., Knight, S. S., and Cooper, C. M. 1995. Use of the Index of Biotic Integrity to assess physical degradation in warmwater streams. *Hydrobiologia* 312:191–208.

Sládeček, V. 1986. Diatoms as indicators of organic pollution. *Acta Hydrochimica et Hydrobiologica* 14:555–66.

Sostoa, A., Caiola, N., and Casals, F. 2004. A new IBI (IBICAT) for the local application of the water framework directive. In *Aquatic habitats: Analysis and restoration*, eds. García-de-Jalón, D., and Vizcaíno-Martínez, P., 187–91. International Association of Hydraulic Engineering and Research.

Steinberg, C., and Schiefele, S. 1988. Biological indication of trophy and pollution of running waters. *Zeit Wasser Abwas Fors* 21:227–34.

Stoddard, J. L., Peck, D. V., Paulsen, S. G., Van Sickle, J., Hawkins, C. P., Herlihy, A. T., Hughes, R. M., Kaufmann, P. R., Larsen, D. P., Lomnicky, G., Olsen, A. R., Peterson, S. A., Ringold, P. L., and Whittier, T. R. 2005. *An ecological assessment of western streams and rivers.* EPA 620/R-05/005, 1–48.

Tison, J., Giraudel, J.-L., and Coste, M. 2008. Evaluating the ecological status of rivers using an index of ecological distance: An application to diatom communities. *Ecological Indicators* 8:285–91.

USEPA. 2009. Water quality standards. http://www.epa.gov/waterscience/standards/

Verdonschot, P. F. M. 2000. Integrated ecological assessment methods as a basis for sustainable catchment management. *Hydrobiologia* 422/423:389–412.

Verdonschot, P. F. M., and Moog, O. 2006. Tools for assessing European streams with macroinvertebrates: Major results and conclusions from the STAR project. *Hydrobiologia* 566:299–309.

Wang, Y.-K., Stevenson, J. R., and Metzmeier, L. 2005. Development and evaluation of a diatom-based Index of Biotic Integrity for the Interior Plateau Ecoregion, USA. *Journal of North American Benthological Society* 24 (4): 990–1008.

Watanabe, T., Asai, K., and Houki, A. 1988. Numerical water quality monitoring of organic pollution using diatom assemblages. In *Proceedings of the Ninth International Diatom Symposium 1986*, ed. Round, F. E., 123–41. Koeltz Scientific Books.

Wright, J. F., Sutcliffe, D. W., and Furse, M. T. 2000. Assessing the biological quality of fresh waters: RIVPACS and other techniques. Freshwater Biological Association, Ambleside, Cumbria, UK. Oxford, UK: The RIVPACS International Workshop, 16–18.

Zelinka, M., and Marvan, P. 1961. Zur präzisierung der biologischen klassifikation der reinheit fliessender gewässer. *Archiv für Hydrogiologie* 57:389–407.

Appendix

TABLE A.1

β-values = Exergy Content Relative to the Exergy of Detritus

Early Organisms	Plants	Animals	β-values
Detritus			1.00
Viroids			1.0004
Virus			1.01
Minimal cell			5.0
Bacteria			8.5
Archaea			13.8
Protists	(Algae)		20
Yeast			17.8
		Mesozoa	33
		Placozoa	33
		Amoebe	39
		Protozoa	39
		Phasmida (stick insects)	43
	Fungi, molds		61
		Nemertina	76
		Cnidaria (corals, sea anemones, jellyfish)	91
	Rhodophyta		92
		Gastroticha	97
		Prolifera, sponges	98
		Brachiopoda	109
		Plathyhalminthes (flatworms)	120
		Nematoda (round worms)	133
		Annelida (leeches)	133
		Gnathostomulida	143
	Mustard weed		143
		Kinorhyncha	165
	Seedless vascula plants		158
		Rotifera (wheel animals)	163
		Entoprocta	164
		Insecta (beetles, flies, bees, wasps, bugs, ants)	167
	Moss		174
		Coleodiea (sea squirt)	191
		Lepidoptera (butterflies)	221
		Crustaceans	232

(continued)

TABLE A.1 (continued)

β-values = Exergy Content Relative to the Exergy of Detritus

Early Organisms	Plants	Animals	β-values
		Chordata	246
	Rice		275
		Gymosperms (incl. pinus)	314
		Mollusca, bivalvia, gastropodea	310
		Mosquito	320
	Flowering plants		393
		Fish	499
		Amphibia	688
		Reptilia	833
		Aves (birds)	980
		Mammalia	2127
		Monkeys	2138
		Anthropoid apes	2145
		Homo sapiens	2173

Note: Values taken from Jørgensen et al. 2005. Calculations of exergy for organisms. *Ecological Modelling* 185:165–75.

Index

Milton Keynes UK
Ingram Content Group UK Ltd.
UKHW021909071024
449327UK00022B/1646